Major Transitions in the Human Life Cycle

The Academy for Educational Development

The Academy for Educational Development is an international, nonprofit planning and research organization that serves schools, colleges, state systems of education, educational organizations, foundations, government agencies, and foreign governments. Founded in 1961, it has served more than 500 clients in the United States and seventy foreign countries.

The Academy prepares long-range educational plans, conducts workshops and seminars, plans and carries out innovative experiments, helps colleges and universities find well-qualified administrators, conducts studies and research on problems involving education, and helps developing countries plan and carry out educational programs at all levels.

Major Transitions in the Human Life Cycle

Edited by
Alvin C. Eurich
Academy for Educational
Development

LexingtonBooks
D.C. Heath and Company
Lexington, Massachusetts
Toronto

Library of Congress Cataloging in Publication Data

Eurich, Alvin Christian, 1902–
  Major transitions in the human life cycle.

  "Based on a series of four annual conferences held at Aspen, Colorado, from 1976 through 1979, under the sponsorship of the Academy for Educational Development"—Pref.
  Includes bibliographies and index.
  1. Life cycle, Human-Congresses. I. Title. [DNLM: 1. Life change events—Congresses. 2. Human development—Congresses. BF 713 M234 $976-79]
HQ799.95.E97   305.2   81–47067
ISBN 0-669-04559-4   AACR2

Copyright © 1981 by D.C. Heath and Company

Published simultaneously in Canada

Printed in the United States of America

International Standard Book Number: 0-669-04559-4

Library of Congress Catalog Card Number: 81-47067

Contents

Contents

Preface

Over the years, the human life cycle has proved to be a challenging subject for scholars, novelists, playwrights, magazine writers, and journalists. Hundreds of books, plays, research monographs, and magazine articles have been written about it. Some have dealt with the life cycle as a whole, some have focused on its various stages, and some have dealt with the transitions from one stage to another.

Nearly all of these scholarly and popular works have been written by a single individual or within the framework of a single scholarly discipline. This book is different from all the others. So far as I know, it represents the first effort to examine the human life cycle, the stages within it, and the critical transitions from one stage to another, from a multidisciplinary point of view. Never before, to my knowledge, have so many scholars and practitioners from so many different disciplines brought their knowledge and experience to bear on the complex, many-faceted subject. They have not only contributed their individual ideas, research findings, and insights to this multidisciplinary effort, but they also have engaged in a lively cross-disciplinary discussion that has added a new dimension to the study.

This book is based on a series of four annual conferences in Aspen, Colorado, from 1976 through 1979, under the sponsorship of the Academy for Educational Development. Funds for the conferences were provided by the Schweppe Research and Education Fund. Each conference lasted five days, and brought together scholars and practitioners from at least twenty different disciplines.

The charge laid upon the participants was to examine the life cycle and its various stages from a multidisciplinary point of view, to identify the special problems and concerns of the people at each stage, to explore the relationships of a particular stage with earlier and larger stages, and to recommend programs of action and research that might help people cope more effectively with the stresses and strains of each stage.

Part I of this book deals broadly with the whole life cycle and reports the discussion from the biological, sociological, psychological, functional-age, and values perspectives. It also identifies critical periods in the life cycle.

Part II deals with the turbulent years, roughly ages 14 through 24. The turbulent years are considered from many angles ranging from the demographic to religious. The discussions terminate by spelling out the priorities for the future.

Part III deals with young adulthood, roughly ages 25 through 40, when most young adults establish families, raise children, develop lifestyles, embark on careers, and plan for the future. They are regarded as years of hopes and lost aspirations. The biological characteristics and changes are given full treatment, and the topics are the problems of stress in family life. Finally priorities for the future are covered.

Part IV summarizes the discussions of one critical period, namely, the middle years. The authors characterize middle age as merely an episode in the life-long development process, a phase of the aging process. The authors consider aging as inevitable, but not immutable. The ways in which people change may be in response to the psychological and sociological environment. Because work and health are central to this age group, careful consideration is given to both. It is noted once more that environment often provides the key to surviving stress successfully. Recommendations are made on various unresolved issues.

Thus, the whole life cycle is reviewed and the various stages examined in detail. Through these chapters it becomes apparent that the family is the center of all stages of development. First, the parents and the children comprise the basic family unit. Second, the children become a part of the work force, establish homes, and bear the grandchildren. When the grandchildren leave home and the nest is empty another stage of life is reached, namely, the aging period. Because of the central place of the family in the human life cycle, these four conferences led to the establishment of a family institute that is undertaking a further analysis of the different periods of the life cycle, and the important role the family plays at each stage.

Acknowledgments

The Academy for Educational Development is deeply indebted to all the individuals who participated in the four conferences. They contributed immeasurably from the knowledge and insights of their respective disciplines. Also special thanks are due to the individuals who summarized the four conferences: Howard Spierer, Joan Waring, and John Scanlon, who also edited the papers in part II. They all did a masterful job. My deepest appreciation, however, is extended to Dr. John and Lydia Schweppe without whose support both financially and intellectually the conferences could not have been held.

Part I
Life Cycle

Howard Spierer

1 Introduction

Human development, the study of man as a whole, appears to be best able to facilitate the unification of the sciences and the humanities and to open the past as well as the future evolution of ourselves and our society—Robert A. Aldrich[a]

What can the biomedical and the social sciences contribute to an understanding of the human life cycle? What are the critical transition periods in the aging process? When do transitional periods and events occur and with what social and biological consequences? How do the different disciplines approach the study of the human life span?

From August 9 through 13, 1976, 15 authorities in the fields of education, sociology, psychology, biology, anthropology, psychiatry, and medicine assembled in Aspen, Colorado, to grapple with these and other significant questions relating to the theme, "Major Transitions in the Human Life Cycle." The conference was conducted at the Given Institute of Pathobiology of the

---

[a]"The Early Antecedents of Aging," a paper prepared for the conference.

University of Colorado Medical Center.

This book summarizes the main themes and issues discussed at the conference. It is based on the oral presentations, discussions between participants, and papers and articles submitted as background reading.

The conference was organized as part of a long-range project on human life-span development. The project calls for interaction between specialists in the biomedical and social sciences to:

- Contribute to the development of constructive attitudes toward, as well as a better understanding of, the aging process.
- Develop a conceptual framework for an interdisciplinary study of the human life span.
- Encourage the study and teaching of the human life span and of critical transition periods.
- Encourage research that could be applied to the solution of individual and societal problems.
- Delineate the characteristics of major transitions in the human life cycle.
- Identify means of helping individuals cope with such transitions in the life cycle.
- Cooperate with activities and centers concerned with the aging process.

The project assumes that the detailed study of the aging process should be one of the nation's priorities. For, the more we know about how physical and social processes interact to affect the development of human beings, the better we can cope with the stresses of modern society.

Further, since the problems confronting us are interrelated, the solutions to these problems require an interdisciplinary approach. Herein lies the promise of closer collaboration between the biomedical and social sciences, both of which are concerned with human growth. This collaboration has the potential to heighten our understanding about what is needed for the human organism to function more effectively at various stages—from infancy to adulthood and old age. The hope is that a merging of techniques and approaches used in the bio-medical and social sciences may create a new way to understand aging better.

The 1976 conference was held at a time when re-search on the characteristics of various stages of adult-hood had increased, with the publication in just two years of at least three handbooks on aging.[1] This re-search is just part of the scrutiny being given to the study of man as a whole and to life-span development theory as an approach to the study of human behavior and aging. Increasing attention is being paid to the problems of older people, as evidenced by the establishment of the National Institute of Aging at the National Institutes of Health, local and state offices of aging, and gerontology centers at postsecondary institutions.[2] And, in the aca-demic community, a controversy has been stirred by Edward O. Wilson's *Sociobiology: The New Synthesis,* a book that raises fundamental questions about the re-lationships between the natural sciences and the social sciences, including the role played by both genetics and environment in the life cycle.

According to a belief that is gaining currency, in-creased knowledge of human development—how people

---

[1] C. Finch and L. Hayflick, eds., *Handbook of the Biology of Aging;* James E. Birren and K. W. Schaie, eds., *Handbook of the Psychology of Aging;* and Robert Binstock and Ethel Shanas, eds., *Handbook of Aging and the Social Sciences.*

[2] If the birthrate stays low, the Bureau of the Census forecasts that the percentage of Americans over 65 will rise from its current level of one in 10 to one in 8.5 by the year 2000 and one in six by 2030. The number of Americans in this age group will climb from 23 million to 31 million by the year 2000 and 52 million by 2030.

grow, change, adapt to circumstances—is a prerequisite to improved management of man's life and his physical, social, and cultural environment.

Much of the excitement of the conference was generated by attempts by scholars in different disciplines to search for a common language, common ground on which to deal with issues. Moreover, their joint quest for answers did not end at Aspen; the conference is but one step in a series of interdisciplinary encounters among scholars.

The 1976 conference sought to identify and to characterize the critical transition periods in the aging process, to assess the present state of knowledge, and to suggest plans for future studies and activities.

2 Different Approaches to the Study of the Human Life Cycle

All the world's a stage,
And all the men and women merely players.
They have their exits and their entrances;
And one man in his time plays many parts.
Shakespeare

Shakespeare, in *As You Like It,* describes the seven ages of man: the infant, the schoolboy, the lover, the soldier, and the justice; and, in the later ages of life, "the lean and slipper'd pantaloon" and, finally, "second childishness and mere oblivion." The aging process concludes with man "sans teeth, sans eyes, sans taste, sans everything." To Shakespeare, then, the human life consisted of moving from role to role, with man assuming the characteristics and physical appearance attributed to each phase of life.

Age, time, and circumstances are major considerations in the study of the human life cycle. Individual researchers may classify time periods into different categories, or may stress different kinds of events, or may believe that one influence (say, genetics) is more significant than another (say, environment). But researchers agree that, over the course of a lifetime, man passes through various periods and experiences critical events.

Regardless of one's approach, two concepts stand out: change and individuality. Change is at the heart of aging. As Matilda White Riley and Joan Waring have observed, "Aging does not mean growing old—only growing older."[1] The process of aging, then, begins at birth and continues until death.

Changes that have important consequences for human behavior are regarded as transitions. These transitions may be due to biological, sociological, environmental, historical, or other phenomena. They may have consequences that are evident now or are manifested at some future date (and thus have "sleeper effects"). They may be evident to friends and to society (going bald, becoming rich, losing a job) or remain unnoticed, although still dramatic, such as losing one's career aspirations. They may be sudden or, more likely, cumulative, as is true, for example, of some diseases.

No two people react to transitions in exactly the same way. Their response is determined by such factors as health, status within society, environmental influences, emotional state, and the values and attitudes of other members of their cohort (people born at roughly the same time, who presumably have shared some similar life experiences).[2]

If the study of the life cycle is the study of the transitions that occur as one grows, then we must closely examine the nature of the transitions and the different

---

[1] Matilda White Riley and Joan Waring, "Age and Aging," p. 358.

[2] A recent book that deals with this aspect of aging is Pat Watters, *The Angry Middle-Aged Man* (New York: Grossman/Viking, 1976).

emphases placed by the various disciplines.

Transitions can be defined in three ways:

1. *By time periods in the life span.*

    According to this view, the life cycle is divided into periods marked by chronological age. Beginning with conception and the prenatal period, man proceeds through birth, infancy, early childhood, adolescence, young adulthood, middle age, old age, and death.

    Each of these periods is distinguished by social and biological characteristics. For example, each period witnesses the impact of particular diseases as well as changes in the endocrine system. Long before adolescence, youngsters develop the ability to conceptualize. Young adults experience difficulty in choosing careers because the results of such decisions are unclear. The middle years produce anxiety in many men as they realize that they will never achieve the aspirations set in earlier years. Old age finds many people losing their reaction speed, but (contrary to popular belief) gaining in intelligence, wisdom, and tolerance.

    A study of the life cycle should encompass an analysis of the characteristics and consequences of each transitional period. Any such analysis would, of course, reveal the relationships between the periods.

2. *By role.*

    Aging consists of learning new roles and relinquishing old ones. Men and women assume a variety of age-related roles that shape their lives—such as mother, father, student, teacher, child, sportsman, businessman, and worker. If one assumes a role valued by society, such as becoming a bride, the adjustment is likely to be smooth. If, however, one shifts from a valued role to one not valued by society, such as a woman losing her husband and becoming a widow, the adjustment may be painful, unless individuals and society provide appropriate support to ease the transition. Assuming a role that clashes with societal expectations also may bring about a troublesome transition, as in the case of becoming a mother of an illegitimate child.

When a person assumes a role is often determined by both biological and social considerations. For example, biology defines the age limits within which women are able to bear children, while society establishes that it is preferable to become a mother when a woman is married and, say, in her 20s rather than in her 40s.

3. *By event.*

Along with transitional periods and transitional roles, there are transitional events. A transitional event is any happening of consequence—including entering school, entering the job market, getting married, becoming parents, and losing one's confidence. Each person's lifetime consists of a series of transformation points: some vivid and dramatic, like divorce; some "unlabeled" but equally powerful, like the gradual decline in aspirational levels or that time in life when we begin to think of years in terms of what remains rather than what has lapsed since birth.

Events differ in content and interest. Biologists, psychologists, sociologists, and anthropologists would produce different lists of key transitional events in the human life cycle, reflecting their different perspectives. Even if the lists contained similar events (such as birth and death), their interpretation as to what is most significant in that event would differ. And, of course, similar events are likely to have a different impact on individuals because of different previous experiences.

There are no absolutes in transitions. That is, although

age is an important factor in most transitions, it is difficult to ascribe a fixed time to any event, period, or role. People confront change at different points in their lives, often going against the accepted norms. For example, society may not deem it appropriate for a woman to marry for the first time in her 40s or for a man to get his first job at age 40. It is thus possible to be "on time" and "off time."

Also, people can experience simultaneous transitions. In early adolescence, for example, one undergoes physical changes (puberty), educational changes (entering junior high school), social changes (assuming a new role), as well as changes in familial relationships.

Furthermore, as Matilda Riley observed in her presentation, societal norms also change, thereby affecting the timing of transitional events. For example, the age of entering school has fluctuated, with the current trend being to begin with nursery school and other forms of pre-elementary education. Young people used to live with their parents until they married; now, many move out before they marry. Today, with the increased independence of young people, the older married couple can look forward to 16 or 17 years of the "empty nest" before either spouse dies. These shifts in age norms indicate that transitions reflect historical contexts.

One such context, which promises to alter the quality of life, the timing of future transitions, and societal norms in general, is the "graying of America." The nation's demographic profile is changing. With the decline in the birthrate—from a postwar high of 3.8 children per woman in 1957 to a low of 1.8 children in 1976—the median age of the population will rise, according to Bureau of Census projections, from 29 in 1975 to 35 in 2000 to over 37 in 2030. If present trends continue, the percentage of Americans 19 years and younger will decline from approximately 35 to 27, while the percentage of Americans 55 years and older will increase from 20 to 27.

It became clear during the conference that age was a powerful variable and provided a useful framework within which to examine human development. The conference participants recognized quite explicitly, however, that chronological age indexed only one aspect of aging. Also to be considered were biological age, psychological age (the capacity to respond to societal pressures and the tasks required of an individual), social age (the extent to which an individual participates in roles assigned by society), and functional age (the ability to function or perform as expected of people in one's age bracket, which, in turn, depends on social, biological, and personality considerations). Just as transitional periods, events, and roles are interrelated, so too are the different aspects of aging.

As might be expected, specialists emphasize the factors associated with their disciplines, study, and research. Their views on aging reflect different approaches to an understanding of the human life cycle. Let us examine their perspectives.

The Biological Approach

Biologists teach us that longevity varies among the species. Turtles may survive up to 150 years, mice only for three or four years. For human beings, the life span for most people is likely to remain between 70 and 100 years—even if cures are found for heart diseases and cancer. The best we can hope for is an improvement in the quality of our physical and mental health so that more people will reach the age of 70 in a condition that permits self-sufficiency and effectiveness.

Donald West King reviewed several theories of aging, acknowledging that some have greater validity than others. One theory is referred to as the genetic theory or programmed death, whereby, after a certain number of generations, embryonic cells fail to divide. Another stresses that our environment is permeated by numerous "agents of injury" of an infectious, chemical, physical, nutritional, and immunological nature; these "agents of injury" produce changes in chromosomes that result in a lack of sufficient protein needed by the body. Still another theory finds oxygen and nutritional substances having difficulty reaching cells; as a result, cells die or become susceptible to injury.

Major diseases are associated with, but are not limited to, various age intervals. As outlined by Dr. King in his presentation on the pathobiology of aging, one possible distribution of diseases by age, though by no means complete, is as follows:

Age Interval	Diseases
0-2	Congenital, infection, allergy, accidents
2-12	Accidents, infection, leukemia
12-20	Accidents, allergy, endocrine imbalance, venereal disease
20-40	Venereal disease, pregnancy problems, ulcers
40-60	Cardiovascular; hypertension; endocrine disorders; ulcers; emphysema; cirrhosis; kidney disorders; cancer of the cervix, breast, lung, and colon
60-80	Continuation of the above, arthritis, deafness, cataracts, demineralization of the bones, cancer of the prostate, cardiovascular collapse

Diseases may occur in any age group and build up over the years, thus hastening the aging process. Cancer is one such example. On the other hand, congenital diseases are not likely to be transmitted to other age groups. The main point is that, although some physical conditions, which are often related to age, are not determined by or limited to age (such as accidents), the response to the diseases or condition has consequences for aging.

In the biological view, every critical period in the life cycle is associated with metabolic and hormone changes, as pointed out by Robert H. Williams, an expert on endocrinology. "The hormones," Dr. Williams said, "should be likened to keys and their respective receptors to locks. Only certain keys fit certain locks; and unless the proper key is provided, certain functions cannot be properly obtained."[3] Dr. Williams provided several illustrations. Puberty and adolescence are associated with increases in the amounts of numerous hormones, including growth hormones, thyroid-stimulating hormones, androgens, estrogen, and progesterone. The female menopause is associated with high levels of gonadotropin and low levels of estrogen. Although males do not have the same menopausal manifestations of the type we see in females, there are gradual decreases in gonadal functions that significantly affect male behavior; the decrease in sex potency and the general loss of strength and stamina tend to lower man's pride, which, in turn, can lead to depression and even suicide. And, in the aged, we find a slowing down in most of the endocrine and metabolic processes.

Hormones affect the size, structure, and function of the cells and, thereby, influence the physical development of an individual. Another characteristic of hormones relating to the aging process is that hormone changes accompany physical and mental illnesses and diseases. Dr. Williams cited glucosteroids and thyroid as examples of hormones that increase during periods of great stress. The greater the discrepancy between an individual's goal and his capacity to attain that goal, the greater the de-

[3] Robert H. Williams, "Hormonal Changes in Association with Aging and Certain Life Crises," a paper prepared for the conference.

pression, anxiety, and tension.

For the biologists, then, the functioning of the hormones and the conditioning of the cells are two significant factors in human development. For, they affect human development at every stage in the aging process and, in turn, are affected by the individual's condition in a holistic sense. Diet, temperature, exercise, and the extent to which we use our muscles, tissues, and organ systems are some of the other factors that have a bearing on aging.

The Sociological Approach

Sociologists, like biologists, believe that an understanding of age-related phenomena is crucial to the study of the human life cycle. They acknowledge that biological factors play an important role in the aging process. But, in their view, many of the problems of age and aging are primarily social phenomena. Matilda Riley and Joan Waring explain the position of the sociologist:

> Only recently, however, have the problems of age and aging been fully recognized as social phenomena: molded by human beings, rooted in human history, and susceptible to social intervention and change. These problems have been taken for granted as immutable, or wholly attributable to biology or "natural law." Yet, to understand their social character is to open all the possibilities of directing social change to improve the quality of life for people at every age. With the assurance that age and aging are mutable, the sociologist now faces a dual challenge: To explore how and to what degree social factors are responsible for personal troubles and social problems related to age; and to find social means for ameliorating or correcting these problems.[4]

---

[4] Riley and Waring, "Age and Aging," p. 357. "Age and Aging" is the source of many of the points made and the specific examples provided by Dr. Riley on the concept of age as a social phenomenon.

Drawing upon her own extensive writings on the aging process and upon the views of her colleagues, Dr. Riley presented two conceptual models for understanding social problems related to aging. The first deals with age stratification. At any moment in life, age places man within two structures: an age structure of *people,* in which man is a member of a cohort, a group of people who were born during the same time interval and, accordingly, who will age together; and an age structure of *roles,* defined by society. The aging process consists of movement within and between these structures as people within one cohort die and new cohorts of people shift from one role to another throughout their lifetimes. ·

Conducting an analysis of cohort characteristics as a means of understanding the concepts of age and aging is an important methodological tool for the sociologist. It is a way of isolating nonbiological and nongenetic considerations. It is also a way of emphasizing that age and aging should be studied "as fundamental aspects of social structure and social dynamics," and not merely "as characteristics of individuals."[5] That individuals in the same cohort may behave or think differently about various issues in no way detracts from the importance of viewing people as members of cohorts.

The stratification of people and roles by age creates problems. Since age is a major (and often arbitrary) criterion for allocating roles in society—such as determining the eligibility for entering military service, receiving social security benefits, running for political office, and retiring from work—it produces inequalities. And these inequalities, in turn, often create hardships for individuals by denying them economic rewards and social standing; and for society at large by creating con-

---

[5] Matilda White Riley, "Age Strata in Social Systems," pp. 189ff.

flicts among groups and by wasting human resources. Then, too, sometimes age inequalities contradict such ideals as equality of opportunity, self-reliance, individualism, and advancement based on merit.

Since roles and circumstances change with history, people in different cohorts often have different values and characteristics and thus age in different ways. For example, recent cohorts have received more formal schooling, better health care, and more job opportunities than older cohorts. Furthermore, age segregates people in many of our social institutions—as in our schools, where young children are discouraged from interacting with older students; and in our retirement homes, where people are isolated from relatives and friends. Such segregation prevents people in different age brackets from providing needed support to one another.

The age structure of roles creates other social difficulties. For example, because of wars, the economy, and other factors, there are imbalances between the number of people in a cohort and the number of people needed to fill certain roles at the appropriate age. Problems have arisen because the baby boom cohort, which created the need for more teachers and schools, was followed by a sharp decline in the birthrate, which has left us with a surplus of educational facilities and personnel.

The result of such interaction between age dynamics, which affects people, and societal dynamics, which affects roles, is an "inherent strain between the continuing flow of new cohorts of people and the societal changes in the roles through which the cohorts flow."[6] It is a wonder that society operates as smoothly as it does.

The second conceptual model was mentioned earlier in this chapter: role transitions. Because of biological considerations and because of the pressure of new cohorts

---

[6] Riley and Waring, "Age and Aging," p. 364.

coming along behind them, people are forced to abandon old roles and assume new roles. These transitions occur throughout the aging process.

If the individual is moving to a valued role that conforms to society's expectations, has the skills and motivation to perform that role, and is sufficiently encouraged, the transition is likely to be smooth. If not—if support mechanisms are not present, if the role is not valued by society—adjustments will be necessary. One possible adjustment is the development of defense mechanisms to resolve the conflict. Old people withdrawing into themselves because of their inability to keep occupied is an example of such behavior.

The Psychological Approach

Other sociologists and psychologists approach the study of the human life cycle from the standpoint of personality development. To these scholars, the traditional periods and roles are at once too confining and too broad to encompass the endless number and kinds of human experiences. Moreover, they believe that much of what passes for life-span development research deals only with a particular period, such as adolescence or early childhood.

What is needed, argued Orville G. Brim, Jr., is a true life-span development perspective. Offering a possible approach to this end, Dr. Brim told the conference participants:

> . . . our strategy should be to look at content—some item, some personality trait, some hormonal secretion, some particular kind of pathology over the life-span and to follow it through its transformations by whatever occurs across the years, and to describe, really, the life course of some human characteristic.[7]

---

[7] Orville G. Brim, Jr., "Personality Development in Life-Span Perspective," a paper prepared for the conference.

Among Dr. Brim's interests are theories about an individual's sense of self. He has looked at various characteristics of self-concept and attempted to study their transformation over the years. To cite two examples: An individual's sense of personal control, his belief that he can control his environment, becomes evident at age four or five, declines steadily until adolescence, rises steadily until age 45 or 50, declines, and then rises in old age. Such beliefs are normally related to age and are continually transformed by experience.

Data on the changes in one's beliefs about physical attractiveness are more limited. We know that four-year-olds can discriminate between attractive and ugly children, and that each child knows where he stands in relation to others. We also know that during adolescence those who mature early or late dramatically alter their sense of attractiveness, one result being that some youngsters who were the idols of their classmates at age 11 or 12 are now viewed as physically undesirable. These beliefs are a result of the interaction between biological and psychological factors and a reflection of cohort values as well as the cultural values of society as a whole.[8]

The Functional Age Approach

If the self-concept approach to the study of the life cycle focuses on what people believe, the functional age approach concentrates on the physical, emotional, and intellectual requirements throughout the aging process. James E. Birren, who has written and published extensively on the biomedical and social aspects of aging, offers the concept of functional age as an alternative to

---

[8] Comparative studies would need to be conducted to determine the extent to which beliefs about physical attractiveness are universal beliefs attributed to specific age groups or characteristic of the United States alone.

chronological age. Although a useful tool in studying aging, chronological age is limited in that it does not tell us which "clusters of functions" are related to age.[9]

At the heart of the functional approach is the theory that every society presents the individual with tasks that need to be accomplished at various stages of the life span —an approach that, in Dr. Birren's view, contains "a hidden biodeterminism." How one reacts to these tasks is determined by biological considerations, the social environment (socioeconomic status, demands of home and job, transportation, recreational opportunities), and one's own personality (ability to adapt, orientation to stability or change).

Among the types of criteria for functional age in adults are: susceptibility to particular diseases, capacity for physical effort, speed and coordination, level of performance on the job, problem solving and reasoning, mental health, morale, peer group identity, effectiveness in interpersonal relationships, and the capacity to process information.

A person's ability to perform certain functions or to exhibit various character traits changes over the years. For example, all behavior that is mediated through the central nervous system slows down with age; older people react more slowly than younger people. By the same token, the older people get, the more likely they are to show improvements in frustration level, patience, experience, knowledge, wisdom, and self-confidence.

Many of these changes are part of an organized, predictable pattern of behavior that is characteristic of aging. As a result of studies of the process of aging in animals, Dr. Birren formulated what he refers to as the counterpart theory, which "views ordered biobehavioral

---

[9] James E. Birren and V. Jayne Renner, "Developments in Research on the Biological and Behavioral Aspects of Aging and Their Implications," p. 2.

changes in the post-reproductive animal as the expression of counterpart characteristics of the animal in earlier development."[10] That is, some traits exhibited by animals in their later years can be attributed to genetic traits acquired at the time of reproduction in earlier years. Applying the counterpart theory to human behavior, one might say, for example, that the favorable complex trait of wisdom is linked in some way to certain superior genetic properties, and that the trait will emerge increasingly with advancing age. Enough time would have elapsed for a person "to learn from others and to live through enough man-made and natural crises to have a large repertory of concepts and responses to events." The trait of wisdom, in turn, brightens the prospects for success in the reproduction of future generations. The counterpart theory, then, suggests a way through which order may be introduced in later life changes.

The concept of functional age could provide the basis for making policy decisions in the areas of reemployment, training, and retirement. Individuals who object to being transferred or retired could, under this concept, request a panel of experts to evaluate their performances and capabilities. In addition, improvements in the delivery of health services, in family life, and in the cultural and educational environments can be expected to raise the functional level of adults.

The study of how people perform also holds the promise of dispelling three prevalent views of intellectual functioning in adulthood and old age—namely, that intellectual decline is continuous and dramatic throughout adulthood, that it is a natural consequence of biological aging, and that it is irreversible. Such was the

---

[10] James E. Birren and V. Jayne Renner, "A Biobehavioral Approach to Theories of Aging," a paper presented at the conference on "Society, Stress and Disease: Aging and Old Age," Stockholm, Sweden, June 1976.

central theme of Paul Baltes. The position taken by
Dr. Baltes and by his research collaborator, K. Warner
Schaie, is clear:

> . . . it is a myth . . . to assume that decline is neces-
> sary, general (across abilities), and universal (across
> persons). . . .
>
> . . . rather than advancing a new myth, we have
> argued for large interindividual differences in adult
> change, multidimensionality, multidirectionality, and
> the import of biocultural, historical change (in addi-
> tion to ontogenetic age changes).[11]

Dr. Baltes cited evidence from studies indicating the
importance of historical and generational influences.[12]
Thus people in the same cohort revealed minor changes
in intellectual performance throughout adulthood compared with people in different cohorts, proving that
cohort differences outweigh age differences. The aged
could be described as intellectually deficient only in
relation to younger, better educated people. Moreover,
studies that plotted intelligence test performances backward from death reported dramatic changes *only* in the
years immediately preceding death.

At the same time, other studies indicated that older
people increase in "crystallized intelligence" (word
meaning, vocabulary) and decline in "fluid intelligence"
(problem-solving tasks). Furthermore, elderly people
trained in thinking through concepts and in problem
solving improved their cognitive performances.

To Baltes and his colleagues, the findings of these
studies proved that there is no monolithic interpretation
of intelligence. There are differences among cohorts and

---

[11] Paul B. Baltes and K. Warner Schaie, "On the Plasticity of
Intelligence in Adulthood and Old Age," pp. 721-722.

[12] See Paul B. Baltes *et al.,* "Operant Analysis of Intellectual
Behavior in Old Age," pp. 259-272; also, Paul B. Baltes *et al.,*
"Recent Findings on Adult and Gerontological Intelligence," pp.
225-236.

within cohorts. Cognitive behavior is subject to change. The failure, then, of older people to perform as well in the cognitive domain is due not to inherent incompetencies, but largely to deficiencies in their environment. Therefore, if older people received appropriate incentives and support (say, social praise or money), they might perform better.

The Values Approach

The importance of environmental influences suggests an approach to the life cycle that focuses on culture, values, attitudes—and the implications these have for the quality of life and for public policy. Particularly germane is the negative attitude that exists in society toward getting old and toward the elderly. This attitude, acquired in childhood and reinforced throughout society, portrays the negative side of growing old and suggests that older people should be isolated from the social mainstream. Such an attitude, Robert Aldrich told the conference participants, creates stress in older people and "actually contributes to the rate of aging of many individuals."[13]

When it comes to the elderly, Dr. Aldrich stated further, we are "suffering from a taboo," one that could be potentially dangerous. For, as older people become an increasing percentage of the population, young people may need to depend on them to join the work force. If that happens, unless current attitudes are reversed, conflicts could arise. We need to apply our knowledge of human development to the task of reversing these negative attitudes, "a task for collaboration between educators, biomedical and social scientists, and the public media."

---

[13] Robert A. Aldrich, "The Early Antecedents of Aging," a paper prepared for the conference.

We have already noted that values change in society. Because of historical and social factors, cohorts may have different attitudes, thereby creating generation gaps. When there are significant differences on major issues, society itself becomes unglued.

In looking at aging from the perspective of values and mores, we become aware of the need to examine our priorities. For example, our current value system would appear to dictate the need for new roles for the elderly, expanded support systems, and new relationships between education, work, and leisure. Opportunities for older Americans to continue their education, pursue hobbies, and participate in athletics need to be expanded. We will best be able to identify these priorities and translate them into courses of action when we understand more about the interrelationships between the biomedical and the social sciences throughout the life cycle.

3

Possibilities for a New Conceptual Framework

My proposal is to create a university which exists only during the summer time (say, eight weeks) which has a small group of regular faculty (say, 12) and a smaller group of visiting faculty that changes each year (say, 4 or 5), with a curriculum that focuses on the issue of aging as a means of integrating the traditional and well-developed domains of human health, medicine, and human society. The Summer University's primary function would be instruction and its students would be drawn from universities all over the world.—Alec J. Kelso[a]

Exploring the possibilities of bringing together individual approaches to the study of the human life cycle into one approach is basic to an interdisciplinary focus.

---

[a]"General Background and Presentation of a Practical Suggestion: A Summer University of Human Health, Medicine, and Human Society," a paper prepared for the conference.

The interaction between biological and social processes can best be examined within a framework that illustrates what we now know about human development and points to future avenues of research.

It will, of course, be some time before we can grasp the big picture. For one thing, there are major gaps of knowledge in the traditional sociological and biological disciplines themselves. For another, we have only recently begun to recognize that other disciplines and professions—such as city planning, engineering, and architecture—have something to contribute to the study of the life cycle because they affect the structure and function of the environment in which we live.

There are several possibilities for a new conceptual framework. The first focuses on how each discipline views a particular issue affecting human development in different transitional periods. Such an approach would permit us to appraise each discipline's contribution. It would enhance our knowledge of what takes place in the various age periods, how each discipline defines human issues, and what changes occur throughout the life cycle. For example, how is cognitive development affected by biological, social, and other considerations throughout the aging process? What do psychology, sociology, biology, psychiatry, and anthropology have to say about the problems that converge in the middle years and their consequences for aging? To be sure, not all disciplines impinge on all aspects of human behavior at all times. Nor are all issues applicable to each transitional period.

There are a number of ways to view the interaction between disciplines and issues over time. The transitional periods could just as easily be described by age intervals (0-10, 11-20, 21-40) as by traditional periods in the life cycle (conception, birth, adolescence, young adulthood, middle years, old age, death).

The issues could be personality traits (aggression,

independence), problems (stress, depression), roles (becoming a mother, student, soldier), or events (marriage, graduation, retirement).

The list of issues to be analyzed could extend indefinitely because each scholar would have his own interests. At the same time, this approach is flexible, allowing us to study not only one variable, but also the relationship between variables of human behavior. For example, one issue might be the juxtaposition between dependence and independence.

The endless possibilities are both an advantage and disadvantage. For some scholars, the idea that a discipline may interact with hundreds of human characteristics and events at various periods renders the concept too diffuse and unmanageable.

The second possibility for constructing a broad conceptual framework within which to study the human life cycle is similar to the first in that it too seeks to understand the relationships between biological and social factors over time. As developed by Robert Aldrich, the model assumes that biological organ systems and social and cultural processes are interdependent and must operate in harmony with one another if people are to function effectively. Interference with one process is likely to have an impact on one or more processes. For example, deterioration of the arterial system in middle age would affect the central nervous system, which, in turn, would affect human behavior. In addition, how well these processes relate to each other is determined by their interaction with critical periods and life events. Changing jobs or one's physical environment may result in a change of values and attitudes, which might have biomedical consequences.

Because the model has been pictured as a long oval that is three dimensional, people have referred to it as the watermelon model, as shown in Figure 1. The horizontal lines represent processes, the vertical lines critical periods and life events from conception (C) through death (D) and perhaps even beyond.

One can use this model to make a more sophisticated analysis of life-span development by isolating major concerns. For example, Dr. Aldrich has emphasized the implications of these interactions for education, genetics, life-styles, and environment. He claims that, by understanding the kinds of changes taking place and the consequences of these changes for society and for the individual, students can gain insight into what can go wrong at various points in life and what action might be taken to alleviate social and biomedical problems. In searching for remedies, students become cognizant of the role of values and ethics.

The watermelon model is flexible. It can be adapted to different cultures. The processes and events studied would also be somewhat different for men and women. And the processes, periods, and events could be extended *ad infinitum*. Although some scholars might single out some events or processes at the expense of others, the main point is that, like the first model, the watermelon model provides a framework for integrating knowledge about the human life cycle.

A concern for values and environmental considerations leads us, according to Dr. Aldrich, to a third approach, one that emphasizes the principles of ekistics, the science of human settlements. At the heart of ekistics, a term invented by C. A. Doxiadis, is the belief that man's environment affects his biological, social, and behavioral growth and development.

Furthermore, different kinds of settlements are most suited to individual growth at different points in the life cycle. For example, one room in one house might meet the requirements of a newborn baby, but not those of a 13-year-old boy.

Figure 1
The Watermelon Theory: A Concept of Human Life-Span Development

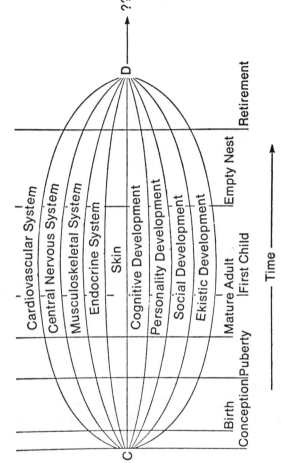

Used with permission of Robert Aldrich.

Doxiadis defined 15 ekistic units, each of which describes a form of human settlement and the number of people who inhabit that space (Figure 2). These units range from anthropos (space for one person) to ecumenopolis (space for 50 billion people).

Figure 2

Ekistic Units

Anthropos	1
Room	2
House	5
Housegroup	40
Small neighborhood	250
Neighborhood	1,500
Small polis	10,000
Polis	75,000
Small metropolis	500,000
Metropolis	4 million
Small megalopolis	25 million
Megalopolis	150 million
Small eperopolis	1 billion
Eperopolis	7.5 billion
Ecumenopolis	50 billion

Source: Redrawn from C.A. Doxiadias, *Anthropopolis, A City for Human Development* (Athens Publishing Center, 1974).

Each of these ekistic units contains five elements that, taken together, produce the environment that characterizes these settlements. Doxiadis and his colleagues referred to these five elements as shown in Figure 3.

Life cycle researchers representing different disciplines can use the principles of ekistics as tools in understanding human development. They can identify the social, biological, and other basic needs of human beings at various intervals in the aging process and then evaluate the degree to which the environments these people inhabit are apt to meet their needs at each time. The optimum condition would be to decide what type of situation is most conducive to human growth and then provide the op-

Figure 3
Elements of Ekistic Units

NATURE(N)
1. Environmental analysis
2. Resource utilization
3. Land use, landscape
4. Recreation areas

ANTHROPOS(A)
1. Physiological needs
2. Safety, security
3. Affection, belonging, esteem
4. Self-realization, knowledge, esthetics

SOCIETY(S)
1. Public administration, participation, and law
2. Social relations, population trends, cultural patterns
3. Urban systems and urban change
4. Economics

SHELLS(Sh)
1. Housing
2. Service facilities: hospitals, fire stations, etc.
3. Shops, offices, factories
4. Cultural and educational units

NETWORKS(Ne)
1. Public utility systems: water, power, sewerage
2. Transportation systems: road, rail, air
3. Personal and mass communication systems
4. Computer and information technology

SYNTHESIS: HUMAN SETTLEMENTS(HS)
1. Physical planning
2. Ekistic theory

Source: Redrawn from C.A. Doxiadias, *Anthropopolis, A City for Human Development* (Athens Publishing Center, 1974).

portunity and suitable environment that would bring about this situation.

The importance of such an environmental perspective cannot be overstated. In contrast to earlier times and simpler forms of human settlement, the modern metropolis, as currently organized, makes human living difficult in a variety of ways. Noise, vehicular traffic, air pollution, overcrowding, and other characteristics present obstacles to social and biological development. Human values are mangled in the process. No wonder that we often hear people question whether our big cities are governable.

Sociobiology represents a fourth attempt at linking biomedical and sociological phenomena. As a result of studies of animal behavior, sociobiologists maintain that the social behavior of animals has evolved over the years in response to genetic influences. In asserting the biological basis for behavior, they believe that one can extrapolate from animals to people and say that we too have inherited genes that influence our behavior—maybe not all of our behavior, but some portion of it. So that our capacity for, let us say, aggression or altruism may be passed on from generation to generation as a result of the workings of the genes.

From the standpoint of understanding the life cycle, the real issue is not so much nature or nurture, heredity or environment, but the influence both have on behavior. As Oscar Hechter, the physiologist, observed, we need to know the limits of genetics—what is it within our genetic background that affects human behavior? By the same token, what are the environmental boundaries within which biological behavior can occur? For example, it is probable that biological influences are predominant during the early stages of development, and environmental influences during later periods in the aging process. Or, to cite another instance, biology, which is a predominant influence in adolescence, reemerges as a strong influence in middle age—but this time, in a somewhat negative way.

To be sure, debate continues on whether sociobiology is useful at all in understanding human behavior. The theories it sets forth, as described most recently by Edward O. Wilson, have been attacked by biologists and environmentalists. Some anthropologists and psychologists assert that, in the long run, we may be able to learn more about biology through culture and sociological principles than the reverse; and that biosociology may have a greater impact than sociobiology.

Clearly, much more research needs to be done on the subject of evolution and human behavior. Nevertheless, because sociobiology involves the application of biological principles and approaches to how man functions in society, it deserves our interest as a growing new discipline.

4 A Closer Look at Selected Transitional Periods and Events

Of all barbarous middle ages, that
Which is most barbarous is the middle age
Of man; it is—I really scarce know what;
But when we hover between fool and sage.—Byron

Since biomedical and social phenomena are present in all transitional periods and events, let us look more closely at three case studies, examples of numerous changes—in physical makeup, social relationships, roles, and attitudes—that take place at about the same time.

Early Adolescence

Of the adolescent period, Beatrix A. Hamburg and David A. Hamburg have written:

> Under modern circumstances, adolescence is the most dramatic example of the complex processes involved in negotiating a critical period of normal development. It illustrates the shifting interplay be-

tween biological, psychological and cultural demands
on the individual.[1]

As a result of these multiple demands, adolescence can
often be a turbulent period.

We are concerned here with early adolescence, the
period between 11 or 12 to 15 years of age, which still
has not received the attention it deserves. This is both
surprising and unfortunate, for, as Beatrix Hamburg explained, early adolescence is a time of rapid and drastic
transitions, a period characterized by delinquency, drug
abuse, teen-age pregnancy, school problems, moodiness,
and aggressive behavior. It is the age for membership in
big city gangs, whose leaders are considered "over the
hill" by the time they reach 18. Early adolescence also
deserves to be studied as a separate and significant period
because some of the beliefs and values adopted determine
behavior and thought patterns expressed in later adolescence and adulthood.

Early adolescence proves to be an unsettling experience because it is a period of simultaneous transitions.
Most significant are the profound changes in biological
status, largely positive changes that are filled with new
strengths and opportunities. Puberty brings about sexual
changes. Girls between the ages of 12 and 13 experience
their initial menstruation, a distinct and easily recorded
event. The only roughly comparable event of sexual
awakening in boys—the initial ejaculation—occurs about
a year or two later. Other sexual developments include
enlargement of the testes in the males and the breasts in
the females, and the growth of pubic hair.

Puberty involves much more than sexual change. Early
adolescence witnesses an increase in height and changes

---

[1] Beatrix A. Hamburg and David A. Hamburg, "Stressful
Transitions of Adolescence—Endocrine and Psychosocial As-
pects," p. 93. This article contains many of the points made by
Beatrix Hamburg at the conference.

in facial appearance, muscular development, and fat distribution.

Males and females mature at different rates. At the time of puberty, females are usually two years younger than males, and they are similarly advanced in language development and certain other kinds of behavior. Indeed, the difference between the early maturing female and the late maturing male may be as great as six years.

How fast a person matures affects his or her psychological development and social acceptance—particularly for males. Studies indicate that males who mature early are considered attractive by their peers and teachers. They are more likely to be superior athletes and leaders in school and to have a higher self-image than late maturing males. Physical attractiveness is shown to be more valued by people in lower socioeconomic groups than by better educated groups, which once again reveals the interaction between biological and sociological influences. The road is not altogether smooth for males who mature early. Some adolescents who have always been looked upon as attractive now have to cope with acne.

The bodily changes create images of self-esteem that persist through the adult years, thereby linking the psychological development of the early adolescent male to his physical development. Since body image is of tremendous concern to early adolescents, many youngsters think poorly of themselves as they believe that their emerging physiques will remain unchanged throughout their adult lives. One result of the convergence of physical and psychological images in slow-developing males is the "small immature boy image in a large well-developed man."

Still another biological consideration in early adolescence is the impact of hormones on human behavior. There is a good chance that changes in sex hormone levels help bring about the unpredictable moodiness, emotional stress, and confrontations with parents that characterize this period.

At about the same time the early adolescent is undergoing biological transitions, he or she is entering junior high school. Like the onset of puberty, entering junior high school is a monumental transitional experience that involves more than just moving from the sixth to the seventh grade. Entering junior high school signifies a change of role and status, from child to adolescent. It represents formal entry into "the teen culture." Furthermore, junior high school usually involves a move from a small and comfortable educational setting to a larger school, different teachers and classes every hour, and a more depersonalized atmosphere. Within such an environment, academic expectations are up, self-esteem down.

Early adolescence is also a time when home relationships are strained. Youths begin to assert their independence, become moody or aggressive, and often look to their peers rather than to their parents for their values. Misunderstandings are likely to arise if parents do not recognize the multiple changes taking place in the early adolescent, or if they believe that the early adolescent is expected to be independent and, therefore, does not require substantial support. Many parents simply do not realize, as the Hamburgs put it, the "heightened need for parental stability and guidance at the time of major biological, school and social discontinuity."[2] The failure of parents to assert their traditional prerogatives may depress adolescents even more than usual.

In addition to serving as an excellent example of how biological and social forces interact to affect human behavior, early adolescence involves transitional events

---

[2] Ibid., p. 99.

whose consequences might be manifested in later phases of the life cycle. These events are said to have "sleeper effects." For example, sexual difficulties encountered in early adolescence may create problems in mid-life.

The Male Mid-Life Period

The menopause has been the subject of much research and controversy. The Aspen conference, therefore, focused on what is known to happen to males in the mid-life period.

Those who believe that nothing much happens to the 25 million American men between the ages of 40 and 60 should consider this description of the mid-life period:

> The hormone production levels are dropping, the head is balding, the sexual vigor is diminishing, the stress is unending, the children are leaving, the parents are dying, the job horizons are narrowing, the friends are having their first heart attacks; the past floats by in a fog of hopes not realized, opportunities not grasped, women not bedded, potentials not fulfilled, and the future is a confrontation with one's own mortality.[3]

Although the mid-life period is not nearly as traumatic for many men as described above, recent studies and reports reveal that important biomedical and sociological changes take place in the middle years. Summarizing the research in his presentation and in his article on "Theories of the Male Mid-Life Crisis," Orville G. Brim, Jr. noted that these changes affect health and physical appearance, family relationships, attitudes toward work, self-concept, and personality. If the events that cause these changes occur simultaneously, a crisis may develop. Otherwise, if the events are stretched out over a period of years, the necessary adjustments will be

---

[3] M. W. Lear, "Is There a Male Menopause?," *New York Times Magazine*, January 28, 1973. Quoted in Orville G. Brim, Jr., "Theories of the Male Mid-Life Crisis," p. 4.

easier and the transition from adulthood to old age will be smoother. Indeed, for many men, the middle years are, for the most part, happy times even while significant personality changes are occurring.

As we have already stated, biology reemerges in the middle years as an important influence—not in the positive sense, as is characteristic of adolescence, but in a negative way. The mid-life period witnesses the gradual decline in the secretion of hormones that affect sexual capacity. Along with a decrease in sexual potency, hormonal changes are also believed to be responsible for an overall decline in physical strength and stamina, which may result in reduced ambition and a general state of despondency. At the same time, middle-aged men may lose, partly or completely, their hair, their teeth, and their hearing and may experience eye trouble—all biological indices of aging, signifying in some instances that life is waning.

For many men, mid-life brings the realization that they will not move any higher in their occupations. They will, therefore, need to scale down their aspirations and anticipated achievement levels. In the process, their confidence in themselves, their feelings of personal control over their own destinies, and their belief in their work competence are likely to suffer. Here, again, some men have little or no trouble facing these realities, while others become depressed at assuming less important roles than they had hoped for.

Even those who accomplished their lifetime work objectives may undergo agonizing reappraisals of their lives. They wonder about the impact of their absence from home on the development of their children. They wonder if they have really achieved all that they could, if the sacrifices they made were worth it, if the future holds anything for them. The pathetic end to such thinking is confrontation with death, acknowledgement that many things will remain undone.

Also possibly contributing to the sense of malaise in the mid-life male is the change in family relationships. The 40-to-60 age period is a time when a man's children go out on their own and when his parents die or become increasingly dependent on him. In addition, husbands and wives may reverse traditional roles: the wives, sometimes seeking a new career, become more aggressive and independent; the husbands, more sensitive and dependent. For some couples, the empty nest period brings tension; for others, contentment.

In discussing the male mid-life period, we have not even touched on the social and biomedical consequences of early retirement. Aside from financial anxieties, early and unanticipated retirement is often accompanied by poor health and by a feeling of being rejected by society, conditions that add to the general strain that characterizes the male mid-life period.

Death

With good reason did John W. Riley, Jr. call his presentation, "Death: The Final Transition." Death represents the culmination of aging. Like most transitions, death is related to age and involves a series of biomedical events and cultural beliefs. It also involves relinquishing old roles (for the dying and their spouses) and assuming new ones (for the surviving widow or widower). Moreover, as death becomes increasingly predictable, we are able to prepare for death and thus do what we can to make the transition as smooth as possible for all concerned.

Although it now seems logical to associate death with old age, such was not always the case. In fact, as Figure 4 on the following page shows, in 1850 only 12 percent of deaths occurred over the age of 65, as compared with 25 percent in 1900 and *63 percent* in 1970.

Figure 4

The Percentage of Deaths by Specified Age Intervals for the Years 1850, 1900, and 1970

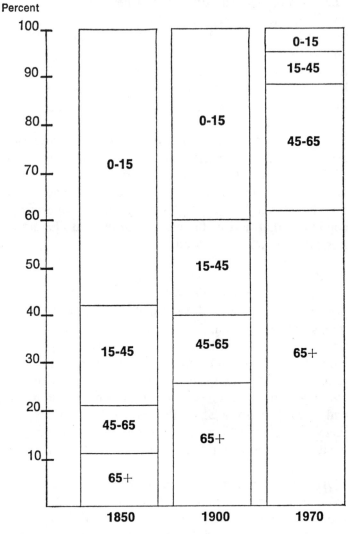

Source: Calculations made by Douglas Ewbank for a meeting sponsored by the Social Science Research Council at Bowdoin College, July 1974. Used with permission.

The result of such a profound change in the distribution of deaths by age is that we are increasingly likely to be able to contemplate the timing of our own deaths. Consequently, we have enlarged opportunities to anticipate death and to prepare for it.

Death, then, should be looked upon as a process. Biomedically, the process may consist of the impact of particular diseases over the entire life cycle finally taking its toll on the human body. We know, too, that an excess or deficiency of certain hormones can cause death. For example, as Robert Williams noted, an excess of insulin frequently causes brain damage and may cause death. On the other hand, a deficiency of biologically effective concentrations of insulin, the main feature of diabetes, can lead to extensive hardening of the arteries, with more than 60 percent of those suffering from atherosclerosis dying of coronary disease.

Medical advances have reduced the threat of early death and of personal physical suffering. People, therefore, feel less threatened by death. Studies supported by the Equitable Life Assurance Society of the United States indicate that societal taboos on death are relaxing.[4] People think more frequently about death than in the past; are likely to maintain that, if a patient is dying, the doctor should so inform him; and are likely to plan for death, including preparing wills, making funeral arrangements, and talking about death with others. The older and more educated a person is, the more likely he is to think about death and take appropriate action.

The increased openness toward death suggests, to Dr. Riley and other sociologists who have studied the meaning of death, the need for a new theory of dying. Dr.

---

[4] For the results of a survey on individual attitudes and orientations toward death, see John W. Riley, Jr., "What People Think About Death," pp. 30-41.

Riley refers to the process of "being dying" as a new form of transition. Once people accept death, they feel free to discuss death, begin to settle their financial affairs, and resolve old disputes.

Through these actions, which may take place over a period of hours, days, months, or even years, people concentrate on the quality of life. Death becomes a celebration of life—a romantic ideal, no doubt, but a trend that may become increasingly evident.

To speak of a new role is not to diminish the importance of traditional roles and behavior associated with death. For the dying person who does not believe in any hereafter, death involves abandoning old roles without assuming new roles. Such a prospect can produce grief, loneliness, and despair both for the dying person and for his spouse. The female survivor, once her husband is dead, ceases being a wife and becomes a widow —a role or condition that may cause feelings of grief, bitterness, and physical illness extending beyond traditional periods of mourning. The degree to which an individual can successfully negotiate this transition and others depends on the support provided by people close to her and by society at large.

5 Coping with Transitions

Despite the omnipresence of threatening experiences, the human species displays remarkable resiliance in adaptation.[1]—David A. Hamburg

Transitions create anxieties in people. Even assuming a role that is valued by society, such as becoming a doctor, involves numerous stressful situations along the way. How much more so the obvious disappointments and difficulties in life, such as suffering a serious illness, losing a job, or becoming a widow.

If, as we have seen clearly, the aging process is filled with transitional periods, roles, and events, then coping successfully with transitions is crucial if we are to function effectively. Coping becomes a mechanism for reducing stress and making transitions less painful. It is essential, therefore, that we search for a theory of coping that can be applied to different kinds of transitions and, in the process, identify the elements of a coping strategy that are likely to produce success.

---

[1] David A. Hamburg, "Coping Behavior in Life-Threatening Circumstances," p. 13. This article includes many of the views presented by Dr. Hamburg at the conference.

Underlying the importance of a coping strategy is the assumption that our present support systems are inadequate in helping people overcome the difficulties involved in transitions.[2] By support systems we mean physical support—facilities, personnel, financial incentives—and emotional support, such as the comfort and encouragement provided by family and friends. Who provides support to the person left to die in a nursing home or hospital? Who provides comradeship to a man who retires after working all his life? Who helps the student making the transition from high school to college, the woman embarking on a second career, the teenager undergoing biological and social change, and the elderly widow trying to overcome her grief and find a place for herself.

Societal myths—particularly regarding older people—are partly responsible for the inadequacy of our support systems. A study conducted for the National Council on the Aging found that people over 65 thought more positively of themselves than did the public at large. For example, whereas 68 percent of older people saw themselves as "very bright and alert," only 29 percent of the public thought older people possessed these admirable qualities. Sixty-three percent of older people maintained that they were "very open-minded and adaptable," in contrast to the public's 21 percent.[3] The problem is that expectations can yield to self-fulfilling prophecies. If enough people believe that older people cannot learn to work or adapt,

---

[2] For a discussion of the family and its support systems, see *Towards a National Policy for Children and Family,* a report of the Committee on Child Development, National Academy of Sciences, National Research Council, December 1976.

[3] *The Myth and Reality of Aging in America,* a study conducted for the National Council on the Aging, Inc., by Louis Harris and Associates, Inc., April 1975, p. 53.

then older people begin to believe that, and act as if they are of little use.

Closely related to the issue of public image is public policy, which, in turn, is one aspect of the broader issue of intervention. Even keeping in mind former Senator Lister Hill's aphorism, that "We must combine the vision of youth with the wisdom of age in making public policy," we still have the questions: Who should intervene in the coping process? When? How? What are the consequences for the individual and for society? With public and private resources growing more scarce, on what basis do we determine priorities, knowing full well that by intervening on behalf of one group we discriminate against another? For example, a lower retirement age benefits younger workers at the expense of the 55- to 65-year-old person. Spending educational dollars on the very young limits the possibilities for those in other age groups.

If we acknowledge that coping, defined by David Hamburg as "efforts at adaptation under difficult circumstances,"[4] involves an appropriate response from the individual as well as from society, we are still left with questions concerning the nature of the response and the kind of behavior that characterizes a successful coping strategy. Dr. Hamburg provides us with some clues based on studies conducted over the last two decades on coping with life-threatening experiences. As a result of observing how patients afflicted with severe burns and severe poliomyelitis and how parents of children with leukemia reacted to these hardships, he suggests a general model on coping.

The model consists of tasks and strategies. Tasks are defined as requirements for effective adaptation, while strategies are the ways in which these requirements can

---

[4] Hamburg, "Coping Behavior in Life-Threatening Circumstances," p. 14.

be met. What makes the tasks difficult to accomplish is that physical disabilities have social and psychological consequences. Patients may believe that they are responsible for their injuries and thus feel guilty. Separation from family and friends creates loneliness, which may lead to depression and self-pity. Sexual difficulties, concern for the future, and feelings of hostility, personal rejection, or inadequacy also may occur. All of which means that just as the transitions themselves involve the interaction of the biomedical and social sciences, the coping process itself is an interdisciplinary phenomenon. As David Hamburg has written, "The study of adaptation links biological sciences, social sciences, and clinical professions."[5]

Dr. Hamburg has identified four requirements of adaptation. The individual under stress needs to accomplish the following tasks:

1.   Contain the distress within tolerable limits.
2.   Maintain self-esteem.
3.   Preserve interpersonal relations.
4.   Meet the conditions of his new environment.

For each of these tasks, multiple strategies are employed. These strategies reflect the different circumstances surrounding each life-threatening situation as well as cultural influences.

Containing the distress within tolerable limits is a gradual process. At first, most patients keep a tight lid on their emotions. They refuse to recognize their illness or, having recognized it, refuse to have any feelings about it. Among those suffering from polio, more than half of the patients initially denied they had polio or were skeptical about the diagnosis. Over a period of days, weeks, or months, the patients progressed from denial to avoidance to acceptance. They sought infor-

---

[5] Ibid., p. 24.

mation about their illness, discovered what needed to be done, and proceeded to act appropriately.

What did they do? Their hopes rose when they observed recovery in others and shared experiences with other patients. Thereafter they provided encouragement to new patients. They maintained their self-esteem by coming around to the belief that they were doing everything feasible to ensure a complete recovery; by thinking of things they did well and for which they were respected; by participating in their own treatment; by developing a sense of identity that rested on nonphysical attributes; by immersing themselves in work and in self-rehabilitation projects; by taking satisfaction in the fact that they, and not their children, were injured; and by strengthening their ties with family members and other people close to them.

The importance of social support cannot be overestimated. Studies show that people under stress seek attachments with groups of people or with individuals. The patients need to be wanted by their families and by the community at large, e.g., to be viewed as a symbol of strength. As a result of their reactions, David Hamburg concludes: "The effectiveness of coping behavior is strongly related to the feeling that one's presence is not only valued by significant other people but is virtually indispensable to them."[6]

Another strategy that proved effective with the polio patients was the formulation of intermediate goals. The most useful goals were those that had a relatively short time frame (weeks or a few months), provided achievable goals, and gave the patients reasons to believe that they could achieve them. As might be expected, the attainment of these goals enhances the patient's self-

---

[6] Ibid., p. 17.

image and, thereby, provides a further stimulus to his or her development.

A study of parents of leukemic children revealed that parents used many of the same strategies in coping with their stress as did the burn and polio victims. Like these victims of serious injury, the parents initially could not believe the diagnosis or the full meaning of the diagnosis. Instead, they looked for solace in other medical opinions and cures. Once parents overcame the hurdle of acceptance, some experienced feelings of guilt, blaming themselves for a late diagnosis, while others searched for a meaning, religious or otherwise, in the tragedy.[7]

Many parents found it reassuring to reminisce about happy experiences. Also helpful were the following approaches: gathering information on the disease, participating in the treatment, drawing closer to parents sharing similar experiences, and focusing on living for today as something they could manage.

The similarity in coping strategies adopted by the victims and the parents led Dr. Hamburg to identify elements in the process of coping with stress. These elements constitute a recommended course of action that will likely take place over a period of months. Although by no means comprehensive, the following list suggests what an individual needs to do to adapt effectively.

1. Regulate the timing and dosage of acceptance, recognizing that movement from avoidance to recognition is a gradual process.
2. Deal with one crisis at a time.
3. Seek information from various sources on the tasks required for effective adaptation.
4. Develop expectations for progress, emphasizing the element of hope.

-----

[7] Many parents of mentally retarded children behave in a similar manner.

5. Formulate goals that are attainable.
6. Rehearse behavior patterns; practice first in safe situations.
7. Test behavior in actual situations, beginning with behavior that involves no more than a moderate risk.
8. Appraise the reaction to this behavior.
9. Try more than one approach; keep options open.
10. Make a commitment to a promising approach and pursue that approach vigorously; prepare a contingency plan as a buffer against disappointment.

The ideal situation would be for the individual to be aware of the possibilities inherent in various coping strategies. Particularly significant is that he realize the benefits to be gained from cultivating personal relationships with family, friends, and others important to him. He needs their guidance and evaluation of progress so that he can set his priorities for the future.

For individuals to acquire the coping skills described above and for society to recognize its obligations in understanding the difficulties involved in transitions and in the coping process, more research is needed on coping behavior in different situations throughout the aging process.

Will such research help? There is good reason to believe so. Most people cope well with adversity. A study of 20 adults suffering from severe burns, conducted from one to five years after initial hospitalization, found that 70 percent had adjusted well.

More than human resiliency is involved, however. We know that family support can make all the difference in the world in effecting a smooth transition. We also know what the absence of such support can do. Matilda Riley and Joan Waring report studies showing that widows "have fewer contacts with children, greater unhappiness, and higher rates of suicide and death than do married persons of similar age."[8]

What gives us hope is that people and societies do change. The impact of transitions is neither inevitable nor irreversible. As we have seen, some of our attitudes toward death have changed for the good. Furthermore, although support systems are lacking to meet many of our needs, the establishment of multipurpose centers for senior citizens, social organizations that help alcoholics and those who are overweight, and lifelong learning opportunities for adults offers some basis for optimism.

The individual coping with stress cannot be successful if he or she must operate in isolation. A person needs help, in the form of support from individuals and from society at large. In addition, differences in coping in a small town *versus* a metropolitan area attest to the importance of the physical environment as a factor in coping successfully.

The nature of outside assistance raises questions about public priorities and intervention in general in the aging process—just two of many issues pertaining to the human life cycle that merit further study.

---

[8] Riley and Waring, "Age and Aging," p. 391.

6 Areas for Further Exploration

The preceding chapters lead us to three conclusions about the study of major transitions in the aging process. First, human development is anything but a smooth process. Second, human development is a complex issue, with biomedical, social, environmental, historical, and other factors affecting our behavior throughout our lifetimes. Consequently, an interdisciplinary approach provides the most appropriate focus to the study of the human life cycle. And third, our knowledge about the aging process and how this knowledge can be used to improve our daily lives is limited. Indeed, on most issues we have barely scratched the surface.

Since the possibilities for future study are really endless, we shall highlight some issues that are basic to an understanding of major transitions.

---

[1] Birren and Renner, "Developments in Research on the Biological and Behavioral Aspects of Aging and Their Implications," p. 49.

Foremost among the possibilities is more detailed study
of transitional periods. We have tried to show that each
transitional period has its own characteristics. We need
more information about the events that take place in each
period, how these events relate to one another, what
constitutes normal behavior and what constitutes deviant
behavior, and the role of values and ethics in helping us
cope with problems. Also helpful would be an analysis
of the relationships between transitional periods and the
events that occur in each period. Is there any relationship, for example, between early puberty and the timing
of menopause? As methodological tools in analyzing
transitions, we could use more studies of cohort behavior
as well as clinical observations of human behavior under
different situations.

The mid-life period—say, the 40 to 65 age bracket—
represents a particularly fruitful area for study. As described briefly in Chapter IV, there is a concatenation of
events in the middle years, including metabolic changes,
the reemergence of biology as a negative influence, retirement, an empty nest situation, a change in aspirational
levels, the reality of death, a decline in a man's sense of
personal control, and a review of one's accomplishments
in life. Moreover, in some ways, the middle years have
different meanings for men and women, with women becoming more assertive and gaining in confidence. In
short, a number of important events and serious problems are associated with this transitional period, which
until now have not received appropriate attention.

We also need case studies of successful transitions and
instances where outside intervention proved decisive.
And, as we noted earlier, an extensive comparative
analysis of coping strategies used in different situations
by different groups of people would be helpful.

The issue of intervention implies continuing to explore
alternatives concerning public policies in the areas of

education, leisure, and work; health, housing, transportation, and educational services for the elderly; and developing new roles for those forced into early retirement. That negative attitudes toward aging are acquired early in life would indicate that a program of intervention in this area should begin in childhood and apply to all ages.

This book has discussed issues related to the human life cycle in the United States in 1976. What can be learned from other cultures about transitional events, periods, and roles and the process of coping?

Increased understanding of the human life cycle could be gained by conducting research on the following specific concerns:

1. What is the proper age for becoming parents? For males and for females?
2. Can the large numbers of elders in our population assist parents in child rearing?
3. What are the requirements of middle-aged men or women for recreation and rest?
4. What should human settlements be like to help people cope successfully with life course transitions?
5. What values established before puberty are changed at maturity and in later life?
6. What are the characteristics that are often stated as "mellowing" in human beings?
7. What is the evidence for believing that a healthy childhood is followed by a healthy middle age and elder years?
8. What does paternal and maternal "bonding" with their infant at birth signify for the family's success?
9. How do conflicting values held by society's institutions affect human cultural adaptation?
10. What new arrangements are needed for the care of the dying?

Insight into the aging process can also be derived from the study of:

- The impact of transitions on different groups within the same cohort, such as the affluent, the poor, and minorities.
- Themes and characteristics that relate to every age interval, such as human suffering, aggression, and moral development.
- "Sleeper effects."

It would be helpful to know the answers to the following questions:

1. What public and private organizations have been especially successful in assisting people in coping with difficult transitions, such as retirement? To what can their success be attributed?
2. What are the implications, for study and research, of classifying the interdisciplinary linkages of the biomedical sciences and the social sciences as subdivisions of an enlarged field of behavioral science?
3. What are the characteristics of transitions in the educational process?
4. What are the most effective means for furthering the general public's understanding of transitions in the human life cycle?

Finally, there is a pressing need for a theoretical approach that will integrate knowledge from all the pertinent disciplines. The 1976 Aspen conference represented only the beginning. Research is needed to identify the analytical tools used by each discipline and their applicability to the study of the aging process.

Selected
Bibliography

1. Ariés, Philippe. *Centuries of Childhood: A Social History of Family Life.* New York: Vintage Books, 1962.

2. Baltes, Paul B. "Prototypical Paradigms and Questions in Life-Span Research on Development and Aging." *Gerontologist* 13 (1973): 458-467.

3. Baltes, Paul B.; Baltes, Margaret M.; Hoyer, William J.; and Labouvie-Vief, Gisela. "Operant Analysis of Intellectual Behavior in Old Age." *Human Development* 17 (1974): 259-272.*

4. Baltes, Paul B.; Barton, Elizabeth M.; Plemons, Judy K.; and Willis, Sherry L. "Recent Findings on Adult and Gerontological Intelligence: Changing a Stereotype of Decline." *American Behavioral Scientist* 19 (1975): 225-236.*

5. Baltes, Paul B., and Schaie, K. W. "On Life-Span Developmental Research Paradigms: Retrospects and Prospects." In *Life-Span Developmental Psychology: Personality and Socialization.* Edited by Paul B. Baltes and K. W. Schaie. New York: Academic Press, 1973.

6. Baltes, Paul B., and Schaie, K. Warner. "On the Plasticity of Intelligence in Adulthood and Old Age: Where Horn and Donaldson Fail." *American Psychologist* 31 (1976): 720-725.

7. Baltes, Paul B., and Willis, Sherry L. "Toward Psychological Theories of Aging and Development." In *Handbook of the Psychology of Aging.* Edited by James E. Birren and K. W. Schaie. New York: Van Nostrand Reinhold Company, 1977.

8. Bellamy, D. "Hormonal Effects in Relation to Aging in Mammals." *Symposia of the Society for Experimental Biology* 21 (1967):427-453.

9. Bengtson, Vern L., and Cutter, Neil E. "Generations and Intergenerational Relations: Perspectives on Age Groups and Social Change." In *Handbook of Aging and the Social Sciences.* Edited by Robert H. Binstock and Ethel Shanas. New York: Van Nostrand Reinhold Company, 1976.

---

* Background reading for the conference

10. Birren, James E., and Renner, V. Jayne. "Developments in Research on the Biological and Behavioral Aspects of Aging and Their Implications." Paper presented at the Institute of Medicine, National Academy of Sciences, May 18, 1976. Mimeographed.*

11. Brim, Orville G., Jr. "Theories of the Male Mid-Life Crisis." *The Counseling Psychologist* 6 (1976): 2-9.*

12. Erikson, Erik H. "Life Cycle." Reprinted from *International Encyclopedia of the Social Sciences.* New York: The Macmillan Company and the Free Press, 1968, pp. 286-292.*

13. ————. "Reflections on Dr. Borg's Life Cycle." Essay presented at the symposium on "Human Values and Aging." Case Western Reserve University, Cleveland, Ohio, November 10, 1975.*

14. Gordon, Chad, and Gaitz, Charles M. "Leisure and Lives: Personal Expressivity Across the Life-Span." In *Handbook of Aging and the Social Sciences.* Edited by Robert H. Binstock and Ethel Shanas. New York: Van Nostrand Reinhold Company, 1976.

15. Gregerman, Robert I., and Bierman, Edwin L. "Aging and Hormones." In *Textbook of Endocrinology.* Edited by Robert H. Williams. Philadelphia: W. B. Saunders, 1974.*

16. Gusseck, D. J. "Endocrine Mechanisms and Aging." *Advances in Gerontological Research* 4 (1972): 105-166.

17. Hamburg, Beatrix A., and Hamburg, David A. "Stressful Transitions of Adolescence: Endocrine and Psychosocial Aspects." In *Society, Stress and Disease.* Edited by Lennart Levi. London: Oxford University Press, 1975.*

18. Hamburg, David A. "Coping Behavior in Life-Threatening Circumstances." *Psychotherapy and Psychosomatics* 23 (1974): 13-25.*

19. Hamburg, David A.; Hamburg, Beatrix A.; and Barchas, Jack D. "Anger and Depression in Perspective of Behavioral Biology." In *Emotions—Their Parameters and Measurement.* Edited by Lennart Levi. New York: Raven Press, 1975.*

---

* Background reading for the conference

20. Huston-Stein, A., and Baltes, P. B. "Theory and Method in Life-Span Developmental Psychology: Implications For Child Development." In *Advances in Child Development and Behavior*. Edited by H. W. Reese and L. P. Lipsitt. Vol. II. New York: Academic Press, 1976.

21. Lerner, R. M., and Ryff, C. D. "Implementation of the Life-Span View of Human Development." In *Life-Span Development and Behavior*. Edited by P. B. Baltes. Vol. I. New York: Academic Press, 1977, in press.

22. Neugarten, Bernice L., and Hagestaat, Gunhild. "Age and the Life Course." In *Handbook of Aging and the Social Sciences*. Edited by Robert H. Binstock and Ethel Shanas. New York: Van Nostrand and Reinhold Company, 1976.

23. Riley, John W., Jr. "What People Think About Death." In *The Dying Patient*. Edited by Orville G. Brim, Jr., Howard E. Freeman, Sol Levine, and Norman A. Scotch. New York: The Russell Sage Foundation, 1970.*

24. —————. "Death and Bereavement." Reprinted from *International Encyclopedia of the Social Sciences*. New York: The Macmillan Company and the Free Press, 1968, pp. 19-26.*

25. Riley, Matilda White. "Age Strata in Social Systems." In *Handbook of Aging and the Social Sciences*. Edited by Robert Binstock and Ethel Shanas. New York: Van Nostrand Reinhold Company, 1976.

26. Riley, Matilda White, and Waring, Joan. "Age and Aging." In *Contemporary Social Problems*, 4th ed. Edited by Robert K. Merton and Robert Nisbet. New York: Harcourt Brace Jovanovich, Inc., 1976.*

27. Sahlins, Marshall. *The Use and Abuse of Biology: An Anthropological Critique of Sociobiology*. Ann Arbor: University of Michigan Press, 1977.

28. Wilson, Edward O. *Sociobiology: The New Synthesis*. Cambridge: Harvard University Press, 1975.

Part II
The Turbulent Years

Edited by
John Scanlon

7

Introduction to Part II

John Scanlon

There are now more than 45 million young people in the United States between the ages of 14 and 24. They are the second largest comparable age group, comprising more than a fifth of the total population. They are really made up of two subgroups, those born toward the end of the unprecedented baby boom that followed World War II, and those born during the "baby bust" period that began in 1961 when the birthrate started to decline.

The life experiences of the two subgroups have been significantly different. The members of the baby boom cohort, by virtue of their sheer numbers, have been in fierce competition with one another for college admission, jobs, housing, and career advancement. The younger members of the age group, those in the "baby bust" cohort, are far fewer in number. They are finding it much easier to get into college, and when they get older, the competition among them for other social and economic benefits will continue to be milder than it has been for the "baby boom" cohort.

The life experiences have also been different in other ways. Many of the older members of the age group were in their teens during the turbulent decade of the 1960's. They were the generation that protested, often violently, against racial prejudice, the depersonalizing aspects of mass education, and the war in Vietnam. Many became disillusioned with their elders and with society. They rebelled against all forms of authority. They turned away from the religions they were brought up in, tuned in on rock music, and "turned on" with drugs and sexual promiscuity. Some of them joined cults in search of the feeling of belonging that they could not find in traditional groups of young people.

Venereal disease is not a new problem among young people in the 14-to-24 age group, but in the 70's it became acute. For example, between 1970 and 1978, the incidence of gonorrhea among female members of the age group more than doubled. The trend is shown in Table 1 on the following page.

63

Table 1
Venereal Disease Among Young People
(1970 to 1978)

Gonorrhea

	Rates per 100,000 People		
Age Group 15 to 19:	Male	Female	Total
1970	960.1	609.5	782.2
1978	977.6	1,481.7	1,228.9
Age Group 20 to 24:			
1970	2,514.5	740.9	1,541.5
1978	2,409.9	1,568.7	1,977.6

Primary and Secondary Syphilis

Age Group 15 to 19:			
1970	18.8	19.8	19.3
1978	16.5	12.7	14.6
Age Group 20 to 24:			
1970	55.6	27.1	40.0
1978	46.0	16.4	30.8

Source: Center for Disease Control, Public Health Service, U.S. Department of Health, Education, and Welfare.

As a result of some of their bizarre behavior, young people were constantly in the news, and the word "teenager" became a symbol of what seemed to be wrong with the younger generation. (It is interesting to note that the term "teenager" did not come into general use until after World War II. It was not included in Webster's Collegiate Dictionary until 1956.) Hundreds of articles were written about young people in the popular magazines, many of them addressed to perplexed parents. One, for example, was entitled "Everything You Want to Know About Teenagers (but Are Afraid to Ask)". The perplexity of parents in dealing with adolescent children is illustrated by the quotation on the following page, and the perplexity of adolescents in coping with the problems of growing up is portrayed in the cartoon on page 66.

In recent years there has been a lively interest among scholars from many disciplines in the problems, concerns, and behavior of youth. Scores of books and journal articles were devoted to such matters.

It cannot be said, therefore, that the participants in the conference on which this report is based were plowing new ground. What made the

Aspen conference different from most other approaches to the subject was its multidisciplinary focus. Scholars and practitioners from more than twenty different disciplines brought their own individual knowledge and experience to bear on the subject, and engaged in a lively discourse that frequently crossed disciplinary lines.

At the outset, the participants were asked to consider the interaction between biomedical, psychological, economic, and social factors on the lives of young people between the ages of puberty and 25. Since most of the Census Bureau data on that stage of the human life cycle

---

Adolescents

- They are said to be "uncommunicative" but it costs me $52 a month for calls on the telephone used by my daughter.
- They complain that they can't get a job if they have long hair. (They should know how hard it is to get one if you have grey hair.)
- They complain and complain even though they can't remember what prices were 20 years ago.
- They are going to have a serious problem some day when they want to tell their children what they had to do without.
- They reject all the older generation's values, beginning with spinach.
- They can't wait to reach the age when they can do as they please, and when they do, they get married.
- There are adolescents so naive that they believe they will never be as naive as they believe their parents to be.
- Adolescents get to be unbearable when they reach 50.

Aldo Cammarota in *Vision,* 16/30 June 1979, p. 60
(translated from the Spanish)

---

covers the age groups 15 to 19 and 20 to 24, these were considered to be the parameters of the demographic data on the age group under discussion, but when some of the participants pointed out that the age of puberty now occurs earlier than it used to, it was agreed that the problems and concerns of 14-year-olds would not be excluded from the discussion.

The overall goal of the conference was to advance continuing communication between disciplines on youth as a stage in the human life cycle. The specific objectives were:

- To identify the major characteristics of, and relationships between,

Emmy Lou

12-12   © 1979 United Feature Syndicate, Inc.

**"You'd think school would have a required course
on how to deal with reality!"**

(Reprinted by permission.)

family, education, work, health, and leisure as these aspects of living affect young people in the age group.

- To identify and gain greater understanding of sex differences, and of important biomedical and socio-behavioral factors that are especially evident in young people, and the relationships between these and other factors evidenced in earlier and later stages of the life cycle.
- To consider and clarify the policy implications arising from a better understanding of the biomedical and social aspects of growing up.
- To suggest programs of action by government agencies, organizations, institutions, individuals, and support systems, that will help young people to cope more effectively with this transitional adolescent period.

The conference opened on Monday, August 6, and continued through Friday, August 10. Some of the speakers came with prepared texts, but others spoke from notes or outlines. At the end of each presentation there was a discussion. All of the proceedings were recorded and transcribed. What follows is an edited version of the transcript, supplemented by the conference reporter's own notes.

— John Scanlon

8 Historical Treatment of the Age Group

John Demos

On occasions such as this one, I think history is supposed to provide a sort of running start, and that's what I will try to do.

At the outset I'd like to make one or two remarks of a prefatory and perhaps even slightly apologetic nature.

First, in trying to range through fifteen centuries of history and covering what is a substantial age group, I will necessarily have to oversimplify along the way. There are many individual experiences of passage through the years 14 to 24 and perhaps many group experiences which I will scarcely be able to touch on at all, and I apologize for that. I think what I'll be trying to do is to trace a kind of mainstream line of change over time even though admittedly the mainstream itself is rather vague and ill-defined at many points along the way.

The second prefatory remark — and it's a kind of apology, I guess — is that I have before me a prepared text which may lend a greater degree of formality to my remarks than I might otherwise wish. This is not my regular mode of operating. However, I think it may prove more economical and efficient and expeditious for me to speak to you from a set of prepared comments, partly because of the enormous scope and diffuseness of this subject and its historical aspect.

Nonetheless, I hope you'll bear with me as I go along in this set piece and I do hope and trust that there will be sufficient time left after I'm done for a good deal of comment and discussion around the table about what it all means.

One more thing. I won't really try to make any significant approach to the present-day situation. I'm assuming that that's what the rest of the conference will be about, but what I am offering will throw some interesting light on the current-day situation.

Okay. Psychologists, psychiatrists, anthropologists, sociologists and all kinds of other ologists, not to mention many among you who are representative of the world of policy and action, have all been accustomed for a long time to take careful account of the factor of age in the stages of life, of developmental patterns and sequences.

Historians like me by contrast have come to appreciate such things only quite recently. Twenty years ago I doubt that you could have found a single certified working historian to talk to you on the subject of adolescence. Indeed, the very suggestion would have seemed bizarre if not ludicrous to all of my historical brethren.

Now you could certainly find quite a number of likely candidates, some of them I would confess better than I. The study of the life course and its constituent parts like adolescence has actually become a rather hot number in contemporary historical scholarship.

The trend was begun with the appearance in 1960 of an extraordinary book by the French cultural historian, Philippe Aries, published in English under the title of *Centuries of Childhood*. I expect some of you know it.

There is no doubt that in terms of popular usage and also in terms of its substantive meaning and connotation, adolescence is peculiarly a twentieth century term. One might even call it a twentieth century discovery.

Aries's work quickly raised up a host of followers among historians on both sides of the Atlantic. By now histories of childhood also have accumulated in fairly large quantities. There is even a journal by the name of *The History of Childhood Quarterly,* and the tide is still flowing.

Meanwhile, some scholars have moved out from childhood to other sectors of the life course. For example, I can alert you to a new and still building miniwave of studies on old age in history. Even the terminal points in life are coming into scholarly vogue. Aries, among others, has written recently on the history of death and just last year there appeared the first full-scale history of childbirth under the catchy title *Lying In*.

Seriously, the one part of this territory which remains largely uncharted is the history of middle age, and I think perhaps that has something to do with where most historians are in their own life course.

If I sound a little ironic in describing these scholarly trends, I don't really mean to disparage them. In fact, I am myself very much caught up in them and more than that I think they do constitute one of the most exciting and challenging branches of current historical scholarship.

One might well say, and I suspect that many of you might well say, it's about time historians noticed the importance of age in human experience.

Our focus today is, of course, not age in general but the age of adolescence in particular. Now I'd like to begin the main part of my remarks with a fragment of a professional autobiography.

I wrote one of my first graduate school seminar papers on the subject of adolescence in American history. That was, I think, 1963. I made one central discovery then that seemed fairly earth-shattering to me at the time, although now it looks obvious, namely the discovery that adolescence as we know it is a relatively new kind of experience.

Correction — I should say that it is a relatively new idea because, in fact, my research up to that point was largely about ideas. Briefly stated, my conclusions were the following: The modern currency of the idea of adolescence owes much to the efforts of one man, Professor Granville Stanley Hall. Hall's work is now largely outdated and forgotten. If he is remembered at all, it's probably for his role as host to Sigmund Freud during Freud's one and only visit to America.

Freud, as some of you may know, came in 1909 to deliver a series of lectures at Clark University in Massachusetts. Hall was the university's president and the person directly responsible for getting up the invitation.

In his own time, Hall was something of an eminence among American psychologists albeit a controversial eminence. Indeed, his reputation went considerably beyond academic circles for he was among other things a remarkably successful popularizer of psychological ideas.

His magnum opus was a vast two-volume study published in 1904 under the imposing title *Adolescence, Its Psychology and Its Relations to Anthropology, Sex, Crime, Religion and Education.*

The title sounds a bit eccentric and the book reads rather strangely today, but at the time it was a blockbuster. It's not too much to say, I think, that the term adolescence and the ideas which Hall and his disciples subsumed under that term achieved a kind of instant celebrity.

It was almost as if Hall's account of adolescence had filled a gap or a need in social experience, so widely and eagerly was it taken up. It's true, of course, that the word itself was not new. One finds scattered —really very scattered—use of it in earlier writings, and I guess it has a certifiable Latin root.

However, there is no doubt that in terms of popular usage and also in terms of its substantive meaning and connotation adolescence is peculiarly a twentieth-century term. One might even call it a twentieth-century discovery.

Adolescence is, first of all, a matter of biology — an intrinsic development process in a physiologic sense. It is also a matter of psychology insofar as it involves the resolution of internal issues around self and others. Finally, it's a matter of social experience, or rather of social context and cultural definition.

I contented myself in my initial study fifteen or sixteen years ago with trying to unravel the main threads of intellectual development in and around the concept, the various influences on Hall and his students, the way their work was understood and so forth.

Subsequently, I went a bit further in an article I wrote together with my wife, in which we proposed a rough little model of social experience to account for the impact of the idea of adolescence during that particular historical moment around the turn of the century.

Looking back it's easy to see the limitations in that early preliminary effort. Like most such models, I guess, ours has proved much too simple to encompass the full range of pertinent facts. Very recent scholarship has greatly extended information about youthful experience in past times. I hasten to add that the newer work is not by and large my own, and what I'll be doing here involves a certain amount of ripping off of professional friends and colleagues.

I think I should particularly acknowledge a new book entitled *Rites of Passage,* by Professor Joseph Kett of the University of Virginia.

In fact, we don't, any of us, have the last word on this subject yet. Puzzles definitely remain and I would be pleased to hear from any and all of you about trying to clear up the cloudier points.

The task as I see it is to understand the changing nature of adolescent experience through three centuries of American history — and let me emphasize the word "experience." The area of ideas—or more broadly, of perceptions — is relatively easy to run through and has, I think, largely now been done.

What is harder to understand is the situation which lies behind the ideas — which even in a sense generates the ideas. More particularly, we would like to know why Americans of our own century have depended so conspicuously on the concept of adolescence as a way of ordering their personal and social experience.

Did earlier generations have roughly the same experience of adolescence, while calling it perhaps by different names? Or is there something intrinsically new and different about modern experience in this connection? Might it even be possible to speak of the creation of a new developmental stage?

One eminent social scientist recently made the prediction that a couple of centuries in the future our young people will be sexually active and reproductively fecund by the age of eight or nine. I'll bet my own scholarly badge that it doesn't happen.

Let me propose, before going any further, that we regard our investigation as having three component strands and as requiring three different analytic perspectives. Adolescence is, first of all, a matter of biology — that is, an intrinsic developmental process in a physiologic sense.

It is also a matter of psychology insofar as it involves the resolution of internal issues around self and others. Finally, it's a matter of social experience or rather of social context and cultural definition. But in a sense, I suppose these three aspects comprise a kind of spectrum from the most nearly universal and unchanging—that is, the biological ones — to the most highly variable and flexible — that is, the sociocultural ones.

Psychology, I think, lies somewhere in between. However, even biology is not entirely impervious to changing historical circumstances.

There is, for example, some reason to think that the attainment of full physical stature comes at a considerably earlier age now than it did in pre-modern times.

According to one estimate, young men in early 19th century America were not fully grown until they were about 25 years old, whereas most of our young people nowadays reach this point by or even before 20.

Similarly, we do have some fairly good data about the age of *menarche* in girls since the middle of the nineteenth century and this material suggests that *menarche,* first menstruation, has been coming along nearly one year earlier with each succeeding generation. That is, from about 15 or 16 in 1850 to somewhere between 12 and 13, I think, nowadays.

For boys, there seems to have been evidence of a parallel decline in the age of puberty from about 16 to about 13 or 14. From this apparent trend, incidentally, one eminent social scientist recently made the prediction that a couple of centuries in the future our young people will be sexually active and reproductively fecund by the age of 8 or 9. I don't know what this would do to the various kinds of work in which many of you are involved, but anyway — I'll bet my own scholarly badge that it doesn't happen.

In Colonial days, the typical little boy was a sort of miniature model of his farmer father, and likewise the girl in relation to her mother.

The point is, of course, that the data on age of puberty covers only a little more than a century—dating back to the middle of the nineteenth century—and that century happens to have been a time of particularly dramatic improvement in nutrition, in health care and in other factors directly related to physical maturation.

If we try to take the story farther back in time, admittedly the data becomes very fragmentary, but I feel quite certain that the age of puberty in, say, 1650, in most of Europe and America anyway, was not appreciably higher than it was in 1850.

These are matters of some consequence to the history of adolescence, and there may now be other elements of human biology that deserve attention within the same general context.

If so, however, I am not the person to draw them out and the remainder of my comments will involve the second and third of the perspectives which I tried to identify a few minutes ago.

Actually, I shall concentrate largely on the socio-cultural picture with occasional glances toward the changing psychology of adolescence. Without further ado, then, let me crank up my time machine and invite you to join in a little journey into the past, seeking at the outset to evaluate the experience of young people during the first century and a half of American history.

This is a period bounded by the settlement of the continent by Europeans at one end and a movement toward national independence at the other. Say—the period from 1620 to 1735. The single most widely salient fact about society in this period was its pre-modern, pre-industrial character.

The population was distributed through hundreds of village communities and in some cases more isolated homesteads. For a good 90 percent of these so-called colonists, day-to-day experience was shaped by the requirements of small-scale agriculture, and the experience specifically of young people embraced themes and tendencies typical of pre-modern farm youth almost everywhere.

It's clear, for example, that children began from an early age to participate in the work routines of their individual households. Seven- or eight-year-old boys and girls would assume simple responsibilities in the care of domestic animals, in gardening, and for such in-house tasks as spinning, candlemaking and food preparation.

As they grew older, the scope of these responsibilities widened, and presumably the details of the process were as variable as the individuals and families directly involved. But the larger meaning for our purposes seems clear. Children moved toward maturity along a smoothly-surfaced path. Their introduction to adult roles began early, and culturally appropriate adult models were present and visible from the start.

Under these conditions the typical little boy was a sort of miniature model of his farmer father and likewise the little girl in relation to her mother.

As I've already suggested, the key to this pattern was the economic organization of the household, but other factors like demographic ones also played a part. Families were very large by our standards — eight to ten children per married couple born over a period of as much as twenty years.

This meant an implicit blurring of intergenerational lines. Many children could envision their own progress toward adulthood by reference to older siblings variously situated along the way.

These patterns were expressed and reinforced in the values, beliefs, laws and social customs of the culture at large. Social and ceremonial activities no less than work ignored distinctions of age. Church-going, family visiting, even education regularly mixed children and adults together.

Legal usage established no single age of majority. Inheritance came most often at age 21 but might also be arranged within individual families across a wide spectrum from, say, 15 to 25. In some communities orphans could choose their own guardians if they were over 12, in others not until 14 or 16.

Youth was seen as a long period of gradual preparation for adult responsibility. There were few sharp twists and turns along the way. The only sort of personal crisis particularly associated with youth was religious conversion.

The age of criminal responsibility was also variously determined. Moreover, the significance of all numerical benchmarks is unclear, since many people in this culture did not know how old they were and others seemed not to have cared.

Now these circumstances imply a rather loose and flexible view of the maturation process, and furthermore, when one looks to literary and documentary materials the same impression holds. When authors considered the life course they did so most often in terms of a four-stage model.

Childhood was customarily understood as lasting from birth to seven or eight years. Youth, which was next, ended at about thirty. Old age began at sixty. Middle age was rarely even mentioned, apparently because the years between thirty and sixty seemed simply the full realization of personhood, rather than another stage of life.

In a sense, childhood and youth on one side, and old age on the other, were deviations from what was seen as the mid-life norm. It is, of course, youth which primarily concerns us here, but the picture of youth which emerges from the literature is short on specifics and vague and lukewarm in tone.

By and large, youth was seen as a long period of gradual preparation for adult responsibility. There were few sharp twists and turns along the way. The only sort of personal crisis particularly associated with youth was religious conversion. Occasionally conversion came around the transition from childhood to youth, as young as seven or eight years old. More commonly it came near the end of youth — that is, in the mid-to-late twenties.

But these connections were very loosely established and the experience of conversion itself was rather hard to pin down. Puberty was not especially noticed, and its connotations were those of gradual accession to power and effectiveness. Nowhere was there any clear sense of youth as an awkward age with its own special confusions and vulnerabilities.

Please consider what has been said so far as a rough overall sketch of young people's experience in the Colonial area and as a backdrop against which to set off the increasingly different experience of nineteenth-century American youth.

After 1800 the time of life previously characterized as youth became increasingly disjunctive and problematic. Social circumstances and I think psychological ones, too, combined to surround the passage from childhood to adult status with new elements of stress.

The nineteenth century can be seen as a time of general confusion or at least of reshuffling in social expectations as to age-appropriate behavior.

The adolescent world of our own times was still some distance off, but it was more and more difficult to sustain the old attitudes of equanimity toward youthful experience. One trend of central importance involved the movement of young persons, especially boys in their teens, in and out of their parental homes.

If we look at autobiographical documents from that period, we find a sequence of comings and goings most often on a seasonal basis. Boys of all ages were expected to be at home during the summers in order to help their families with the heavy work of farming. But in winter and spring they might well go off to look for work elsewhere in activities such as lumbering, fishing, or clerking in the commercial towns that were growing so rapidly during this period.

Sometimes too they would go to school for a while, though education was for most people a very irregular process.

The result of all this was a new and complex web of youthful experience — in part autonomous and self-contained and in part still bound by taut cords of blood and community.

One scholar has recently characterized this trend by the term "semi-dependency." From the standpoint of the youth themselves the change must have been very unsettling. For a long period, a young man was virtually on his own, obliged simply to send back money to his family. Yet, the old lines of familial subordination still applied. The young were expected, in all things, to defer to their elders and especially to their parents.

Let me illustrate the substance of these norms with a single quotation taken from a letter by a father in a small town in Maine to his twenty-five-year-old son who was then living temporarily in Boston:

> I am afraid that you do not exercise enough to keep you in good health and spirits, and to remedy that difficulty as much as possible I should like to have you retire to rest by nine and rise certainly by five in the morning, and take a long walk in the morning air at least one hour every morning when it is suitable weather.
>
> Don't neglect this because I think this is most important.

Such instructions might, I think, have been quite plausible in a different cultural setting, basically an earlier one, even for twenty-five-year-old sons. But in this case, the young man had been living on his own for a good long time.

In fact, the period of the nineteenth century can be seen as a time of general confusion or at least of reshuffling in social expectations as to age-appropriate behavior.

In some respects there were new elements of youthful precocity. The average age of religious conversion, for example, seems now to have begun a steep decline. In the excitement of recurrent religious revival there was pressure on the young, especially on boys and girls in their teens, to declare a personal experience of salvation.

Politics was another form of excitement in the life of the new nation. By the 1820's and '30's, political debate and struggle had become major preoccupations for Americans everywhere, and if we look to political activity on the local level we often find children and teenagers present as spectators or even as participants in limited ways.

Moreover, we know that the young continued to be exposed to the emotional concerns of their elders. Autobiographical accounts show their attendance, for example, at various deathbed scenes, and here is a brief representative case.

> Then my kind uncle opened up one of the coffins and let me see how decayed the body had become, and told me that my brother Edward's body

was going to decay in a similar manner and at last become like the dust of the earth.

This may seem shocking to us but, in fact, I think it serves vividly to measure the distance in certain respects between their world and our own.

Now in a setting where four-year-olds were encouraged to examine decaying corpses while twenty-five-year-olds were advised by their fathers as to the minute particulars of their daily experience — in such a setting we may feel that age norms of all kinds were rather loosely maintained, and this impression certainly has much truth in it.

It was in education that the tensions and conflicts around issues of authority over youth revealed themselves most fully. In schools and colleges there was a minute concern with rules and regulations.

The pattern — or rather the non-pattern — can be seen with particular clarity in the records of school attendance. In virtually all the schools of the period students of widely different ages were mixed up together, usually in a single classroom. For example, a little academy in Massachusetts in 1831 reported a total enrollment of 20 pupils and gave their ages as follows: One student each of 12, 13, 14 and 15 years old. Two of 16, three of 17, one each of 18 and 19, two each of 20 and 21, one of 23 and one of 25.

There is no record of the teacher's age in this case but given what is known of similar schools it could well have been as low as 20. We can imagine that the regular maintenance of order and authority might have become quite problematic under such conditions.

In fact, it was in the field of education that the tensions and conflicts around issues of authority over youth revealed themselves most fully. On the one hand we find in the schools, and especially in the colleges, of this period a minute concern with rules and regulations. Most colleges put out elaborate manuals of conduct which students were expected to learn by heart.

These manuals covered a broad range of topics: study hours, dress, forms and places of recreation and above all, attitudes of deference towards authority.

At Yale, for example, the rules said that students must not wear hats within five rods distance of tutors and eight rods of professors.

Everywhere there was an elaborate system of punishment. Yet, it was one thing to declare the rules and even to punish infractions of them, but it was quite another thing to secure from the students underlying attitudes of compliance. And, when we look at the record of student life and behavior in these colleges, we gain a very different impression.

The early nineteenth century was the great age of student rebellions — the tumultuous events of the 1960's notwithstanding. Fistfights between students and professors, riots over food and lodging conditions, not to mention frequent duels among the students themselves— such troubles became a commonplace of nineteenth century college life.

In fact, as early as the 1760's, Yale had been wracked by years of continuous rebellion which eventually forced the college president to resign. At Brown University in the 1820's students were given to stoning the president's house on a nightly basis.

By the middle of the nineteenth century, urban youth was increasingly characterized by the not young as a social problem.

At the University of Virginia in the 1830's there was a particularly long and violent insurrection culminating in the murder by students of a senior professor. The list could easily be lengthened if time and interest permitted, but the basic point should already be clear. The strenuous disposition to force compliance on the one side, and the urge to rebel on the other, suggests that all authority relations were somehow undermined.

Circumstances had reduced the capacity of the older generation to provide the young with leadership, with guidance, with help of various kinds. The young, while still expected in theory to follow their elders, were in fact faced with new and difficult choices to be made largely on their own.

This factor of choice was now emerging as essential to a broad range of youthful experience —.choice of occupation, choice of residence, choice of values, of friends, of spouse. Questions which in traditional communities had been more or less decided *for* young people were increasingly matters of individual consideration.

And, where the alternatives were immediately juxtaposed, such choices and questions could be painful indeed. The predicament of a boy named Smith, age 15, during a series of intense revivals in upstate New York was fairly typical; again I quote:

> The whole country seemed affected by religious commotion and great multitudes united themselves with the different religious parties which created no small stir among the people, some crying, "Lo, here," and others, "Lo, there." Some were contending for the Methodist faith, some for the Presbyterians and some for the Baptists.

Now the boy who recalled and wrote about this experience later on felt great uneasiness and confusion and spent many hours in the woods near his home praying for guidance.

I've said that the predicament was typical, but its resolution in this case was rather unusual. Eventually the boy received his answer. An angel appeared instructing him not to join any of the competing sects but instead to found a newer and truer religion of his own.

And thus was born that quintessentially American church known to later generations as the Mormons, with young Joseph Smith in the role of founding prophet.

By the middle of the nineteenth century it was clear that these various pressures bearing in on the lives of youth had a specific locus, the new and bewildering environment of the city. Here the choices, the temptations all converged, and urban youth was increasingly characterized by the not-young as a social problem.

A famous book about children in the 1850's and '60's was entitled significantly, *The Dangerous Classes of New York.* But not all of the young people of the city were openly dangerous. Many were themselves in danger, and in direct response there sprang up a whole new genre of sermons, lectures, published books and essays, the genre of advice to youth.

These materials, whose remnants now lie gathering dust on hundreds of shelves in old libraries across the country, epitomized both the problem of mid-century youth and the dominant cultural strategy for responding to that problem.

The problem in a nutshell was how to choose well and wisely from among the infinite possibilities thrown up by urban life. The response was moral integrity and above all what counselors call "decision of character." Uncertainty and vacillation must give way to willed decision. Here was the central message, and if this message now sounds a little thin and trite, it must then have struck some resonant chords, for it was widely and endlessly repeated.

We come now to the final part of my little historical model, a period covering the last three or four decades of the nineteenth century. This in fact takes us right up to the climax of G. Stanley Hall's career and the dawning of the modern concept of adolescence.

I will speak only briefly about this period for my time is growing short, and the crux of the story is, I think, already clear.

Many sectors of American life in the late nineteenth century were characterized by what one scholar has called "a search for order," and this same impulse did, I think, directly affect the experience of youth. The effort was made in many ways to channel and organize the jarring influence of a generation or two earlier.

Consider, for example, the field of education. It was in this period that our nationwide system of public schools assumed definitive shape. It was then and only then that primary schools and high schools emerged as distinct institutional entities.

What this meant among other things was an end to the old pattern of mixing ages in the schools and henceforth the young would be separated into grades as they passed through their student years.

Henceforth, development was subject to a kind of social processing, with schools as the key institutional device and with other institutions also playing a part.

But the efforts of schools in this connection were complemented by trends internal to families. The earlier forms of semi-dependency whereby youth moved back and forth between family settings and the wider community gradually disappeared.

Henceforth youth was to be a time of fuller and fuller and longer and longer dependency. Demographic change also played a role. The birth rate had been falling throughout the century so that the average size of family was reduced by about half.

The result was a new style of careful, highly self-conscious parenting. Another scholar finds in this period a whole ethos of "intensive family life." Mothers, in particular, were expected to direct all their energies to individualized nurture of their children.

The larger impulse which underlies all this activity, whether in domestic life or education or whatever, was to create a systematically planned environment for the young. Youth might then progress in a more orderly fashion toward the great goal of adulthood, bypassing insofar as possible the perils and pitfalls along the way.

In a sense the outcome of these trends was a broad-gauged standardization of youthful experience. Henceforth development was subject to a kind of social processing with schools as the key institutional device and with other institutions also playing a part.

There were, for example, new organizations devoted to the improvement of both morals and practical skills in the young—the Boy Scouts, the Girl Scouts, the YMCA, the YWCA and so forth.

Adolescent fashions in dress, in recreation, in speech and social mannerisms reflected the same sort of trend. Indeed, it is to this period around the turn of the century that we can trace the appearance of a "youth subculture" with features that are familiar to all of us today.

The effort to standardize the life of young people was ambivalent in motive and complex in result. On the one hand there was some intention to constrain precisely those elements of experience which seemed most youthful — the adventuresomeness, the opportunity to sample a wide range of behavior styles and beliefs and values.

On the other hand, there was implicit recognition that adolescence was a bona fide life stage with its own particular requirements; and insofar as the standardizing tendency brought young people together in specially contrived settings of one sort or another, it greatly accentuated peer-group consciousness.

To be an adolescent was to share with others of similar age not only a developmental position but also a social status. This is to say, in other words, that many adolescents found themselves caught up in what Erikson calls "identity diffusion" of a sort hardly conceivable during earlier periods. The process of deciding who one was, what one wanted to be and how one's particular character would intersect with the social order — here was a labyrinth of personal and social complexities. Perhaps there are other psychological perspectives that one can also bring to bear on this and I'm sure some of you are better qualified than I to canvass them. For the moment, I would simply like to conclude by suggesting what, I suppose, in the broader sense is the moral of this somewhat long and tortuous story, namely, that the experience of people passing through this life stage, ages 14 to 24 — as indeed is true for people in all life stages — is itself in motion through time; historical circumstance, historical pressures, the whole momentum of history in one way or another are likely to change the contours within which each one of us as an individual is able to experience the passage through any particular part of the life course, including this one.

Maybe the term adolescence, and all the connotations the term has for us today, will no longer seem sufficient as time goes on.

One of the things that I have tried to present for you in these remarks is the way in which the concept of adolescence arose, and I leave it as a somewhat less clear, but I still think basically supportable proposition, that not only the concept but in many respects the experience of adolescence was something new that began let's say a century or a century and a half ago. This raises the further question whether, as we move into the future, we may come to still newer and different ways of experiencing passage through this same chronological portion of the life span.

Indeed, the term adolescence is, I suspect, a term that would come most readily to mind for dealing with this period in the life span today. Maybe that term, maybe all the conceptual apparatus, maybe all the connotations the term has for us, will no longer seem sufficient as time goes on. I am reminded in this connection—I wish I had been reminded a few days ago and I would have done some more homework—I am reminded of the essay by Ken Keniston whom some of you may know. Some years ago, it appeared originally in the *American Scholar*. It was called "Youth as a New Stage of Life." In that essay, Keniston, in fact, revived the old term, youth, the term that was, as I said, customary in talking about this time of life in the colonial period, but he revived it in a rather special context. He was suggesting, in effect, that we really need to conceptualize another stage of experience between what we think of as adolescence and adulthood proper.

He developed a rather interesting list of some of the central characteristics, issues, and so forth, of what he called this new stage of youth. Speaking as a historian, I feel that one of the things that seems to happen as history moves along over a long span of time is that our sense of the life course and indeed perhaps of our experience of the life course becomes increasingly differentiated. Where before there were four stages, now there are six or seven or eight. And maybe the number will continue to increase. It seems likely that the process cannot continue indefinitely, but in the near future, anyway, it may be that we will need new ways to think about our experience passing through these years.

Discussion

Leonard Duhl: One thing that's fascinating to me is that I am not sure that all of the young people I have been dealing with have had the same history that you are talking about. As you pointed out, you are talking about middle-class American youth. We often maintain that our history has been an attempt to standardize the experience of youth so that we can bring people from multiple cultures into a relatively standard culture. And yet, as I look at all of the youth that I know now, whether in college, in the streets of Oakland or any other place, I find entirely different histories. The history of their adolescence, whether they are American Indians or blacks or what have you, seems to be very different. I haven't been able to find, anyplace, good histories of any of the other cultures except modern American middle-class. I would like to know where some of this history might be.

Demos: Well, it doesn't exist in printed form where you could readily put your hands on it. One of the difficulties a historian faces is getting all the materials. One thing about middle-class people is that they write, and they have for a long time. They tell you some of what is going on in their lives. They have the time, the money, the leisure, and the access to printing presses to produce the documents on which, alas, most historical scholarship is based. This is not to say, however, that groups a little apart from the middle class are not beginning to be heard from. Recently, for example, history of black family experiences has become a very lively and interesting vein of historical scholarship. There have been several big new books dealing, at least in part, with the black family. I have looked so far without much result for ideas in those books about the life cycle in general and adolescent experience in particular. It's not really there, but I would expect that once the groundwork of understanding about black family life in general is laid down, then we may get some refinements in the direction of studies of things like adolescence.

Alfred Kahn: John, I wanted to raise a similar question. Apart from the ethnicity issue, is there not a social class issue? I have had the privilege of reading Keniston and some of your work. I appreciate the beautiful job you did in summarizing it for us. But I have also read *Youth and History*, by John R. Gillis. There seems to be a story there that has to do with social class and stratification, at least in Europe. Is that applicable here?

Demos: Well, I think there is definitely a sense in which a large part of the impetus to channel and organize and standardize youthful experience began in the late nineteenth century. In short, the founders of the Boy Scouts and the Girl Scouts, for instance, were certainly concerned not only to kind of control and constrain the lives of their own middle-class young people, but also, and probably more particularly, to somehow bring into the fold and reduce to some sort of standardized pattern, the young people of other ethnic, racial, and class groups, too.

I do think that one of the interesting things about the nineteenth century part of the story is the way the middle-class youth in itself seemed dangerous and potentially out of control. But then, I think as the story moved along, the effort became not only to control the middle-class kids but also to find ways to embrace them under the same broad umbrella of organizations—ethnic and lower class youth, and so on.

Paul Bohannan: While we're at this, John, let's broaden our perspective a little bit further, because your description of the early seventeenth century and eighteenth century situation is a variation of a very broad pattern which is recognizable from peasant societies in Africa in the twentieth century, and it may have some resemblance to what it was like to grow up in a hunting and gathering society. Unfortunately, we have very little information on this. However, the Harvard people who are studying the Kalahari Bushmen have made some interesting observations about them. Growing up in that kind of society was, again, very very different if for no other reason than that they lacked peer groups. The type of community—and hence the context for human development—underwent a sea change at the time of the agricultural revolution and another at the time of the industrial revolution. We have a whole variety of patterns in today's world, associated not only with different geographically distinct societies, but with ethnic groups and other culturally distinct part-groups in complex societies.

John Schweppe: John, may I make a comment on your reference to the trend, over the past two centuries, toward earlier maturation? When I went through the Tower of London, the armor I saw there would fit a man about five feet four inches tall. The men who wore the armor were really very small. And I think we can say that in the last 50 or 75 years our general population has increased considerably in stature. One wonders why this has happened. Is it a genetic manifestation? Is it purely a nutritional event? I don't think anybody can answer for sure, because we weren't doing that kind of research back in the 1850's, but I think we can say that nutrition certainly must have played a pretty prominent role.

One other thing you said raises some questions. Many of the families you mentioned had eight or nine children. They would end up having 50 percent or less by the time of puberty, for the simple reason that pneumonia and other diseases killed off a lot of children. Today we are interfering with Darwinian Law in the sense that by the use of modern antibiotics and modern means of medicine, we are preserving children who otherwise would not be able to survive; we are changing the patterns of adaptation. I just leave that open for whatever one wants to think about it, but really, medicine, in a sense, has changed the patterns of Darwinian development of the human being for the foreseeable future, and will continue to do so.

June Christmas: A couple of thoughts occur to me in relation to eighteenth and nineteenth century African-American adolescence. One, even though there aren't many first-hand reports by people in that stage of life, we do have a number of reports of older people who described what life was like and what children and adolescents did. And I wonder if we might not be able, in reviewing some of them—particularly those that appear in written documents and those that came at the end of the nineteenth century in the form of first-hand oral histories—to get some sense of what the historical determinants of slavery have been on present-day African-American life.

The other point that struck me is that the adolescents John Demos talked about were able to make some kind of work choice, but work for the adolescent slave was a given. It did not necessarily provide a sense of accomplishment. It had, in fact, quite a reverse connotation. I also found myself thinking that today there are many adolescents who cannot visualize work as a possibility for the future. It simply isn't available to them. That is a part of present history just as condemnation to work for somebody else was part and parcel of the history of adolescent slaves.

Demos: Yes. You have suggested that the change in the fundamentals of experience for black youth over the past 150 years, from slavery time to the present — the sheer amount of change — must have been fully as great although quite different in its particulars, as has been the case of middle-class youth coming from the small farms to the city or suburban environments, and I think you are absolutely right. It's only recently that historical scholars have begun to discover and use the wealth of material of one sort or another about the black experience during slavery. It's only a beginning, but I hope there will be more of it.

Sidney Werkman: One of the themes that was most powerful to me in the excellent history you gave us, John, was the introduction of order and standardization in the nineteenth century. That seemed to be concurrent with a decrease in any concern with the essence of man. One of the things that is most fascinating to me about our increasingly pluralistic society is what indeed are the deeply shared ideas that allow it to function. It does sound as though the nineteenth century, and perhaps the early twentieth, with their tremendous velocity of change — the scientific revolution, industrialization and all the rest — left us without that long and powerful sense that history has given us of what human nature is. It's as if with the nineteenth century and the early twentieth century we stopped being moralists. Instead, a lot of other people came in — psychiatrists, psychologists, and so forth — to ask questions that had been essentially moral questions from the beginning of civilization.

Demos: I think that the beginning of the story of standardization, which is so central to our society today, is really found way back in the nineteenth century. In fact, I think it could even be argued that the second half of the nineteenth century was the time of greatest pluralistic development in this country — that to a degree, anyway, the various efforts and mechanisms of standardization which were put into place in the late nineteenth century plus the development of a mass-production, mass-consumption system in the twentieth century have narrowed the range of pluralism, at least in contrast to what it was earlier. The mid-nineteenth century was a time when in some very deep-seated way, Americans felt that their world was flying apart in all directions and that all order had broken up, the moral order particularly, as you said. Although this was perceived as a very liberating kind of situation, opening up a sort of success myth, it was also very frightening.

Since that time, I think you could read a lot of things around the theme of an attempt to put some limits on experience. Much of the concern with pluralism — with infinite choice and so on — has always zeroed in on the age group we are talking about.

Terry Saario: Could you expand on that, John, and talk a bit about sexual expectations and whether they became more rigid between the eighteenth and nineteenth centuries or whether they became a little more elastic?

Demos: I think that in the nineteenth century there was a massive increase in sexual and—what sociologists call sex-role—stereotyping. During the period I'm most familiar with—the seventeenth and eighteenth centuries — sex roles, while they certainly existed, were not a matter of terribly rigid or self-conscious experience for most people. Male and female roles, broadly speaking, overlapped at many points. Work was, in many sectors, shared. Social experience was substantially shared and so forth and so on. I'm struck by some of the little things. For example, in seventeenth century court cases you will often find assault and battery—cases of open interpersonal conflict between male and female. A woman fights with a man, over a cow, where part of a stone fence has been knocked down.

When you get to the nineteenth century, all that has suddenly become inconceivable. From that point on, men and women are thought to occupy entirely different spheres, which is really saying they have entirely different characters, that they are different species of being, and all that stuff.

I think there has been some retreat from the extreme sex-role stereotyping of the nineteenth century for a good two or three generations now, and that it has accelerated over the last fifteen years. But there was certainly a massive change in the nineteenth century.

Ernest Bartell: One of the things that occurred to me, John, was that many of the changes in the life experiences of young people during the eighteenth and nineteenth centuries coincided with changes in our economy. In Colonial times, for example, we had a self-sufficient, agrarian economy, and dependency was confined largely within the family. With the growth of exchange and marketing, trade was growing so the dependency shifted somewhat from dependency in the household to dependency on an interdependent kind of market economy, and the youth got caught a little bit in that growth period—that is, they went out of the home to earn a cash income, but still returned some of the income home. That was part of the growth of market specialization. Now, this type of market evolution leads to a certain degree of standardization, too, because the technology that came along then was designed for economies of scale, that is, to realize the most gains from larger-scale production and trade. Some of the social standardizing was quite consistent with the standardizing of jobs to achieve the economies of larger-scale production, which was centered in cities. There was a certain standardizing of economic roles all along which I suspect contributed to the social standardization. In the economic development theory, you move finally to the wonderful age of mass consumption.

Presumably, in a finely tuned, well-working interdependent kind of market economy where all the economies of scale have been achieved, one finally moves to great freedom of choice, and that should restore autonomy at the consumption level. We are seeing some of that, I think, in the kind of recent youth culture at the consumption level. The trouble is that at the producer's level, the dependency thing became very difficult as we moved into a less labor-intensive kind of economy. It was hard to find jobs for young people, so in more recent times, we have had an economic tension between the dependency of young people who are not being productive — not getting into the productive process — while the consumption possibilities for a while, were very, very broad and very great. This left young people facing an identity confusion in finding themselves. So there is a certain economic foundation to some changes in the life experience of young people. Maybe I haven't made it clear, but it's there.

Robert Aldrich: I just wanted to flag an issue that is going to be talked about a great deal more, and that's the issue of cities. I think one of the things that's happened is that most of the American cities were designed by white, middle-class males for the kinds of people who were going to live in them during the latter part of the nineteenth century. A misfit now exists in trying to accommodate the female of the species — in trying to make the city function for her or for blacks or for American Indians or Chicano populations. The misfit for many of our cities is profound. When you stop and think about the capital investment in these misfits and how much it would cost to make changes, it poses an issue of such size that it almost fits in the energy crisis category, it's that big.

Later on, I'll talk a little bit about some of the things they are doing in Toronto, which I think are at least a step in the appropriate direction, but I wanted to just flag this issue so we can come back to it, because it really affects this age group profoundly.

Demos: I suppose the further irony is that the cities that were designed in a mistaken way by white, middle-class males have now been deserted by many of the same white, middle-class people.

9 Demographic Considerations

Arthur J. Norton

Now we move from the historical perspective to a demographic perspective, and it should be understood that demographic considerations reflect social, economic, psychological, technological and political developments, because the disposition of people to behave in certain ways relative to what I am going to be principally concentrating on — marriage, fertility, education and the labor force—is unalterably bound to the social and cultural sanctions that are imposed by the institutions that I just mentioned.

I could describe the 14-to-24 year age span as the age of preparation for labor-force entry, the age of transition from youth to adult, and the age of sexual differentiation — the age when diverging routes are taken by young men and women regarding education, marriage, career, and so on.

I would not presume—no sane person would—to be able to speak intelligently on cause and effect relationships of these many diverse interacting forces. I merely want to call your attention to the fact that a series of complex interactions do create behavioral patterns that provide interesting data for social and demographic analysis. I always find it necessary to qualify my remarks in such a general manner before

91

launching into a monologue on the demographic developments influencing any age-group segment of the human life course.

Last year we considered the 25-to-40 year segment, which I cleverly described as the age of application. Well, with equal cleverness, I could describe the 14-to-24-year age span as the age of preparation for labor-force entry, the age of transition from youth to adult, and the age of sexual differentiation — the age when diverging routes are taken by young men and women regarding education, marriage, career, and so on.

What I would like to do very briefly is trace relevant demographic developments from the entry into and the exit from the age range of 14 to 24. That range, by the way, is the one the Census Bureau uses in gathering and reporting data on the young people we are talking about at this conference.

John Demos covered three centuries in his remarks, but I am going to confine myself to developments since 1960. But don't feel disappointed. There have been many dramatic developments since that time, and certain demographic patterns for the future have begun to emerge.

Let me begin by giving you some numbers. There are 45 million young people in this age group, and they represent 21 percent of the total U.S. population. About 37 percent of them are under 18, and about 67 percent are from 18 to 24. By the year 2000 this age group will diminish in importance in terms of numbers to a little more than 15 percent of the total population. So, we are talking about a straddling age group. The older members are the tail end of the "baby boom" cohort, and the younger ones represent the beginning of what is called the "baby bust" cohort.

Table 2 on the following page, and the chart on page 94, show the relative size of the 14-24 age group in relation to the rest of the U.S. population.

Historical contingencies have an important impact on the life experiences of people in all age groups, and those in the 14-to-24 age group are no exception. The older members, because of their sheer numbers, are relatively disadvantaged. They are in fierce competition with one another for education, jobs, and other opportunities for advancement. The younger members, because they are fewer in number, are relatively better off.

Since an individual's sense of worth, security, and self-prognosis for success may influence him or her to act in a certain manner, at least

Table 2

Age Structure of the U.S. Population, July 1, 1979
(Includes Armed Forces Overseas)

Age Group	Number	Percent Distribution
Under 5	15,649,000	7.1%
5 to 13	30,647,000	13.9
14 to 17	16,276,000	7.4
18 to 24	29,285,000	13.3
25 to 34	35,024,000	15.9
35 to 44	25,136,000	11.4
45 to 54	22,957,000	10.4
55 to 64	20,952,000	9.5
65 years and over	24,658,000	11.2
Total	220,584,000	100.0%

Source: U.S. Bureau of the Census

some of the actions of young people in recent years may be a result of the failure of society's institutions to satisfy the needs of an unusually large age group. A general sense of insecurity among the "baby boom" members of the age group may very well be one reason for the recent rise in crime, victimization, and divorce rates.

An increasing number of young people are experimenting with what might be called "independent living." This may be seen as a transitional phase, and it may become the norm for young adults in the future.

Some new young adult behavior patterns which may last and be beneficial to society have also occurred, again in part because of the unique situation of the "baby boom" cohort. Included among recent changes have been a delay in the age at which young adults marry for the first time and an increase in the incidence of young adults leaving the parental home to establish non-family living arrangements prior to entering into family responsibilities of their own. This independent living phase may be seen as transitional and may become the norm for young adults in the future.

Distribution of the Total U.S. Population
By Age and Sex
(1970 and 1978)

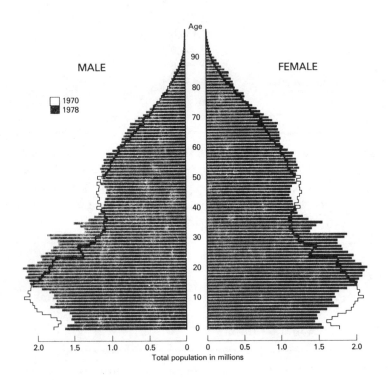

Source: *Current Population Report.* Series P-25, No. 800, Bureau of the
Census, April 1979.

Back to the demography of the 14-to-24-year olds. Young people at ages 14–17 are relatively homogeneous with respect to generally descriptive demographic characteristics. Ninety-four percent are enrolled in school; 98 percent have never married; 99 percent live in families — usually in a child status. Labor force and fertility activity are negligible in an aggregate sense.

As we move to a consideration of the next segment (18–24 years), changes begin to become apparent. School enrollment drops and other pursuits signal the movement to what we call adulthood. Forty-five percent of 18 and 19 year olds are in school, a level which drops to 22 percent for the 20-to-24 year olds; in 1978, 80 percent of the 18-to-24 group were high school graduates with 30 percent of the graduates enrolled in college. Men and women are enrolled at equal rates compared with a 60/40 overrepresentation of men among college students a decade ago.

These developments, along with the delay in marriage and the increase in independent living for young adults, seem to indicate that these people and their successors may be better equipped for later life challenges than their predecessors were.

There has been a lessening of distinction between men and women with respect to their social and economic roles, and this is another in a series of crucial considerations that must be taken into account when reflecting upon implications for the future.

One of the most important demographic developments of recent years has been a decline in the fertility and birth expectations of women 18 to 24 years old. Data on labor force participation give another indication of change; 76 percent of men and 62 percent of women 16 to 24 were in the labor force in 1978. The rate for men has remained unchanged during the last decade, while that for women has increased by 47 percent. Also, for women, going to school has replaced keeping house as the main reason for not working.

Other important statistics reflecting the current status of 14-to-24-year-olds are: 12 percent live in poverty; 13 percent are black, and close to 30 percent of the blacks are in poverty.

However, the very basic changes I have described regarding marriage, living arrangements, fertility, education, and labor force activity have occurred among the various social and economic strata. These changes indicate that a new transition system—or rite of passage—is emerging, enabling young adults to more adequately prepare for greater responsibilities, family and otherwise. Since there is less pressure to conform to social pressures, there is more time for selectivity, experimentation, and rational planning for future roles.

Speaking of roles, one of the interesting features of the changes I have described is that there has been a lessening of distinction between men and women with respect to their social and economic roles — another in a series of crucial considerations that must be taken into account when reflecting upon implications for the future. Indeed, the pattern for much of the future — individually and collectively — is established during this developmental phase of the life span.

Discussion

Robert Aldrich: Do you have any recent statistics on single-parent families?

Norton: Yes, we do. Some of them were published recently in the journal of the Elementary School Principals.

Aldrich: The reason I asked is that the needs of people in single-parent families are being very badly met. We don't have very many models on how to do this reasonably well, and it seems to me that this is an area that we should again flag here. There is need for research as well as direct action.

Norton: Well, about 11 million children live in single-parent family situations. That's about two out of every ten children under the age of 18. For blacks it's more like five out of ten. I should point out that these are what we call period statistics — that is, statistics that come from a snapshot of the population taken at a given point in time. They show highly transitional statuses.

Willard Wirtz: What were those figures ten years ago?

Norton: As I recall, the figures for 1970 were 5.6 million single-parent families, with probably 7 million children.

Alfred Kahn: It's interesting to me that we have 11 million children in single-parent families, and we have about 7¼ or 7½ million children

— with about 3 million mothers — receiving Federal Aid to Families with Dependent Children. That suggests that we do have the model Bob Aldrich was talking about. It's AFDC.

June Christmas: The aggregate statistics you gave us about children living in one-parent families were dramatic. But the difference between the overall population and the black population is even more dramatic. It seems to me that every time we hear statistics that are overall we don't really get enough of a picture to allow those of us who are interested in policy and in programs to get a true sense of what it means to be in certain social and economic situations. For example, there are still black male college graduates who earn only what white male high school graduates earn. That is very revealing. I think it would be very helpful if we could get more data broken down in a way that allows us to see what the reality of life is for a lot of people. It doesn't come across as vividly when you look at the overall statistics.

Norton: Such data are available in a very descriptive way, but there is no question about the differentials you mentioned. Unfortunately, we haven't really devoted much time and effort to gathering qualitative data that would highlight the differences. We just haven't had the money to do it.

John Demos: One point specifically. What is the new work that is just now getting started? Can you describe it for us?

Norton: It's called the Survey of Income and Program Participation. It's being funded by HEW and we're currently pretesting it on 12,000 households. It's a longitudinal survey that will follow the same individuals for five consecutive calendar quarters collecting social and economic information at each of the five interviews.

Sidney Werkman: Getting back to single-parent families — are they unusually mobile?

Norton: Yes, but they don't tend to move very far. They are mobile within a small geographic area.

Leonard Duhl: Do you have any data on the legal and illegal migration from Mexico into the United States, and particularly into California? There have been estimates that there will be 30 million Mexicans in this country within a short time, which is more than there are blacks at present. The governor of California claims that there will be a Mexican governor in California within ten years.

Norton: We don't have that kind of data. We at the Census Bureau don't differentiate between so-called legal and illegal people. We try to count everybody. We go to great extremes to avoid prying. We just want to know how many people there are at a given point in time at a given place.

Duhl: At Berkeley we're beginning to see a shift in admissions. In searching for quality admissions at the University it is now easier to get qualified Chicanos than it is to get qualified blacks. That represents a rapid shift in the last ten years, and it's going to change a lot of things.

Vivian McCoy: Do you have any data about the career expectations of people in this age group?

Norton: To my knowledge, we have very little information on expectations, but we do have data on labor force participation among young couples. It's about 75 percent joint participation.

Philip Blumstein: Can you say anything about the data you have on household arrangements — specifically focusing on heterosexual couples living together without marriage?

Norton: Not what you would call hard data. We simply go and ask who's living here. Sometimes we find a household that contains two adults who are not related to each other and not married to each other. And there the matter rests. However, the number of heterosexuals living together has doubled since 1970—from about 500,000 in 1970, to about 1.1 million in 1978. So there has unquestionably been an increase in cohabiting couples. Sometimes it's economically based and sometimes it's not. If we look at the data within that 1.1 million couples there are something on the order of 200,000 to 300,000 who have children under 14 living with them.

Blumstein: You don't ask about maternity of those children?

Norton: Not specifically.

John Cannon: Do you ask questions about religion?

Norton: No.

Willard Wirtz: Let me ask a question in two forms and you take your choice. They both involve the degree of confidence you feel in the data on employment and unemployment. In one form the question would be how much confidence do you feel in the monthly figure which reports unemployment among black 16-to-19-year olds? The broader form of the question would be how much in general can we trust the aggregation of any of the figures on employment between the ages of 16 and 24?

Norton: That's a tough question, and it's a little bit outside of my bailiwick because the data you mentioned are analyzed and published by the Department of Labor. However, I'll try to answer it as best I can. My guess is that in terms of measuring change, the employment-unemployment figures are pretty good, but in terms of measuring what the situation actually is at a given point in time they're not so good. My hunch is that this also applies to the age, sex, and racial breakdowns.

Wirtz: But even in terms of measuring change, isn't the sample so small that the change itself becomes suspect?

Norton: You may be right. I just don't know. But the sample has to be small in order to make monthly reporting possible.

McCoy: Do you have any data on students in the 18-to-24 age group who are married?

Norton: Yes. Each October we take a survey of school enrollment by age, by race, and by whether the students are married or not married. I'm not sure of the exact figures, but the number of married students is declining.

Demos: But so is the number of marriages.

Norton: True.

Demos: If my sample of my own graduate students is any good, I would say there's been a very major change between a time as recently as ten years ago, when the majority of them were married, to the present, when virtually none of them are.

Norton: That's an interesting point. There have been all kinds of dramatic changes in our demographic profile over the past ten years.

Demos: I would like to press you on the "transitional independent living" trend you mentioned. I think you suggested rather briefly that maybe there were some good things about this new pattern. You talked about options. Perhaps ultimately it may be a better way of preparing for what lies ahead at the next stage of life, but I wonder whether this independent living pattern is, in fact, transitional. Is it really intended to prepare people for what lies ahead, or is that merely what they tell you?

Norton: Let me start out by saying that the one thing that's held up over time in terms of being highly correlated with eventual divorce is early age of marriage. Income, education, and other factors don't hold up. So, for a growing number of young people, living together before marriage is a way of finding out what marriage is like before committing themselves to the real thing. But, we don't have any data on how many of these couples eventually get married to each other.

Sheila Kamerman: If 20 to 24 is the most typical age of divorce for women, and there's a delay in marriage, are we seeing any delay in the time at which women are likely to get divorced?

Norton: Not yet, but the duration of marriage before divorce has been declining.

Donald King: Is there any way to do a longitudinal study on the 1.1 million unmarried people who are living together? For example, among those who eventually married, did the divorce rate go down because they had more experience?

Norton: That's a good subject for further research. We just haven't had the money to pursue it.

Blumstein: I'd like to respond to Don King's question about the 1.1 million who are living together. I've done research in a related field and it was inconclusive because we couldn't get adequate data. It seems to me that the Bureau of the Census is in a unique position to gather it. The census people have the legal power to intrude on people's privacy. They certainly have the moral authority to require people to respond to their questions.

Norton: We don't have the legal authority. We are mandated by other federal agencies and sometimes by legislative acts to collect certain kinds of information. Generally, respondents are not legally required to cooperate in surveys. But we can't go beyond the seven items asked of 100 percent of the population unless we have a special mandate to require compulsory response.

Cannon: Do you have with you any family income data that would shed some light on the whole question of dependency? For example, how dependent—or independent—is a young person who works part time while going to school or who works only intermittently?

Norton: Unfortunately, I don't.

Ernest Bartell: What about income data for the unmarried people who are living together? Do you have any data about the extent to which they are dependent on each other's income?

Norton: I have no real information on dependency.

Cannon: One question that really interests me is when is our social/economic system ready to adjust to the young adult? Another is what happens when the young person suddenly leaves the fold, and has to adjust to a lower standard of living?

10 Young People as Individuals and as Family Members: The Implications for Public Policy

Sheila B. Kamerman

There are two things you ought to know before I begin my remarks. The first is that I'm basically a social policy analyst, which means that my approach is multidisciplinary and my primary interest is in questions of public policy. The second is that my major focus is on youth in western industrialized societies, including, but not limited to, the United States.

I don't have a prepared text, but for the benefit of those of you who like a neat and orderly outline, let me say that I will be making three main points. The first is the lack of clarity regarding the parameters of this age group that we're talking about. People define "youth" very differently, using very different ages to denote the beginning and end of this period. The second point that I'm going to discuss is the need for us to think about every cohort of youth in its own historical time. When we talk about the current 14-to-24 age group, we must remember that the young people in it are far different from the young people of a generation ago. And the third thing that I'll talk about is the need to begin to interject some family-related considerations into our discussion in order to be able to raise some very significant questions for public policy.

Let me illustrate some of the problems in defining the parameters of the youth population. I'll begin with the Census Bureau definition: youth is the age group from age 14 to 24. Parenthetically, I might mention that even in Census data it depends upon which current population series you look at. Some publications refer to those aged 15 to 24. The bureau's latest statistics define youth as the group from 16 to 24, subdivided into two groups, 16–20 and 20–24. Legal majority in this country varies among the states. Usually it's either 18 or 21. Voting age, needless to say, is 18. If you look at other specific items of federal legislation there are even more variations. The Runaway Youth Act defines youth as those under 18. The Youth Employment Demonstration Act applies to youths up to 21. The Adolescent Pregnancy Act covers youths of 17 and under. If you read articles by experts on the teenage pregnancy problem, some people talk about girls under the age of 19 but others talk about those under the age of 15.

The three most significant tasks of youth are leaving the parental home, entering the job market on a regular and permanent basis, and forming a new family. Although these tasks tend to be completed at different times by different people, they tend to be completed by the time young people reach the age of 25.

Clearly we're not addressing the whole question of the onset of puberty; that usually happens before the age of 14. Webster's Dictionary defines youth as "that part of life which follows childhood and precedes maturity." In social policy analysis we usually find it easier to identify a chronological age, simply because legislation has to be specific about who is eligible for whatever it is that is going to be provided.

One of the most consistent criteria that can be used for indicating the beginning of youth is the school-leaving age. In the United States it's 16 and in most European countries it's either 16 or 15. But when we turn to the question of when youth ends, the task is far more complicated, because that varies enormously among different countries.

Basically, the tasks of youth have to do with completing certain transitions. The three most significant ones are leaving the parental home, entering the labor force on a regular and permanent basis, and forming a new family. In general, although these tasks are completed at different times by different people they tend to be completed by the time young people become 25. In effect, however, there is no consistent definition of what youth is, but for policy purposes we can probably assume that the group 16 to 24 is the central age group that we're talking about.

Now we come to the second point that I wanted to make — the need to remember that just as individuals have their own individual life experiences within a historical context, so does a cohort.

Very briefly I would like to identify what I think are some of the most significant aspects of the historical and social context in which the current population of youth has grown up. As Arthur Norton pointed out, some of the young people in this cohort are members of the "baby boom" cohort and others are members of the "baby bust" cohort. There are major implications to this and I'll return to them in a moment. First let me talk about the historical and social context.

This is a group that has grown up in one of the most acutely stressful periods of history. It's also the first age group that has grown up in an intensively automated and computerized society.

More than 90 percent of today's young people were born in families with native-born parents. More than 90 percent lived in non-farm settings. More than 81 percent completed high school. More than 95 percent had both parents survive their childhood and more than half had three or fewer siblings. This is a group that has grown up in one of the most acutely stressful periods of history. Through the media or television, it has experienced the Vietnam war, the civil rights movement, the riots and the student uprisings of the 1960's, the Middle East war, Watergate, the first resignation of a President in U.S. history.

In addition, of course, it's also probably the first age group that's grown up in an extensively automated and computerized society. In a whole series of ways youth are experiencing a kind of technology that was not known before. They have also witnessed major changes in the roles of men and women — major variations in the nature of their role models and in the kinds of models that are available to them. Among other things the group that we're talking about is more likely than previous groups of young people to have had working mothers, to have experienced divorce within their own family, to have lived with a member of the opposite sex before marriage, to have experienced divorce themselves, and to have had the experience of being unemployed. All of these are particularly characteristic of the age group we are discussing.

As students and job-seekers, the older members of the age group were in oversupply. In effect, they were sellers in a buyers' market and they're still suffering as a consequence. They grew up in a period in which the so-called problems of youth—that is, the physical manifestations of what has been defined as unacceptable behavior—received an extraordinary amount of attention and left something of a stigma on youth in general. In contrast, the younger members of the age group —those born since the beginning of the "baby bust" period—are now in short supply. In each case there are important policy implications.

Maybe we can begin to get some additional insights and raise some new questions for policy purposes if we begin to look at the family aspects of youth.

Now for my third, and major point. Given the various experiences of this particular age cohort as well as our consciousness of the fact that subsequent cohorts will have different historical and life experiences, how do we begin to think about youth?

One of the things that I have discovered over the last few years since I've been working in the field of family policy is that there are two age groups which rarely get mentioned when one talks about government policies with respect to families. One of those groups is the elderly and the second is the group that we call youth. In the process of preparing for my comments today it occurred to me that there is a growing tendency to focus on youth as independent and autonomous individuals and to support them as such. Less attention is paid to the fact that young people also have families. Maybe we can begin to get some additional insights and raise some new questions for policy purposes if we begin to look at the family aspects of youth.

In order to explain the implications of what I'm saying, I'd like to arbitrarily select a few illustrations. In general, when we talk about the group aged 16 to 24 or 16 to 25 we're talking about a group which experiences the most significant social and personal transitions of any age group in society within a very limited period of time. These transitions are critical for what happens subsequently to individuals in adulthood and during the rest of life. As I mentioned, these changes include leaving school, leaving the parental home, entering the labor force on a permanent basis, and forming new families. One of my hypotheses is that all of these transitions occur in a family context. They occur while young people are still living within their own families —their parental families—or they occur while they are living in the new families they form. The tasks they accomplish may vary by race or class, but essentially they are tasks which must be completed in our society and similar societies by the time the young people are approximately 25. To illustrate the implications of employing a family perspective, I will focus on what I think are the three most essential transitions — leaving home, entering the labor force on a permanent basis, and forming a new family.

As Arthur Norton pointed out earlier, the vast majority of American young people live in families. That is, they either live in their parents' families or they live in the families they're forming themselves. If one looks at the younger group, those aged 14 to 17, it's about 98 percent. Even when we look at the group aged 18 to 24, 88 percent lived in families in 1977, and interestingly enough, more than 90 percent of the black youth this age. More specifically, for the 18-to-24-year-olds: 24 percent of the males lived in husband/wife families and 62 percent with their own parents, while 38 percent of the females lived in husband/ wife families and 46 percent with their own parents. Clearly, the group that lives alone is a relatively small group, one which included only 13 percent of the males and 11 percent of the females in 1977. Although this group has grown rapidly in the past decade, we really know very little about how they live. For example, we have no idea whether these youths are really separated from their families, or whether they are actually in very close contact with their parents and see them frequently.

Another interesting question — and I'm delighted that the Census Bureau is going to conduct a survey that will give us this kind of information—is how much parents contribute in money to the children. To what extent do intrafamily transfers influence the way young people are living? If they play a significant role, what are the consequences —for the parents as well as for the children? For example, when young people move out of their parents' homes does that mean that the parents are going to have less income? Two possibilities could lead to lower family income: (1) the withdrawal of financial contributions by youth who now require these funds to support themselves; or (2) increased parental contributions to support youth now living in a separate setting.

Although some families may be better-off financially if the same income is shared among one family member less, we know very little about what the actual consequences — and patterns — are.

When young people decide to live together is that a new form of what used to be called engagement? To what extent does it influence divorce rates? We just don't know.

Another interesting aspect of the trend towards living together has to do with the question of whether what we now have is what the President of the French National Union of Family Organizations refers to as "a new form of engagement." He says that all this living together really isn't anything new. Instead of going through a formal "engagement," where a boy gives a girl a ring, a young couple now simply shares an apartment or a house together. He suggests that, ultimately, it makes very little difference which procedure is followed. It is merely a matter of semantics. If that's the case — if living together is just a form of surrogate marriage — is there any influence on divorce rates later? This is another example of the type of questions that need answering if we are to understand the implications of this living-together trend.

With regard to labor-market entry there's another kind of question that needs to be raised. Indeed, there are a series of questions. The development of successive generations of workers basically takes part through the socialization of young people before entering into the labor market. The family and the education system provide nearly all of this socialization. The role of the school has been fairly extensively studied, but the role of the family has been given very little attention in previous research. Although there were some studies in the 50's that looked at the whole question of the role of families in influencing occupational choice, there are a whole series of other questions that have never been raised in regard to the socialization of young people for entry into the labor market. For example, what sex role norms are used? What occupational aspirations and expectations are instilled? What very specific labor market knowledge and skills do parents impart to their children? And in what ways do these influence career choices and achievements?

In some families, labor market socialization is done either poorly or not at all. For example, I read an article by Mary Keiperling recently which described what happened when she asked a group of 10-year-old girls how many of them expected to be working when they were 40 years old. Only about ten percent of them raised their hands. Mary was horrified. Personally, so was I. The fact is that at the present time over 60 percent of the women who are 40 are in the labor force. Within the next few years it will probably be 70 percent, and by the time those 10-year-olds get to be 40 it may be still higher.

We are moving to a period of time in which nearly all children will have both parents in the labor force.

I think we have to recognize that parental example is the most important influence on the young's attitudes toward work. Maybe it helps to explain the very extraordinary growth in the female labor force participation rates over the last 15 years. Research studies of the National Longitudinal Study survey data reveal that women whose mothers worked are more likely to be in the labor force when they grow up, and that they are more likely to have more positive attitudes toward work when they grow up. Research in West Germany and Sweden reached the same conclusion. Lois Hoffman's recent review of the research on the effects of maternal employment on children underscored these earlier findings. She reported that both the sons and daughters of working mothers "showed better social and personality adjustment, had a greater sense of personal worth, more sense of belonging, better family relations, and interpersonal relations at school."

From a policy perspective, this country does absolutely nothing to support family formation by young people.

Since most of the women that were studied in the series of research projects reviewed by Hoffman had worked for some time before their children had reached adolescence, she concluded that the positive benefits of mothers working were cumulative, and that they surfaced in the high school years. I mention this not only because it has significance for the age group we are talking about, but also because it will have even greater significance for the cohorts who are about to enter adolescence. In effect, we are moving to a period of time in which nearly all children will have both parents in the labor force. We have very little understanding of what the implications of this may be for socializing young people in their later attitudes towards work.

The final point I want to make has to do with family formation. I would like to remind you, that although about half of the young people in this age group are not married by the time they reach 25, about half of the women have already had their first child. Clearly, new family formation is an extremely important fact for this age group. Yet from a policy perspective, this country does absolutely nothing to support family formation. I stress this because, having worked for years in Europe, I am acutely conscious of the different values placed on family formation in the U.S. as contrasted with other countries.

The United States is unique among 65 countries in providing no family or child allowances. We are unique among about 45 industrialized countries in providing nothing in the way of statutory maternity benefits, paid maternity leaves, let alone parental leaves. Finally, we are unique among a still larger number of countries in not providing maternal or child health care programs. In effect, we do nothing to facilitate family formation by young people in this particular age group or by those who are slightly older.

Although the probability of divorce and of being a single parent are relatively high among women in this age group, we are doing relatively little to facilitate their entry into the labor force.

Some of the reasons behind later marriage and delayed child bearing may have to do with economic factors and the growing stress on young people to be able to manage having children at the very time they have to cope with beginning their regular participation in the labor force. At present, I would stress, we are doing nothing, also, to address the whole question of how a young couple, both of whom are in the labor force, can cope with having children.

Now, from a very different perspective, I want to mention two other kinds of family-related issues. First of all, as I said during the discussion of Arthur Norton's remarks, 20 to 24 is the most typical age for divorce among young women. For some years afterwards, therefore, these young women are going to have to support themselves. And many of them have children. Furthermore, 16 percent of the black women in this age group are already single mothers, heading their own households. This means that the probability of divorce and of being a single parent are relatively high among women in this age group. Yet we are doing relatively little to facilitate their entry into the labor force.

Finally, let me say that although I have stressed the need to pay more attention to young people as family members, I do not mean by this that we should underplay the significance of policies focused on youth as individuals. Quite the contrary. I wanted to suggest rather, that we need to have a somewhat broader perspective as we begin to explore who these young people are and what is needed in the way of policies concerning them.

Discussion

Robert Aldrich: Sheila, I have a question. I have been very disturbed in the last six months by the appearance of a very well-organized anti-child force in this country. I recently read a newspaper article containing quotes from letters put out by this group which said essentially "Down with children. Let's not have any more. If you have children it will spoil your fun." That kind of approach. I wonder if anyone else in this room has seen any of their literature? The interesting thing about it is that it's fairly recent. I was driving from Portland, Oregon, to Seattle a few days ago, and just outside of Centralia, Washington, I saw a billboard which said essentially the same thing. Don't support the International Year of the Child and so forth and so on.

Leonard Duhl: There's another kind of development that I find somewhat more disturbing. It's the kind of thing that I've noticed over the last five years in California — the growing number of apartment houses and condominiums that exclude children under the age of 15.

Kamerman: I've spoken in situations in which I've been heckled by members of some of these groups. One thing that's usually very effective is to remind them that when they get to be 65 they're not going to have any Social Security whatsoever if there aren't any young people around who are going to pay for their Social Security pension. My guess is, by the way, that the concern with having children is short term. There is a real anxiety that's sweeping much of Europe at present and that I think will be here in another five to ten years, and that's the worry about too few children being born! That's the opposite of the present concern in this country.

Duhl: I wonder if this problem isn't part of the general fragmentation that's taking place — the segmenting of society by age groups and by interest groups all the way across the board. There's been some feeling in California that the kids who felt completely superfluous and the old people who felt superfluous were the people who voted in Proposition 13. They did not feel that they are getting a piece of the action.

I think we are approaching a point in the history of the world when children and youth will be superfluous. Work by people is no longer important in our society, because we are designing a society which makes large segments of our population superfluous, and thus there is no longer a need for children. At times I feel the reason our society gets screwed up is not because we can't keep it sane and working, but because a screwed up world gives jobs to people like us who have to deal with the results of the screwed up society.

Anyway, I think these are the kinds of problems that are not youth problems at all. They are societal problems. What kind of society do you have where meaningful work has no role? A lot of the jobs in the labor market are meaningless jobs. And you get kids asking themselves: "Do I really want to work in that kind of job? Doing that kind of thing?" They're starting to raise a whole lot of qualitative questions about work and family. I think these questions are much more significant than labor force statistics, because we don't really give a damn about kids.

Kamerman: Before we go on to anybody else, I'd like to respond briefly to what Len has just said. First of all, I don't think work is unimportant in our society. I think it's overwhelmingly important. It's important to the society and it's important to the individual. It's important both to those who are working and those who are not working. The problem may be a lack of work in certain circumstances, but I have no question about the significance of the role of work.

My second comment has to do with the question of children being superfluous. What I'd like to suggest is that in an industrialized society such as this, children—not one child in one family to one set of parents, but children as a group—are economically essential to the adults in the society as a whole, because without their productive labor as they enter the work force there is no support for the base of the economy. I also think that part of what's coloring our perspective right now is the fact that we are at the tail end of the baby boom. We have been threatened by an enormous number of young people entering the labor market at one particular point in time and we haven't handled this well. I have a feeling that ten years from now we will have a very different perspective.

A. J. Kelso: When we talk about broadening our perspective, it seems to me that it's also important to realize that we're talking about a very small segment of the past—the last couple of decades. At other times in our history the situation was quite different.

I also think we ought to broaden our perspective geographically. Whatever modest experience I have had with people from other cultural backgrounds suggests that for them the process of growing up is not a process of becoming independent; it's a process of becoming interdependent, which is very different from the kind of process that we expect.

Kamerman: I think you're quite right. In the late 1950's, a lot of people I knew were convinced that there was going to be massive unemployment — that the labor force was essentially much too large and that a very small percentage of the people were going to be needed to actually work. Then, in the 1960's, the labor market expanded rapidly in both numbers and types of jobs available. In effect, one possibility is that work is so important in our society and societies like ours that as one form of work ends other forms of work will be created. Very often, as I'm sure several of you will agree, it is very hard to make a distinction between what is defined as "work" and what is not work. We usually distinguish *paid* work from unpaid, but apart from that, there's a very fine line between the two.

John Schweppe: I remember about four or five years ago Senator Kennedy suggested that at some point between the ages of 18 and 20 every young person — male or female — should be required to devote one year to some form of national service, either military service or something like the Civilian Conservation Corps of the Depression era. I think that was a good idea. It would give our young people a better perspective on life and work. Instead of losing a year or two they might gain a year.

Willard Wirtz: Let me say this. I don't think we're going to get through the next five years without the development of a substantial youth service program. We're that close to it. Jobs in the traditional sense are drying up, and I think the emergence of a significant-service component in this preparation period is almost imminent.

Schweppe: How much would it cost?

Wirtz: The current assessment is about $7,000 to $10,000 a head. That would be far beyond any conceivable budgeting in the next ten years, but I think those figures are completely unrealistic. It would be more realistic to assume that such a program will cover one million kids the first year, maybe two the second year, then three and so on, and that the right per capita cost will be about two to three thousand dollars. That involves getting it out of our heads that we should pay the minimum wage for the service program. We're talking about national service, not work in the job-market sense. Therefore, it shouldn't be covered by the minimum wage. If you get rid of that nonsense and if you decentralize it to the local community — get it out of Washington — I think the cost would be about two to three thousand dollars per head, per year.

John Demos: I personally think that this service component idea is tremendously important and I hope it will be one of the recommendations of this conference.

I think it's important for a lot of reasons beyond the fact that it deals with problems of youth unemployment. One of them is that in some broad, and somewhat intangible, but nonetheless perhaps very important way, it might move us in the direction of sort of bringing children and youth back into some larger social network. And this really gets me to the reason I wanted to put in a comment here — the whole broad question of the lack of support for children and youth and family in this society. It can be looked at in the broadest way of all, I think, in terms of a certain kind of anti-child sentiment that may run quite deep and go back quite far in our history. But there's another, sharper aspect to it, and that, I think, is the developing sense over a century or two of our history that, essentially, children are the sole responsibility of their own parents. "You take care of your kids, buster, and I'll take care of mine." That's what it comes down to. The whole issue is really whether, how much, and in what ways we care for other people's children. I think that's different from the way it is in other countries, and I think it's different from the way it was in this country two and three hundred years ago. I wish that we could come up with some set of devices — prescriptions almost — for starting to move in the other direction.

I want to make one further point. In terms of policy making, it seems to me to be of some interest that people in positions of high public responsibility, both in government and the private sector, no longer need, as they used to, to have a family life that is recognizable in conventional terms. I think it's a really significant point that for the first time in American history the governor of the largest state and the mayor of the biggest city are both single people. I've even heard that in the business world it is thought to be an advantage for some people to not be encumbered by family ties, so they can be moved about more easily, and special demands can be made on them in terms of travel and the length of their working day. I think one would have to be very cautious about assuming that such nonfamily people are sympathetic to family problems and issues. The biggest difference among the kinds of people I know is that between those who have or have had children and those who have not. In part it's an age difference, but it's more than that, and it may ultimately have some bearing on this whole policy issue.

June Christmas: I can see four or five positives that might come out of the youth service corps idea, but I have a question as to how it might ultimately relate to the problem of long-term unemployment. I can see that if some of these public service activities were human services, or public service, or conservation, that youth might learn to understand certain kinds of social problems and social needs — especially if the program were operated on the local level. They might develop some sense of community. I can also see that if the youth service corps brought together not only the disadvantaged, but those who are well off, this might begin to break down some of the barriers that separate us economically, ethnically, and so forth. Finally, I think that society might come to value youth a little more, and that for some people community service could be a way of building better work habits. These to me are all positives. But I'm wondering, as the youth service corps idea has been conceptualized, whether it really would contribute to the long-range solution of preparing people for entry into the labor force.

Kamerman: That's a good question, and we ought to deal with it when we talk later about careers and vocations.

Duhl: I think the policy question is how can we create social institutions that provide not only job opportunities for young people, but also an opportunity to reconceptualize their role in society. The Peace Corps has been very successful in doing that.

11 Youth: Stage or Problem?

Alfred J. Kahn

To those who make public policy in the United States, youth is a problem. It's not stage, it's not status, it's almost a diagnosis. This policy orientation is manifested in all of our legal tradition and in our legislation for youth. We have youth *service* programs, youth *treatment* programs, youth *rehabilitation* programs, and youth *control* programs.

These are the major government youth programs. If you ask somebody on the White House domestic policy staff to give you a list of government youth programs, you'll find that almost all of those listed deal with youth as a problem. Except for the recent youth employment program, these are very small programs, in terms of expenditure, but there is one big exception — education. Our national education budget for the last fiscal year amounted to about $166 billion — federal, state, local, public, private. But the strange thing is that education is not thought of in the context of youth policy. It is something apart.

Youth has been described as the preparation for the next stage in the human life cycle. I'd like to argue that it should have its own validity — that there are things about it that deserve their own recognition.

I am not saying that education should or should not be the anchor point of our national youth policy. Instead, I'm suggesting that it would be a good idea to explore whether we could refocus the core of federal youth policy on normalcy, on development, on positive aspects of the stage of life we are talking about here.

115

Maybe our approach ought to be to face the fact that in a society of 215 million, there are 45 million young people who are different from the ones who are younger and the ones who are older. Maybe we ought to recognize that life consists of stages and phases, each of which is different and each of which has its own validity.

Youth has been described as the preparation for the next stage in the human life cycle. I would like to argue that youth should have its own validity and that there are things about it that deserve their own recognition. Maybe young people aren't very important in the marketplace, but maybe they do bring social benefit to the larger society in a way that validates their role. I'd like to suggest that.

In any case, my role here is to attack the notion that the sum total of youth policy is rehabilitation, therapy and doing something special about minority youth. My approach has its difficulties, because it is legitimate to ask: Can we focus on development, on the majority, on normalcy, and not give short shrift to minority students who are having such a difficult time in the kind of society that we have today? Would they be helped by this approach, or would we be creaming off the resources of the society in a way that doesn't address their needs as well? That's one of the questions that I'm preoccupied with. I can argue the case that, in general, universalistic social strategies help people who are poor and who are at the margins better than selective social strategies; that is, programs that are means tested and focused on specific groups. However, I also recognize that one cannot be sure just what the results would be for a given group at a given time.

The other problem with the current approach is: Can this youth stage really be conceptualized as having standing of its own, and be accepted as having standing of its own? Those are two different things. You can conceptualize it and develop a point of view of how it would be, but I don't know if you can sell it politically. By way of example, let me cite the efforts of the Carter Administration to develop something called a "Youth Policy."

As you know, the Carter Administration began with large ambitions for the reorganization of the federal government and the improvement of the services it provides. What is relevant to our discussion is a report the President's Reorganization Project produced in July of 1978. It was publicly circulated, and there were hearings on it. The report dealt with a problem which in one way or another has been written about and talked about since the early 1950's — that is, the fact that we have very very loose boundaries in human services, enormous numbers of programs, ten federal departments and agencies, 100 programs, $23 billion and (when you put it all together) imprecise federal policies and goals, unclear assignments of responsibility for the delivery of services, lack of effective citizen involvement despite a lot of milling around, uncoordinated service delivery at the city and county levels, complicated administrative requirements, inaccurate information about services available, weak federal enforcement in monitoring the policies and programs, and poor coordination of the federal agencies. The report laid out alternative ways of dealing with each of these problems, but by the time it was published, the political mandate for a major initiative was lost (except for the recommendation for a Department of Education). The report therefore offers only the terms for a debate.

Shortly thereafter (August 1978) an initiative within the Department of Health, Education and Welfare (Office of Human Development Services) produced proposals for a possible youth initiative. The National Institute of Education also issued several relevant RFP's ("Requests For Proposals") in an atmosphere in which major initiatives would not be forthcoming. The former called for "demonstration" projects to help communities cope with youth and problem-oriented service demonstrations. The latter sought "youth budget" data, understanding of teenage pregnancy, and the history of school social services. All three of these reports and proposals saw the differences between therapeutic and prevention efforts and wondered about the nature of this age cohort. Was it a stage? What were its characteristics?

The OHDS document said:

> What this statement suggests is that the troubles facing today's youth may be far more significant and long-term than the immediate problems of the runaway, delinquent or pregnant teenager, and may in fact affect all youth. These troubles seem to relate to the very process of adolescent transition to adulthood. While the problems of runaway youth, juvenile delinquents, alcohol and drug abusers and teenage mothers are widespread and, in some cases, reaching crisis proportions, many experts are beginning to explain such behavior, not as individual maladjustment to a well-ordered society, but rather as understandable, but perverse adaptations which occur when social arrangements make healthy transition to adulthood difficult or impossible. If youth find few healthy avenues to adulthood open, the natural drive for self-assertion increasingly expresses itself in aggressive delinquent behavior or in resigned, apathetic escapism.

Where are we? Most of the programs that provide services to youth do so as part of broader programmatic objectives, such as the National

Institute of Drug Abuse. We have a series of small, problem-related so-called categorical youth programs. Runaway Youth is in the Office of Human Development Services, which is a service unit in HEW. The Juvenile Justice and Delinquency Prevention Program is in the Department of Justice, or Law Enforcement Assistance Administration. ACTION, which has something called Service Learning Programs, and National Youth Sports Force and Summer Youth Recreation programs is oriented to minority youth and their problems. These are relatively small, and they are located in four different departments. The reorganization staff wondered: should we pull these together, keeping them as individual programs but reaping the benefit of a streamlined program structure?

The reorganization staff—as indicated—offered some possibilities: one could attempt to create policy and service coherence by: a) creating a youth agency within the Education department, recognizing the centrality of education as "the major life activity of most youth", but also assigning that agency some responsibility for the problem-oriented social services; b) consolidating all social service programs for youth (now dispersed throughout government, but especially within at least four departments) in HEW's social service agency, Office Of Human Development Services. The choices refer to agencies; the differences are between policy leadership from a developmental or from a problem-oriented focus. If the focus is developmental, can it be based in education? If the focus is problem-oriented, is an age category a good principle for organization? And what happens to the rest of youth policy?

Apart from political obstacles, the first alternative ignores the statistics that have been given to us at this conference, showing that there is an increasingly large number of youth in the age cohort we are talking about who are no longer related to education as an institution.

The problem with the second alternative, as most of you know, is that OHDS has a very weak political standing within HEW, an even weaker political standing in government generally, and thus far has failed in its efforts to integrate human services programs. In fact, it has met constant defeat at the hands of Congressional committees that are backing categorical service programs. (That, by the way, points up another problem. The Congressional committees themselves have become categorical, and so have the interest groups that press for legislation. In fact, the entire legislative process has become categorical.)

But even if one could consolidate service programs, they are problem-oriented, not universal, separate from education and not related to work-transition for most youth. So the problem remains: how to create a center for coherent consideration of youth trends, problems, needs — while recognizing the centrality of education, work, for all youth — and yet meeting some functional principles for the organization of government generally (health, personal social services, education, etc.) and desperately needing a broad view of youth.

In any event, the Carter Administration, after a lot of milling about, has achieved very little in the way of a coordinated, streamlined policy for dealing with youth as a whole.

We need to promote youth policy for the majority over youth policy only for those having trouble.

My argument is that we need a focus at the federal level of policy development for the nation, for coordination of services and research. And we need long-term overall federal policy goals. Maybe this conference can make a contribution to the debate. But even the question of where to locate the debate on a national scene is a tough one. In the White House? In the Office of Manpower, in Budget, in HEW? How do you get Congressional committees to do for youth what they've already done only for the aging — establish the Select Committee on Aging and develop a point of coherence which was insisted upon because the political pressure became so great?

The issue of how to go about helping young people is a long story, and I hope it's going to be part of our discussion. We never do well if we think that we are dealing only with people who cause trouble. That's also a false perception of the issue that we're talking about. So I favor universalism over particularism. I think we will do better for everybody with that approach.

What I have been trying to suggest is that maybe the real direction in policy is to look at youth as a stage, not as a problem. We need to promote youth policy for the majority over youth policy for only those having trouble. Such policy will do better, as well, for those of the margins.

Discussion

Robert Aldrich: Al, I really enjoyed your very quick run-through of some awfully important issues, and I'd like to draw everyone's attention to what's been done in Canada. We don't have to go very far to find some effective grappling with some of the things you were talking about. It's going on right next door to us.

In 1961, I participated in a national conference called Mid View of 1970, regionalized all across Canada. It was an effort to find out what Canadians wanted for their youth. About 50 percent of the participants were young people, and much of the research that was done and presented was done by young people — some as young as 13. It was an incredible experience. I could count the number of people in the United States—and I know a lot of people—who even heard of the conference. But I think it would be a useful thing to resurrect the conference reports, and then to look at what's gone on subsequently.

The children and youth program is very powerful in Canada. They now have developed probably the best North American model of how to deal with human civilization on a city-wide basis. I don't want to give you the details, but the leadership that has arisen in Canada for children and youth in cities and small towns, and the policies that are being developed by the Canadian government have had from the very beginning strong participation by youth. I'm tremendously impressed with what they are accomplishing. When I compare it with the chaos in our country, it makes me damn mad, because I think we could do at least as good a job if we made the effort.

Maybe it's sort of un-American to say that we can learn from anybody else, but I think the time has come when we damn well better start looking at what other people are doing and are doing in real good style.

June Christmas: Al, I've often said that I'm the Commissioner of mental illness rather than mental health, because much of what I have to do with regard to service has not been directed toward helping people grow up healthy — learning how to cope and adapt and to develop — but rather getting people when they are casualties, and perhaps those who are the most trouble to society, or perhaps the most ill.

Philosophically, I would say that if we addressed youth as a whole, we would not be overlooking those who are troubled and those who are in the minority. That would be fine if we were starting at point zero in designing youth programs. But how would you, with the philosophy that you've espoused, deal with the fact that there are some young people who do need more help than others, whether it's more money or more attention?

Kahn: I do recognize that fact. There are two types of problems, one practical and the other almost an issue of morality. I believe that we somehow have to deal with the larger policy question without slighting the troubled groups. The problem is that we deal with the second in terms that almost negate the first and create a distortion about what the youth policy question is. We act as though the institutions are fine, and that all we have to do is to get these other kids to fly right. I'm arguing that the institutions aren't fine.

Margaret Hastings: I don't have the answers but I do have a few observations about youth policy that stem from my own experience in working on legislation dealing with education.

Policy requires consensus. From what I have observed, I would say that our typical way of making policy in this country goes something like this: You pass a bill which creates a program, and then you fund it. Pretty soon it delivers service. Sometimes it dies out, and you create another one.

In my estimation, the annual budget process is antithetical to a comprehensive approach to policy making. In the long run, nothing really happens because what is really required is the bringing together of the major power interests in our country to see if they can agree on a policy they are really committed to. Frankly, I'm not sure that this is even possible in this country. What I'm saying is that I think the mess we're in stems from the lack of real national commitment to youth as they really are.

Kahn: Yes, we're in our present predicament because of the way our process works. The issue is how do we break out of it?

Leonard Duhl: I think we're really talking about the need for a national dialogue. When Bob Aldrich talks about Canada, what intrigues me is that Canada went into a national dialogue. It was not a dialogue about expenditures on youth services. It was a national dialogue about Canada as a place to live, as a community, over the next ten years. In order to break out of the situation we're stuck in — all these fragmented programs—we really need to start talking about what kind of society we want to have. It's that level of discourse that we have to start moving to, instead of arguing over the allocation of funds.

One other thing the Canadians did proved effective. They started a multiplicity of small functional programs by giving anybody with a good idea a few thousand dollars in seed money. The people that they've supported over the past 16 to 20 years are now the leaders of Canada —not on the national level yet, but on the local level. So in a way we have to start making a shift, not about youth, but about how we start really discussing where the hell we're going as a society on many many different levels. Maybe the theme should be: "Let a thousand flowers bloom." Let's try thousands of programs all the way across the board, and let the local communities start emerging with answers rather than HEW.

Kahn: I think we're going to have to do both. It's that kind of a problem. We need local initiative, but we also need a national policy.

Aldrich: I want to just say one other thing because I think it illustrates very clearly what's happening in another culture which is just as urbanized and industrialized as we are. I was lecturing in Japan not long ago, and one of the public lectures I gave dealt with effects of urbanization on the family. The Tokyo papers must have reported it, because a couple of days later the Prime Minister called me at my hotel to ask if I would come and give the same lecture to his staff. I agreed, and we had a delightful discussion afterward. Then, a few days later, I spent another evening with the man who is responsible for police service throughout Japan, the chief planner of highways, and a number of other key people.

I mention this because it illustrates what a heck of a high priority the Japanese government places on family and children. For example, they're taking very stiff action to stop the dissolution of small neighborhoods. In Tokyo, it's practically impossible for a developer to tear one of these down and put in high rises. I think we have a lot to learn from Japan as well as from Canada and some of the European countries.

Kahn: I think I've left a number of people depressed. What I really wanted to do was to challenge all of us, in the rest of the discussion, to think in larger terms about youth and youth policy.

12 The Challenge of Shaping a Personal Identity

Edmund D. Pellegrino, M.D.

My intent is to initiate a discussion of one of the more urgent challenges of the age group we are considering in this conference. I refer to the challenge of fashioning a personal identity—the process of selective interaction between our internal demands and desires and those of the external social, natural and divine worlds which present themselves with increasing force as we emerge from childhood. The decisions to accept some values, to reject others and modify still others ultimately give us our identity.

The task of definition is peculiarly human, for only humans need to know who they are and what they can become; only humans can shape their identities consciously. The task is incomplete even at death, when the final adjustment is demanded. It is too intimate to yield to a general formula or chronology. Nor is it special to any one stage of life.

Nonetheless, the years 14 to 24 represent in our culture the time when we face most directly and acutely the conscious decisions which begin to set us apart as definable persons, with our own configuration of values, beliefs and principles—those we choose to hold and those we choose to reject or ignore.

With full cognizance of the dangers of isolating any one set of influences from the others, I will concentrate on the intellectual features of identity formation for the following reasons:

- The formation of an identity involves choices which are ultimately rational.
- Certain tools of learning are needed if the choices required are to be consciously and confidently made.

123

- These intellectual tools must be designed to deal with value choices.
- This is precisely the dimension most notably lacking in collegiate and university education in recent decades.
- As an educator, and as a physician, I am impressed repeatedly with the distress and the perturbations of personal growth that ensue from a defective education of the intellect in making value decisions.

Though he lives in a nexus of emotional and social forces and is delimited by the richness or poverty of his genetic endowment, man perceives himself a thinking being. He is the only creature who needs to know who he is and what he can become, who demands meaning of his life and of events, and who insists on making his own choices. Many things and events determine the constraints which an individual life must face. These are the cards fate deals us, but it is our free choices that determine how we shall play the cards.

While the process of identity formation continues throughout life and cannot be completed until death, there is a convergence during youth of a number of factors which make this period of life a most crucial one.

It is this ultimate grounding of identity in an intellectual process that will engage my attention. I would like to address the subject under the following headings:

- What a mature intellect is, and what it is not.
- The process of identity formation and the anatomy of its development.
- What special tools of intellect are required in this process.
- What is the role of the college and the university and what obstacles are encountered.

An essential task of adolescence is to begin to form one's own identity, to make active choices among the many values presented — those choices which will define who we are, what we will become and what meaning we give to what happens to us.

While the process of identity formation continues throughout life and cannot be completed until death, there is a convergence during youth of a number of factors which make this period of life a most crucial one. For one thing, the adolescent is beginning to emerge from the influence of family, church and school and becoming exposed to the multiple and often conflicting value systems of his peer groups, of the youth counter-culture, and of cultures foreign to his own. It is at this time also that the intellect is ready for the questioning of received opinion and is, as Piaget has shown, ready to deal with abstract questions, to fomulate answers, and to pass from a stage of passive acceptance of knowledge to the skeptical examination of what is put before it. This is a time also when the creative assimilation and synthesis of ideas and thoughts begins to occur. During this period, the young person devotes the major part of his time to the advanced stages of education and is forced to make choices about a job, the formation of a family and the establishment of independent relationships with other persons. These latter decision points all involve a comprehension of values and choices among them.

In primitive societies, the transition from childhood to adulthood is more formalized and structured. Young people must undergo certain well-prescribed rites of passage. If they navigate those successfully, they are accepted as mature adults and are immediately a part of grown-up society.

In the more sophisticated or the more complexly organized societies, the passage from childhood to adult life is much looser and depends more upon independent choice. There is no formal time of entry and no set of rituals which upon completion assure young people that they have become adults.

In modern society the dilemma of passage is greatly complicated by a number of socio-cultural factors. As a result of the loss of a homogeneous value system, widely conflicting values and countercultures present themselves with equal force. The influence of all authority structures such as the family, the school, the church and government has been weakened. Young people face the paradox of technological capability which promises a material paradise on earth on the one hand, and a threat to their survival as human beings on the other. A growing and overt disparity is constantly displayed on television and in the media between the stated goals and values of our society and the performance and behavior of public figures. Those who happen to be taken as models — athletes, entertainers, and the whole panorama of youth heroes from rock artists to race car drivers—often exhibit values counter to the traditional ones. There is a general loss of faith in the value of rationality itself with a strong pull to intuitive, romantic and mystical modes of knowing.

These difficulties notwithstanding, in our Western society at least, each person must ultimately fashion an identity, a self which is definable and justifiable on some rational grounds. This process will ordinarily require a maturation of the intellect as an instrument of decision and choice. Let us look, then, at the characteristics of the mature intellect. I will address this question from a special point of view and this must be kept clearly in mind as the rest of the discussion proceeds. I will define the mature intellect in humanistic terms — that is, in terms of those characteristics which are most closely related to being human and being part of humanity, because the process of identity formation is distinctly human. Animals do not emerge as distinct individuals except in their psychological behavior and physical characteristics. We have little evidence that animals will deviate from the set patterns of behavior of their species although they may manifest differences within a very limited set of behavioral responses.

Intellectual maturity is the very antithesis of some of the major characteristics that social commentators have described in today's youth culture.

The mature intellect manifests itself in a capacity to locate itself with respect to conflicting claims, values, beliefs and principles, encountered in the interaction with society, with nature or with the divine. A person with a mature intellect, in this sense, is one who has made his or her own selection of ways of living, of seeing and being. He or she has actively assimilated a set of values and made them part of himself or herself so that he or she can explain and justify his or her actions. The person with a mature intellect is willing to subject his/her choices to criticism and to opposing viewpoints without yielding to every challenge, yet being open to change, when facts or arguments demand. The mature intellect has respect for the choices of others and can see its own choices in relationship to them; it understands that its own identity is always undergoing re-examination and that the process will never be completed. The mature intellect is capable of dealing with questions of ends and purposes in relationship to some concept of the good life and the good society.

Intellectual maturity is the very antithesis of some of the major characteristics that social commentators have described in today's youth culture. For example, the mature intellect would be inconsistent with the idea of the "dislocated man" so well described by Robert Lifton. The "dislocated" man is disconnected with the past, feels no sense of responsibility for the future, sees himself as isolated in time; he has no allegiance to, or belief in, what has gone before, no hope of modifying what happens in the future. The mature intellect is inconsistent with the feelings of uncontrollable drift described by many young people who feel that their ideas are not their own and who feel manipulated by powerful forces. The immature intellect confuses fantasy and reality — a phenomenon in no small part engendered by constant exposure to television. To see the horrors of the Vietnam War — to see a human being murdered by deliberate gunshot to the head on television — in sequence with equally violent events in the regular "shows" is surely to raise questions about the value of human life. The mature intellect is inconsistent with the inability to know when to say "yes" and "no" and the failure to see the boundaries between good and evil, just and unjust.

If maturity of the intellect is the antithesis of "dislocation" it must also be distinguished from overconfident, static stances and unrealistic certitudes. The mature intellect is therefore not to be confused with the strong adherence to a value system with no comprehension of the possibility of validity in its counter-views or without the capacity to engage intellectually with those of opposing views. A mind which has conclusively but passively located itself on all the major issues of life without active examination and assimilation cannot be considered mature.

The mature intellect is not equated with "feeling good" about values unless that feeling has been subjected to critical reflection.

Maturity of intellect is not to be confused with professional or technical expertise or scholarship in any of the usual human disciplines. Specialized study enables one to solve individual problems in a particular realm of thought or action. Specialized education, however, does not fit one for asking or examining the value questions which transcend a specialty. Indeed, narrowness of focus essential to scholarship impedes the broader vision needed to define goals and purposes. Education and scholarship are not the same thing. It is possible for a specialist to locate his specialty within the larger framework of human experience. But only when that happens does the expert become an educated person.

The mature mind has no relationship to age. Some never advance beyond the earliest stages in the development of the intellect, while others may reach the stage of maturity in their thinking relatively early.

The mature intellect is not equated with "feeling good" about values unless that feeling has been subjected to critical reflection. Many people experience successive moments of exhilaration, "getting psyched up" as they encounter some new idea, some exotic experience or some unusual person. The transitory exhilaration is usually followed by disappointment and rejection until the appearance of another new experience or idea starts the cycle again.

One popular way to avoid consciously making one's own choices is to join the youth culture. This is another mark of the immature mind. Here a pre-set system of values and a lifestyle can be appropriated and justified simply by peer preference and the sharp contrast with the "straight" world of the adult. The characteristics of the youth and countercultures of our time have been too well delineated by sociologists and psychologists to be repeated here. The countercultures usually, however, are anti-rational in spirit. They emphasize "experience" engendered by the simultaneous stimulation of the senses by music, visual images, drugs, in combinations that the adult finds ludicrous or physiologically unbearable.

The above are all examples of refusal or inability to engage in the process of selective interaction with the claims and values set before us in admittedly confusing disarray in a pluralistic society. They are all mechanisms for avoiding the difficult critical intellectual process of value choice either by rejecting all choice or assuming passively some preformed system.

We do not become fully ethical beings until we are clear about our values.

Let us now turn to the process involved in identity formation. That process itself is a selective organization of values, ideals and principles that bring the inner demands of our nature into some form of confrontation and resolution with the demands external to us in our environment. The key words are "selective" and "interaction," a set of decisions to accept some of the values offered, to reject others, and to modify still others. This selective interaction is an active, conscious process which we can justify and explicate.

The selections we make determine those values which will become realities for us and which become the standards by which we judge ourselves and expect others to judge us as success or failure. Our choices give meaning to what happens to us. It is the configuration of those choices, taken as a whole, that in large part defines us as persons. We begin to have a "stance," to stand for some things and not for ot!..... and thereby we achieve an identity. As William James says, "We belong in the most intimate sense wherever our ideals belong." Choosing values of our own from the multitudinous possibilities and claims before us converts us therefore into "fighters for ends" (again William James' phraseology). All the other choices which occur at any time of life, and particularly in adolescence — namely choosing a job, a religion, a life's companion, a political party — all are related to the values with which we choose to identify. We do not become fully ethical beings until we are clear about our values, because only by the selective interaction to which I have been alluding can we develop standards of morality — benchmarks, as it were, against which to evaluate our own ends and purposes and those of other human beings.

There is a rough anatomy of development of the capacity for selective interaction. With no pretensions to the sophisticated formulations of the development of personality and intellect like those of Piaget or Erikson, I would like to propose the following stages:

- **First is the stage of passive reception of external values.** Starting with the earliest years of consciousness, the young person is exposed to the values of school, church and family. These values are, at first, taken in unselectively and parroted back at appropriate times and on appropriate occasions. This passive reception of values corresponds to the major mode of learning in the early years of life when things are learned by rote and images; stories and poems are easily memorized, as are historical dates and the names of objects, and people. Multitudinous questions are posed to satisfy the enormous curiosity of the young child. But the answers are not subjected to tests of validity or verifiability. Everything is regarded as information without the selective interposition of critical questioning.
- **The second stage is that of beginning skepticism.** The child begins to appreciate that there may be a number of different answers to the same question coming from different people. He/she begins to see

also that there is a world of fantasy that does not square with the world of experience— Santa Claus, for example. In this stage, we see the beginning of the use of logic to monitor the information contained in memory, to question our observations and to initiate debates to resolve doubt. Negative views of what is offered are common.

In this stage the intellect begins to be used as a critical testing instrument. The opinions and behavior of parents, friends, public figures come under question as do even the received religious beliefs and practices. The need for a definition of terms and for clarity of statements is realized. The child or young adolescent will not at this point have formulated his or her own stance, but he or she nevertheless begins to move in that direction by using the "via negativa" of doubt, skepticism and attempting to disprove what is claimed to be true.

- **The third stage is that of selective affirmation.** Here the process of active assimilation and incorporation of values begins to occur. The selection of a particular configuration of values and beliefs begins to define the individual as this person and the self begins to take on a more distinctive shape. The person begins to appreciate that there are indeed different answers to the same question and that these express different values which each of us must evaluate and on which we must take some position. The self then begins to determine its own relationship to others, to the world, and thus to emerge as a recognizable entity. This is the stage when the intellect begins to mature.

- **The most advanced stage of maturity is that of creative assimilation.** Here the person who develops to this stage not only uses the usual methods of deductive and inductive logic but also supplements, expands and extends the possibilities of a personal identity by creating new forms, modifying others, and adding original insight into the synthesis of his or her ethical and moral stance. There is a closer identity here between the self and some creative work of the imagination in which personal expression and identity are greatly enhanced. Each of us at this stage shapes his/her identity as a work of art, and with personal style — a stage of sophistication not open to all.

The correlation or lack of correlation with Piaget's classification of the stages of cognitive growth in children is not the essential point here. I mean only to indicate that cognition and the capacities of the intellect do exhibit a developmental anatomy which has a rough correlation with age. Important for our discussion here is Piaget's observation that it is in early adolescence that young people begin to develop the capacity to deal with formal logic and symbolic forms, and abstract entities. These are the intellectual capabilities required in making value judgments. They are essential to the selective interaction between internal and external demands, desires and beliefs out of which an identity can be formed.

There are people who have never progressed beyond the first stage of passive reception of values. . . . There are others who become arrested at the stage of skepticism.

There is also a pathology corresponding to each of the developmental stages I have described. That pathology manifests itself in an arrest of growth and development at one stage or another. Thus, there are people who have never progressed beyond the first stage of passive reception of values; they have identified themselves with a value system but have not actively and critically examined that system.

There are others who become arrested at the stage of skepticism. They lose faith in reason entirely and become convinced that there are no correct or possible answers to the fundamental questions about life and existence, or if there are answers, they cannot be ascertained by the use of reason. Such persons may become complete cynics, taking a negative stance toward every claim and proposition concerning values, the good, the true, the beautiful. Their skepticism makes them unable to formulate any alternative of their own.

There is a pathology associated even with the stage of selective interaction. One may overemphasize the process of continual change and be too ready to yield to counter propositions. The true meaning of tolerance, for example, is to differ with someone, to hold to one's view, and yet to respect the person and the opinion of those with whom we disagree. Tolerance does not demand absolute relativism of value systems. Rather, it rests in the recognition of differences and the willingness to give them fair hearing. The pathology of the last stage is usually manifest as an overemphasis on the limitations of reason. The artistic, imaginative, creative, romantic, intuitive elements become overvalued. Personal style is confused with critical assimilation and substituted for the hard and undramatic work of reason. There is a kind of reversion to the first stage of passive inhibition—this time of one's own creativity rather than someone else's values.

For most of us, the slow and vexing process of identity formation demands some careful, often error-ridden, selective interaction with the values presented to us from a myriad of sources.

Clearly the developmental stages described, and their disorders, overlap often. They follow no fixed chronology in any particular person. Arrest of development can occur at any stage. Sometimes, all the stages may be telescoped, and confusion about identity and values may be suddenly resolved with the emergence of the "new man" in a true mystical experience. (We may think of Paul on the way to Damascus, or St. Augustine in the garden.) These are rare events. Too many young people, and adults as well, hope for the blinding light of inspiration from God, or from a drug, or transcendental meditation, or even the Devil. For most of us the rather more mundane, slow and vexing process of identity formation demands some careful, often error-ridden, selective interaction with the values presented to us from a myriad of sources.

It is essential here to re-emphasize that the psychosocial and cultural influences are powerful indeed, and that my emphasis on the cognitive elements in no way denies their great significance. However, I do hold that the process of selection does finally depend upon the use of the usual rules of deductive and inductive logic, of *post facto* ratiocination and pre-logical suppositions. There is a logic, an epistemology and a metaphysics of value choices, and to dissect these entirely free from the psychosocial and cultural is not possible in the unity we call man. Suffice it to say, that as humans, we do explain, justify, criticize, judge and reject or accept value claims. It is my assignment here to describe the development of these critical facilities and this requires some violence to the complex unity we call a person.

The development of critical faculties of mind coincides with the sociocultural need to form an identity, and these in turn both coincide with the process of formal education which is the major occupation of youth in the years 14 to 24. Let us turn then to the function of the universities and colleges in the process of intellectual maturation.

Manifestly this process of identity formation takes place in all of our life experiences and at all levels of education—primary, secondary and higher. However, in our Western society, the major part of the preparation of the intellect for this process—at least in the most formal and most sophisticated sense — occurs during the college and university years.

Having said this, I wish to underscore the fact that those who do not have the privilege of attending the university must not thereby be deprived of the opportunity to develop a critical intellect. Education of the intellect and wisdom does occur outside the university, reluctant though the academicians may be to see this. Indeed, as I will suggest later, the university as it is constituted today may in some cases inhibit the development of a mature intellect.

The required attitudes of mind are those traditionally inculcated by the liberal arts.

Nonetheless, the university does have a major function in initiating the process of selective interaction and its major responsibility is to provide the tools of intellect required in the process. The university cannot guarantee that the student will graduate with a mature intellect but it must initiate the process, and above all, restore and sustain the student's faith in the feasibility of the enterprise as an exercise of human reason.

What tools are requisite for the process of selective interaction?

The required attitudes of mind are those traditionally inculcated by the liberal arts. I refer here specifically to those tools of learning and of the intellect which are most characteristically human—the capacity to reason logically and critically, to engage in dialectic with conflicting claims and ideas, to engage in moral discourse, to understand the structure of language, to be able to define terms, to use words effectively and with style—that is, with individual and personal nuances. Included also in the liberal arts are the capacity to understand the continuity of human experience and to see the relationship of past, present and future developments in ideas and culture, to be able to judge beauty and to arrive at some personal conception of a good creative effort in literature, music and the visual arts.

The topic of liberal education has been under discussion in the Western world for 2,500 years. The tools required for the enterprise of identity formation are the tools of the liberally educated mind as set forth in Plato and Aristotle and further refined and enriched by Cicero, Quintillian, the Renaissance humanists and in more modern times by John Henry Cardinal Newman. Among these tools are the capacity to disentangle fact from opinion, the proven from the merely plausible, to identify and address another person's line of argument and to see the relevance of our own subject or discipline to the culture in which it relates and to be able to tackle a subject on our own and to learn without a teacher hovering over us.

These are, in fact, the liberal arts. They are "liberal" by virtue of their capacity to "free" us from the potential tyranny of others. This freedom is crucial to the process of identity formation and value selection because in fact every statement of values, beliefs, principles about life and the world, is a claim in the philosophical sense. We are asked either directly or indirectly to give assent to a statement that this or that matter is so important that it should modify our behavior and indeed become an overriding factor in decision making. The mature intellect has the capacity to free itself to the highest degree possible from passive assent and from being overwhelmed by the claims and opinions of others.

If we keep this idea of freedom clearly in our minds, we will not confuse the liberal arts with the humanities. The humanities are disciplines and therefore specialties. They are taught today as disciplines for their content and their methodology. It does not follow therefore that anyone who is a scholar or expert in a field in the humanities is necessarily liberally educated.

What is important, however, is that the humanities are the preferred tools for teaching the liberal arts. It is the philosopher who can teach us the art of dialectic, the rules of deductive and inductive logic, the evaluation of evidence and the kind of knowledge we have, and it is the philosopher who can expose for us the metaphysical foundations of our beliefs and practices. If philosophy is taught this way, then it is a tool for the advancement of the liberal arts. If it is taught for the content of philosophical systems or the history of philosophy or even for the generation of a new system it may not be effective in teaching the liberal attitudes of mind noted above.

The same reservations and distinctions must be made in the teaching of literature, history, and language — the other three disciplines classically included among the liberal arts. The historian can use history to teach how to think critically, how to evaluate evidence, how to understand and comprehend language and culture. If, however, he teaches his discipline merely for its content or method then he, too, defects somewhat from the purpose of history as one of the tools of the liberal arts. I hardly need repeat that literature and language can suffer the same disability.

The major obstacle to the teaching of the liberal arts today lies in the confusion of scholarship and technical expertise with education and the equating of simple exposure to the humanities with the teaching of the liberal arts. Scholarship by definition requires intense concentration on some aspect of a discipline—on an organized field of knowledge and method. It requires clearly the use of the intellect to evaluate the data within the field, but the mere teaching of the field will not assure that the use of these tools of mind are communicated to the student. Those attitudes of mind must be addressed explicitly.

Two things are required for the development of a mature intellect. One is the liberal arts as the tools of learning. The other is a set of beliefs, a set of principles and values that can serve as benchmarks against which conflicting value claims can be measured.

Today our emphasis on pre-professional and professional education, while understandable and commendable, tends to depreciate the liberal arts. It is for this reason that I have worked two decades in an attempt to inculcate the teaching of the liberal arts within professional medical education, within the education of the other health professions, and in the continuing education of physicians. Here the subject matter of medicine is used as the material on which to sharpen and test the tools of the intellect—that is, of the liberal arts.

The college and the university, but especially the college, is charged in Western society with two things — first to prepare the minds of young people to engage the world of ideas, concepts and principles as free minds, and second to prepare young people for specific positions in society. The first task is best achieved by the inculcation of the liberal arts and the second by the specific disciplines and the content of the professions.

Two things are in fact required for the development of mature intellect. One clearly is the liberal arts as the tools of learning, and this we have discussed. The second is a set of beliefs, a set of principles and values that can serve as benchmarks against which conflicting value claims can be measured. The importance of a set of beliefs and values is that it provides a necessary focus for the selective interaction between internal desires and demands and the external claims of society. An educational institution which provides only for the development of the tools of intellect and does not provide a set of values on which these tools can be exercised by each student in a personal quest for identity does not fulfill its total educational responsibility. A belief system presents itself as an organized, coherent, integrated reconstruction of all the dimensions of life and experience, which gives them meaning.

Much of the confusion in all age groups, much of the uncertainty about identity and the process of identify formation, may be related to the absence of creditably articulated and defended belief systems during the crucial formative years of college.

There are practical difficulties in fulfilling this second requirement for the maturation of an intellect in a democratic pluralistic society. In the case of educational institutions under religious auspices the problem is simpler. The Catholic, Protestant or Jewish institution stands for an explicit belief system which can and should be clearly articulated and presented to the world and to its students as a base and a stance from which the crucial issues confronting mankind can be measured and judged. Even church-related institutions in the past two decades have often defected and defaulted from a clear statement of their own belief systems. They have reflected the confusion created even among believers by the tumultuous upheaval of value systems that afflicted the universities of the late 60's. There is evidence now, however, of the beginning of the return of confidence in religiously sponsored institutions. A renaissance of education under church sponsorship may be in the offing. Church-related institutions do not fulfill their responsibilities as universities unless they do attend to the inculcation of the liberal arts against the background of a clearly articulated, intellectually and critically justified set of beliefs for which they stand.

In the case of the public and secular institutions the problem is more difficult. We have traditionally in this country forbidden teachers to proselytize, and this prohibition should not be relaxed. However, the interdiction of the proselytization has been equated with a stance of moral neutrality with respect to the major issues of truth, justice and the good society. Teachers have prided themselves on the suspension of personal moral judgment in the classroom. The dominant emphasis has been on the analytic dimensions of their teaching to the exclusion of any attempt to snythesize or to give meaning to the phenomena under consideration. Indeed, one of the most lasting memories I have of my conversations with the troubled students of the late 60's was their great disappointment, disdain and disaffection for many of their faculty members who maintained moral neutrality in the face of such obvious discrepancies as racism, poverty, the Vietnam War or the dangerous potentials of unrestrained technology. Students then, as students now, have always wanted to know the answer to the question: "What do *you* think?" "How do *you* feel?"

Much of the confusion in all age groups, much of the uncertainty about identity and the process of identity formation may be related to the absence of creditably articulated and defended belief systems during the crucial formative years of college.

I believe the problem of providing a benchmark set of beliefs in public institutions is resolvable if we can become sufficiently mature intellectually as a society to understand the supreme social importance of mature intellects in our midst. When we gain that maturity, I can conceive of three ways in which secular and public institutions could initiate the critical discussion of values in a framework of stated belief systems:

● The first would be to encourage each teacher to be more open about his own value system. As crucial questions rise in class a teacher should be prepared to explain and justify his own point of view as long as he makes it clear that it is his own point of view and not the point of view of the school, or the society in general. As long as the teacher stimulates critical examination of his position there should be no real danger of a "takeover" of one ideology. Students expect teachers to have positions; they are disappointed and shortchanged when a teacher does not illustrate that it is part of an education to be able to take a position and to defend it. A set of beliefs is also essential because a student needs to have a stance and some benchmarks before he can examine some of the more esoteric, bizarre, and far-out value systems that are being offered to all of our society today. There is no reason why an adolescent should not examine the claims of any or all value systems — especially when they violate reason and depend largely upon intuitive and immediate knowledge. To examine these positions it is necessary to have a stance from which to examine them. We cannot protect students against exposure to these systems; the best we can hope for is to prepare them to deal with them critically.

- A second would be to propound the secular philosophy, since in many cases this is the philosophy implicit in our public educational system anyway. If the secular philosophy is propounded responsibly and with the same attitude of critical examination, then I see no reason why it cannot also serve as a foil for discussion, a subject for examination, and a stimulus for the student's own thinking. As long as we permit rigorous and critical examination of the claims of the secular philosophy then theoretically we should be able to use it as a means of inducing each student to take his or her own position in the matter.

- A third way, and I think this would be the preferable way, would be to define the common set of beliefs we do hold in our democratic, pluralistic society. The ideals of equal rights, of tolerance, of justice, of truth, of mutual assistance, of common concern — all of which are in the Western tradition — certainly form a belief system. The uniformity of this set of beliefs would to a certain extent be mitigated by the obvious differences of opinion among the viewpoints which make up our pluralistic democracy.

By taking one of these three routes and preferably the last, even a secular or public institution could provide a set of beliefs as a template or as a set of benchmarks for the student.

Up to this point I have emphasized largely the interests of individual members of our society and particularly the individual quest for identity. There is, of course, very great social significance in what I have been saying. A mature society requires that as many of its citizens as possible possess a mature intellect and have a clear view of their own position and stance. It is difficult to think of the preservation of a democratic society and a democratic world order without individuals in that society capable of examining the questions of value and purpose in a critical fashion. Much of the difficulty in our nation today in arriving at national policy for health, social welfare, security, environment, and so forth, is related to our incapacity to define what we mean by a "good society." We have not debated the question or reached a consensus on what goals and ends we wish to achieve and what kind of society we wish to foster.

We need more citizens who have mature intellects and who have a set of beliefs which provide them with a stance from which they can engage in conversation with their fellow citizens. It is only a society with individuals whose own identity is clear that can provide the conditions for the compromises that are necessary in any common societal and political effort. A human being can compromise only when he knows precisely what his own values are, and at what point he may yield in the interest of a common goal. Without knowing the limits of good and evil or the limits of justice or equality we cannot know if we have compromised these values irreversibly.

It is imperative that we do not confine the process of selective interaction and the preparation of the intellect for this end to the university. There is a grave danger in assigning this vital task exclusively to any institution.

The mature mind is also the only defense members of our society will have against a technocracy — that is, rule by experts. Every field of special knowledge is possessed of a special language and special methodology which is closed to the non-expert. We are all at the mercy of experts in environment, in health, in nuclear armaments, in space travel, in finance, and so on. Unless we preserve our intellectual capacity to judge the values and purposes to which they wish to put their technology we will be their unwilling subjects. Once we equate expertise in a special field with expertise in the values pertinent to that field then we have lost our freedom as a society. We are subject then to the kind of society designed for us by the particular technocrats whose knowledge and whose methodology is dominant at any particular moment in history. Science and technology become an ideology and a political system. The dangers are all too obvious.

Therefore, if we are to preserve our liberty and still make use of the capabilities of technology, if we are to vote intelligently and participate in public policy debates, if we are to retain those elements of life which are most distinctly human, it is important that we foster the healthy, mature development of personal identity and of intellects capable of arriving at that identity in a reasonable and rational way.

It is imperative that we do not confine the process of selective interaction and the preparation of the intellect for this end to the university. There is a grave danger in assigning this vital task exclusively to any institution. The values of that institution too easily become the official position or the only respectable one. The general public must be exposed to the same questions; the process is, after all, a lifelong one. Mere exposure for a period to the universities will not suffice.

We must find means to raise the fundamental questions in a variety of places — in community organizations, in local government, in the media, and I believe in a more formal and organized program of continuing education in the humanities conducted by colleges and universities for graduates and non-graduates as well.

We must overcome the fear of adults that the young will be seduced to error if they are permitted to examine all the facets of a belief system, critically and openly.

A formidable obstacle will be the disinclination of Americans to debate value or "gut" issues. We are not accustomed to the rules of polite disagreement about belief systems. We are fearful of offending, and we have yet to learn how to distinguish between an idea with which we disagree and the person with whom we disagree.

To achieve these goals we must overcome the current loss of faith in the use of reason to decide questions of value; to restore the liberal arts as tools of the intellect and make them essential parts of the educational process both within and without universities. Above all, we must prepare faculty members who themselves are capable of teaching in this fashion. This may be the gravest obstacle we have to overcome.

Lastly, we must overcome the fear of adults that the young will be seduced to error if they are permitted to examine all the facets of a belief system, critically and openly. It is understandable that adults wish to replicate their own world and their own values. If we follow this policy we can only expect a static society which would never enrich, expand, and advance in achieving its main purposes. This fear of seducing the young to error very often betrays an underlying lack of faith in the values we hold. That, in turn, may well be due to our own incapacity to develop a stance of our own which we can justify and explain to others. I can only remind you that it was this charge of seducing the young when he was trying to develop their critical intelligence that brought Socrates to his sad, but nonetheless dignified end, 2,500 years ago. The tragedy is that the death of Socrates continues to be enacted in a thousand different ways up to our day.

Discussion

John Schweppe: How can colleges and universities help young people define themselves? I'm thinking now in terms of such questions as: "Who am I?" and "Where am I going?"

Pellegrino: As I indicated, by providing them with the tools of the intellect, through the liberal arts, and by providing them with a set of values to examine critically. Professors should not hesitate to express their point of view on such questions.

Alvin Eurich: I was interested in the distinction you made between the liberal arts and the humanities.

Pellegrino: John Demos gave us an excellent example of how the content of the humanities — in his case, history — can be used as an intellectual tool. As I indicated, however, the purpose of the liberal arts is to free us from the potential tyranny of the opinions of others.

Harold Visotsky: I have a pragmatic problem with what you said about identity formation being an intellectual process. A lot of young people look for their identity in peer groups.

Pellegrino: I think you're right, but I don't think that all young people who join a peer group really want to belong to that group. Some are either pressured into joining, or they go along because they don't have a value system of their own.

Paul Bohannan: When it comes to helping young people find themselves, I don't think we should rely exclusively on the university. We ought to develop "back-up" institutions. For example, I could make a case for the family being the starting point in the identity-formation process.

Pellegrino: I thoroughly agree with you.

A. J. Kelso: Do you make a distinction between "teaching" and "professing?"

Pellegrino: Yes. To profess something is merely to make a claim. Not all professors teach.

13 The Evolution of Human Genders

A.J. Kelso

During the past few decades, the path our ancestors followed in the course of human evolution has become much clearer to us. The major landmarks, from the adoption of erect posture to the appearance of *homo sapiens,* are well known and the pattern is so familiar it seems simple. But the results of human evolution, the variety of contemporary human beings, serve as reminders that human evolution is not as simple as it seems. Change within has not been uniform. As long as our attention is directed to the whole species, we may miss some of the complexities within.

Males and females illustrate the unevenness of human evolution. Of course, the genders evolved together, yet across the species and within local populations males and females faced separate environmental stresses, particularly severe on sexual and reproductive characteristics, and in responding to these each gender has changed in different ways. Within our species evolutionary change has been relative to the adaptive problems to be solved. Females have changed more than males and the difference suggests a stage in human evolution which has been neglected, a stage possibly as significant as erect posture or the acquisition of culture.

For a long time, human sexuality was thought to be basically similar to primate sexuality in general. We can take little comfort from realizing that this view was based on a comparison of ourselves with animals caged in zoos. But about fifteen years ago Jane Lancaster and Richard Lee, basing their argument on free-ranging monkeys and apes compared with a contemporary hunter-gatherer group, concluded that human beings, in reproductive features, are decidedly different, "qualitatively different," from other primates. This helped change our outlook; human beings were not simply sexual apes, they approached sexual and reproductive matters in their own way, the human way. As other investigators sought to identify the specific characteristics of the human way, it became clear that the distinctive features are to be found in human females. The most striking differences are: 1) continuous ability of females to become sexually aroused; 2) increased reproductive capacity; and 3) a long post-reproductive life. Compared with other primates, human females stand out as conspicuously different, whereas males are similar to other primates, especially our closest living relatives.

Continuous Ability To Become Sexually Aroused. This feature is usually called loss of estrous periodicity and, of course, refers specifically to females. Sexual activity in mammals is usually confined to a brief period lasting perhaps only a few days out of a year. In primates the cycles are more rapid and the receptive periods — that is, estrous periods—are signalled and controlled by hormonal changes in females, any one of whom may enter estrous at any time.

There is a kind of continuum in sexual receptivity from African apes to human females ... This is both evidence of our common ancestry and a measure of the extent to which human females have differentiated in the hominid line.

The general primate picture is, then, one of primate males continuously capable of sexual arousal living in groups with females who are discontinuously arousable. In extending the comparison with human beings, it will be helpful to focus on our closest relatives, chimpanzees and gorillas, and, as much as possible, to use data from free-ranging groups. Sexual behavior is distorted in captive animals, and, because it is rarely observed in free-ranging animals, many have concluded that chimps and gorillas are not especially sexy. But we do not know what, if any, difference a human observer makes and, of course, we don't know what goes on in the forest after the observer has gone.

African ape females are sexually receptive only a few days of their reproductive cycle. Gorilla and chimpanzee males appear prepared for arousal of sexual interest at almost any time. Adult females in the group become receptive at different times and as they do the adult males compete with one another for sexual access to the estrous female. Among chimpanzees an estrous female is treated more deferentially than a non-receptive female and, if meat is available, the former receives the greater share. In African ape communities it is far more common for the group to be without any estrous females than it is to have one or more. The African apes, in other words, are typical of the primates in general, discontinuously receptive females with continuously arousable males. The duration of receptive intervals is longer among chimpanzees (6 to 8 days), and among them it is even longer in those groups living more widely spread out in open country (8 to 10 days). Gorilla females are receptive for three or four days. The cycle recurs in both gorillas and chimpanzees approximately every 35 days. A few months after an ape female conceives, her receptive periods cease altogether for three or four years until the infant is nearly weaned.

In contrast human females can become sexually aroused at any time, they remain sexually competent throughout pregnancy, and recover their capability for coitus within a few weeks and their ability to conceive within a few months after the child is born. To be sure, there is a kind of continuum in receptivity from African apes to human females. On one end, Jane Goodall notes a few instances (less than one percent) of female chimpanzees who, even though they were not in estrous, were mounted by males. On the other end, Udry and Morris have shown human females to be most responsive sexually midway between menstrual periods and least responsive four days before the onset of menstruation, though they continue to be responsive throughout the cycle.

To Robin Fox the evidence for a continuum is sufficient ground to disagree. Fox claims that the estrous cycle still operates at a very profound level in the human female. Recognizing a continuum, however, should not obscure the real differences between human and African ape females. The continuity from apes to human beings is both evidence of our common ancestry and a measure of the extent to which human females have differentiated in the hominid line. In effect, the changes have synchronized male and female sexualities more closely in men and women. The synchrony is far from perfect, but much closer than between the genders in the African apes.

Chimpanzee females mark their receptive periods by a pronounced swelling of their genital regions. The swelling is much less pronounced in gorillas. Mature human females, capable of continuous sexual arousal, display attributes continuously attractive to the sexual interest of males — well developed breasts, everted lips, rounded hips and buttocks. Desmond Morris discussed these and their evolutionary implications in his widely read book *The Naked Ape*. Since these female features become fully developed at sexual maturity, there is a possibility they may be the continuous morphological expressions of the underlying continuous capability for sexual arousal in females.

Increased Reproductive Efficiency. Thirty years ago G. G. Simpson observed that human reproductive efficiency is the highest in the animal kingdom. Basically, this means human beings do not waste their reproductive potential; they are capable of producing more babies than gorillas and chimpanzees, and infants are more protected and cared for and so are more likely to survive infancy. Reproductive efficiency, therefore, consists of two related components: reproductive capacity and ability to care for the young once they are born.

An increased reproductive capacity may have been an adaptive solution to the problem of high mortality which existed in early hominid populations.... But it may be an adaptation to conditions that no longer exist. Yesterday's solution thus becomes today's problem.

In regard to their capacity for reproducing offspring, women exceed gorilla and chimpanzee females in all significant measures; they live longer reproductive lives, they ovulate more frequently and ovulation resumes more quickly after a child is born. Human females are capable, under optimum circumstances, of bearing an infant every fifteen months. Over a reproductive life of approximately thirty-seven years, from age 15 to 47, a woman is therefore *capable* of bearing twenty-five children even without twins or triplets. Fortunately, optimum circumstances are rare and few women actually reach their full reproductive capacity. Among modern day hunter-gatherers, fertile women who survive their full reproductive life usually bear fewer than seven children. Contemporary hunter-gatherers, however, live in difficult marginal subsistence settings and for this reason fail to represent accurately the human capacity. Hutterite women, generally well nourished and actively seeking to have as many children as possible, average between ten and eleven children.

By comparison, chimpanzee and gorilla females are estimated as able to bear approximately six or seven infants. In fertile females, capacity is probably close to reality. The estimate is not secure, however, since it is based on very little data. No one, to emphasize the point, has ever observed the birth of a gorilla or chimp infant in the wild. But if the estimate is even close, then human females are able to produce many more offspring than African ape females can produce. In this regard —the capacity to produce infants—human females are reproductively more efficient.

An increased reproductive capacity may be an adaptive solution to the problem of high mortality which existed in early hominid populations. The most compelling evidence for our improved reproductive capacity comes indirectly from the explosive population increase in recent decades following a marked decline in mortality rates. The recent increase illustrates the capacity of our species to produce more of its kind and the growth has taken place even though all cultures have one or more constraints on unrestricted fertility — for example, marriage regulations, intercourse taboos, sanctions favoring infanticide, abortion, celibacy, chastity, virginity, homosexuality, masturbation, surgical sterilization, and recently, the availability of a multitude of pills, potions, devices, gels and foams which inhibit conception.

The overall effect of culture on human reproduction may be more insidious, for culture defines the very nature of the relationship between the sexes. According to Columbia University anthropologist Robert Murphy, culture defines the relationship in a way which reverses the biological reality. Citing the work of Masters and Johnson in his support, Murphy argues that the common cultural beliefs about human sexuality express views which in fact are the opposite of the biological reality. Females are in fact sexually more competent than males, their need for sexual arousal less than males, their capacity for sexual gratification longer lasting and of finer quality than males. Overall, the effect of the cultural reversal of these beliefs represses female sexuality and thereby constrains reproductive capacity.

So improved reproductive capacity may be the powerful remnant of an adaptation to conditions which no longer exist. Yesterday's solution thus becomes today's problem. Improved reproductive capacity was not a complete solution even yesterday because of the immediate effect it had on child care and nurturing, the other aspect of reproductive efficiency. More babies per mother meant more infant care for survival. In the economics of evolutionary biology, survival is always contingent upon the successful completion of a businesslike transaction across generations. Sociobiologists have developed this marketplace analogy to enable them to measure several aspects of behavior in adaptation.

Infant care is a way one generation protects its investment in the next. Though at a very general level one may see social organization as a protection of the mother-infant relationship, in most mammals, the tasks of infant care are performed primarily by mothers, and in primates the responsibility is more exclusive since mothers rarely bear more than one infant. Among the gorillas and chimpanzees observed in the wild, mothers regularly cared for but one infant at a time. Among chimpanzees the care is extended over a long time and reports are common of a chimpanzee mother carrying an infant on her belly as a juvenile rides her back. Yet even among chimpanzees, mothers rarely care for more than one infant.

Human females are unique among the primates in having a long post-reproductive sexually active stage in their life cycle.

In human evolution, with multiple infants per mother relatively common, primary child care tasks extended to juveniles, adult males and childless females. The early hominids protected their generational investment by involving all other members of the group more directly in the nurturing of children. In human groups child care is the direct concern of all individuals; it is imbedded in affective ties, sharing networks and reciprocity relationships, an entire social reticule expressed in the kinship systems of preliterate societies. *Homo sapiens* is a species in which both genders and all life stages beyond infancy have been selected in part for their ability to contribute directly to the care of infants.

Long Post-Reproductive Life. Human females continue active sexual lives long after their reproductive lives are over. There are no data on gorillas, but chimpanzee females, according to a recent study from the Yerkes Primate Research Center, continue to ovulate throughout their lives. Human females are unique among the primates in having a long post-reproductive sexually active stage in their life cycle; there is nothing comparable in apes, nothing comparable in men. The value of experienced females in stabilizing and nurturing small populations with multiple infants per mother is easy to imagine, though very difficult to measure. Grandmothering augments parenting and thereby increases the opportunity for selection to improve on reproductive efficiency.

The post-reproductive stage is not merely the addition of years under favorable conditions, for it occurs as well under adverse conditions. Nor is a long post-reproductive life understandable as simply another step in an irrevocable ontogenetic program. The post-menopausal life stage, currently more than one third of a Western woman's life, is understandable, to return to the business metaphor, as an advantage to the individuals of one generation enhancing the likelihood that the investment in the next generation will be successful. Mothers may have several infants at once, but a long post-reproductive life assures that the proportion of adult females to infants remains high in human populations.

The patterns of female sexuality and reproduction are distinctive in human beings—patterns which took shape after our ancestors separated from the lines which led to modern gorillas and chimpanzees. As mentioned earlier, the patterns reveal that evolution within our lineage has not been uniform. The differentiation of females has been so strikingly different from males, one can hardly escape the implication that the changes must have been of fundamental importance in human evolution. This brings us to the question of how the changes influenced the course of human evolution. What was their significance? Erect posture, for example, freed the hands for carrying and for making tools. But what role did the differentiation of female sexuality play in human evolution? One fact, itself an enigma, offers a clue: our ancestors spread across the earth very rapidly. Changes in female sexuality may have enabled that explosive dispersion to take place.

Whatever the significance, females changed in more basic ways during the course of human evolution than males changed.

If the biochemical evidence is accurate, our lineage had separated from the African apes by from four to six million years ago. As recently as two million years ago, the hominids were still confined to the tropics. One million years ago there is evidence of our ancestors beginning to move out of the tropics and into temperate climates in the Old World. Within the last million years the hominids spread around the world. The dispersal took place well after the first recorded fossil evidence of erect posture, and well before evidence of the earliest significant cultural achievements. For these reasons there is reason to believe the spread of our species is somewhat independent of the adoption of erect posture and acquisition of culture.

Without the hard evidence of fossils, one can only speculate on the sequence of and interrelationships between events. Our ancestors were colonizers. A colonizing species constantly faces high mortality as its populations encounter new adaptive challenges. High mortality selects for improved reproductive efficiency. Once adapted, populations faced the problems of improved reproductive efficiency which, in turn, created again the conditions calling for more colonizing. As Duane Quiatt argues in a forthcoming publication, improved reproductive efficiency in females provided individual males with an opportunity to improve on their reproductive fitness by affiliating more or less permanently with one or more females. The synchronizing of male and female sexualities rendered these affiliations attractive and at the same time facilitated colonizing by enabling very small groups to move out and occupy new territories. Even a pair of individuals could become a colony. Thus, improved reproductive efficiency with continuous capability for sexual arousal could have worked together in enabling our ancestors to spread rapidly into new regions. Extending primary child care to all members of the group, to post-reproductive females in particular, represented an investment in parenting which further secured the chances of survival in the subsequent generation.

Whatever the significance, females changed in more basic ways during the course of human evolution than males changed. While this paper deals with three of the most distinctive changes, others have been proposed. Human females regularly copulate from a ventral position; African ape females regularly in a dorsal position. Human females are capable of orgasm during intercourse; ape females rarely if ever experience organsm in intercourse. Human females masturbate as a means of sexual gratification; ape females rarely masturbate and it is possible that in the wild they never do. Sexual activity continues in ovariectomized women; it ceases when the ovaries are removed from primate females.

Some of these apparent distinctions may disappear as more data become available on primates living in their natural settings. At present they add strength to the impression of basic changes having taken place in females during human evolution.

There is a familiar evolutionary precedent in the evolution of mammals. The differentiation of the mammals involved fundamental changes in the sexual and reproductive characteristics of females, changes not paralleled in mammalian males. In many ways the distinctive features of mammalian reproduction — internal fertilization with implantation, placentation, parturition, lactation — are the distinctive features of mammalian females. Similarly, the distinctive features of human sexuality and reproduction are the distinctive features of women.

The turbulence of adolescence and early adult life in contemporary society is thus likely to be felt more keenly by females than males. The human female during those years is adapted to a close network of support from others, particularly from older post-menopause coresidents. In place of the intimacy and support of kinfolk, a contemporary woman faces alone the critical decisions accompanying sexual maturity, the early pregnancies, childbearing and responsibility for child care. There is no biological precedent for such isolation among our primate relatives or for that matter even our closest human ancestors. Living alone, or nearly so, is an unnatural act.

Men have the largest penises and are alone among the apes in lacking a penis bone. These features deserve more serious attention than they have been given, but apart from these there is only one other thing one can say about the sexuality of men that one cannot also say about ape males: Men talk a lot more about sex. A good place to stop.

14 Chemical Dependence and Drugs

Donald West King

The age group we are discussing is a very healthy age group. The childhood diseases have pretty well run their course, and the death rate is extremely low. Most of the deaths are caused by accidents, homicides, and suicides. (See Table 3 on the following page.) There is, however, a relatively heavy dependence on drugs of various kinds. Most of my remarks will deal with that problem, but I also want to talk briefly about mental illness and the growing problem of environmental pollution.

Therapeutic Drugs in Excess. Most of the discussion we hear about the drug problem among young people has to do with marijuana and the so-called "hard" drugs, but there are lots of other drugs which, if taken in excess, can be dangerous—not only to adolescents but also to older people. Even an overdose of certain vitamins, such as A, D, E, C, and B_6 can produce harmful side effects. There is also a broad spectrum of antibiotics, some of which affect certain people in unexpected ways. Penicillin is a good example. Some people just can't take it because they're allergic to it. Unfortunately, there are certain chemotherapeutic drugs that are extremely useful in treating certain diseases, but some of the people who take them develop secondary malignancies.

Even aspirin can be dangerous. Many young people in the age group we are discussing take aspirin in large quantities, and they develop severe stomach and pancreatic ulcers. Cortisone, a very useful anti-inflammatory drug, sometimes produces insensitivity to pain, which can be quite dangerous. Antihistamines, when taken in large doses, have serious toxic ramifications.

Table 3
Ten Leading Causes of Death
Among People 14 to 24
(1977)

	Rate per 100,000
Accidents: Motor vehicle 44.1 All other 18.4	62.5
Suicide	13.6
Homicide	12.7
Cancer	6.5
Heart disease	2.5
Congenital anomalies	1.6
Influenza and pneumonia	1.3
Cerebrovascular diseases	1.2
Anemias	0.4
Diabetes	0.4
All other causes	14.5

Source: Mortality Statistics Branch, Division of Vital Statistics, National Center for Health Statistics, DHEW, Hyattsville, Md.

Now we come to contraceptive drugs, which more and more young women are taking. The oral contraceptives usually contain estrogen or progestin, or combinations of the two. Some are long acting, and taken only once a month. Others are taken daily, beginning about five or six days after menstruation. Both of these drugs have harmful side effects.

Estrogen produces nausea, dizziness, and vomiting; fluid retention or edema formation; painful breast engorgement, hypertension; mild to severe headache; blood clotting disorders; alteration in liver function tests and jaundice; changes in blood glucose levels, and increases in hormone binding proteins in the blood.

Progestin causes fatigue, mental depression, and lack of motivation; decreased menstrual flow; changes in liver function; acne, weight gain, and hirsutism.

Women under 50 rarely have coronaries—they seem to be protected from them until the beginning of the menopause—but some preliminary studies indicate that women who have taken contraceptives for at least ten years have a higher rate of coronaries than those who have not. So maybe in a few years we'll see an increase in coronaries among women under 50.

Caffeine, Nicotine, and Alcohol. These are the drugs that are most widely used by adolescents and older people. Caffeine, of course, is present in Coca Cola, coffee, and tea. Most people think it's not important, that it isn't anything to worry about. But caffeine is an important stimulant. It increases the response of the reflexes. It also decreases the appetite, and some people use it to lose weight, but if taken in large doses over long periods of time it can be harmful. People who use it that way do so at their own peril.

The adverse effects of nicotine are so widely know that I don't have to go into them here, but the tragedy is that many young people start smoking at an early age.

Teenagers are also beginning to use alcohol at earlier ages than was the case 20 or 30 years ago. When used in moderation, alcohol is not a particularly harmful drug, but when used to excess over long periods of time it can produce acute intoxication, delirium tremens, hepatitis and cirrhosis of the liver.

The question that's usually asked about alcohol is how much is too much? A friend of mine has an answer to that. I don't know whether it's right or wrong, but I'll pass it along to you anyway. He claims that if you have eight ounces of 50-proof alcohol a day for ten years you are going to find yourself in trouble. That's about four martinis a day. When you take a drinking history from patients you sometimes get four bottles of wine a day, or two bottles of whiskey. Those people are in real trouble already. I guess the best answer is that the ability to handle alcohol varies widely from individual to individual, but if you drink more than you can handle over a long period of time you will develop acute chronic alcoholism.

Hallucinogenics. The one that has received most attention is LSD, which young people call "acid." There's also mescaline, which is present in cactus buttons. These drugs precipitate dizziness and nausea, and they have psychic effects which may be pleasurable for a while, but they can be extremely dangerous.

Some people include marijuana among the hallucinogenic drugs, but other people claim it is not hallucinogenic. The number of people throughout the world who use marijuana is probably around 300 million. Here in the United States it's about 20 to 30 million, and my guess is that maybe half of the 45 million people in the age group we are discussing here have tried marijuana at one time or another. We know it produces euphoria, sleepiness, and a dream-like state, but we don't know much more than that about it. In some states it's outlawed, but interestingly enough in a few other states it has been given legal status for experimental use in the treatment of glaucoma.

Tons and tons of anti-anxiety drugs are prescribed with little or no justification.

Depressants. The depressants are a large group of drugs that include the barbiturates and the so-called anti-anxiety agents. Among the longer-acting barbiturates are barbital, which is sold under the trade name Veronal, and phenobarbital, which has the trade name Luminal. The most commonly used intermediate-acting depressants bear the trade names Nembutal and Seconal; they induce sleep. Many people take Nembutal or Seconal every night of their lives.

The anti-anxiety agents — such as Miltown, Equanil, Librium, and Valium — are the most widely used prescription drugs in the United States. Tons and tons of these drugs are prescribed with little or no justification. Their toxic and psychological effects can be extremely harmful.

Narcotics. These are the so-called "hard" drugs. They include heroin, cocaine, methadone, morphine, and opium. All of them are covered by the Controlled Substances Act of 1970. The granddaddy of them all, of course, is opium. It's the substance from which most other narcotics are made, and it's been around for hundreds of years. After the Civil War, for example, it was estimated that about four percent of the population was taking opium in one form or another.

Heroin, a widely used derivative of opium, is the drug most commonly sold on the streets. Twelve kilograms of opium will make only one kilogram of heroin. Then it's diluted as much as 20 or 30 times with quinine, barbiturates, or lactose. Although heroin is a tremendous respiratory depressant, it also has certain other side effects caused by the contaminants that are put into the final mixture. One of these side effects is hepatitis, which is transmitted by contaminated syringes.

The heroin problem, although still serious, has been diminishing in the past four or five years. In 1972 it was estimated that there were about 700,000 heroin addicts in the United States. This figure dropped to about 400,000 in 1974, and of those approximately 80,000 were on methadone maintenance. Methadone itself is an addictive drug. It's not anywhere near as dangerous as heroin, and you can maintain people on it, but eventually you have to wean them away from it just as you have to do with heroin.

Cocaine seems to be growing in popularity. A lot of people seem to be sniffing it these days. Its abuse potential is high, and it can have severe physical and psychological effects. Morphine is not as widely used as it was around the turn of the century.

Deaths from narcotic overdoses seem to be on the decline. In New York City last year, the number was about 800, almost half of what it was in 1966. That's not too bad when you consider that there were 1,700 murders in New York last year—and New York is not the murder capital of the world. There are several cities in the Southwest that have a higher murder rate, per capita, than New York.

Mental Illness. There are so many experts on mental illness in this room that I don't have to dwell on that subject except to say that about 60 percent of the 48,000 deaths in the 14-to-24 age group are caused by mental illness. In addition, there are 5,600 suicides, 4,600 homicides, and 1,700 fatal accidents. Two things about these figures are interesting to me. First, mental illness must have been a contributory cause in some of the deaths by suicide, homicide, and fatal accidents. Second, I would guess that about half of the deaths by suicide resulted from an overdose of drugs.

Environmental Pollution. We're all aware of the tremendous array of contaminants that pollute the air we breathe, the water we drink, and the food we eat. Most of them are carcinogens. The air pollutants are probably the most pervasive. For example, there is one county in New Jersey that at one time had the highest incidence of cancer in the country, largely because of its heavy concentration of industry. We've all heard about the smog in the Los Angeles area, produced by the exhaust from automobiles. And then, of course, we're beginning to be conscious of the fact that smokers not only injure themselves but also pollute the air that others breathe.

Pesticides and industrial waste are the principal causes of water pollution. It took us a long time to recognize the dangers of DDT. It has not only contaminated the water we drink, but it has also gotten into mother's milk. It's now banned in this country, but in several other countries it's still being used as a pesticide. Love Canal in upper New York state brought home to us the dangers of industrial waste, but there are probably about 600 Love Canals around the country. Asbestos is still being dumped into Lake Superior.

Rural areas are thought to be healthier than highly industralized cities, and that may be true, but we've all heard about the cattle feed contaminated by fertilizer and the migrant workers in California being sprayed with pesticides.

In recent years we have become increasingly concerned about food additives. Some are harmful, but others are not. Cyclamates and saccharine are probably not carcinogenic, but they have been largely pushed off the market by a recent amendment to the Food and Drug law, which says that any drug that causes cancer in any species in any dosage is banned.

The fact of the matter is that we are surrounded by toxic substances, and there's no way to ban them all. We're going to have to live with them.

Discussion

Leonard Duhl: Several of my colleagues and I have been gathering data on toxic substances in the equipment and materials used for construction. We have found that buildings constructed with sealed windows and built-in air conditioning systems have a higher moribidity and mortality rate than other buildings. That applies to homes as well as office buildings. And with the energy crisis and the slowing down of air conditioning, there has been a tremendous buildup of all the toxic stuff in the rooms.

I mention this in reference to your comments because it seems to me that what we've got here is a situation where we spend most of our lifetime trying to control the environment, and each time we control the environment with one thing we end up having to control the thing we controlled, and then we are in an endless process of controlling the controller. Isn't it about time to start raising questions about this whole philosophy? Maybe we ought to shift gears. This is something the adolescents are raising questions about all the time. Do you have to live in a system that controls everything or, as they say, do you take to the hills?

Robert Aldrich: I can add one other example. In the March issue of the *Journal of Epidemiology,* there's a good article by someone in our department showing that the occurrence of carcinoma is directly related to how close the individual lives to power substations, telephone poles, and transmission lines. It's very sophisticated.

Sidney Werkman: Don, I was curious about why you said that tons of anti-anxiety agents are given to patients without justification.

King: Maybe because of my personal experience.

A. J. Kelso: I think a lot of people use them because they're so easily available

Harold Visotsky: That's right. People get upset about something and they reach for some Valium. Why should they do that?

Werkman: Why shouldn't they take something to relieve their anxiety? There are 25 million people in this country who are chemical copers, which means they regularly use agents of the kind Don was referring to, but that doesn't mean we are an over-medicated society.

Edmund Pellegrino: I think we are. We spend 18 billion dollars a year on drugs in this country. We've abandoned our critical faculties when it comes to the ingestion of drugs. We have an insatiable itch to take something.

John Schweppe: I'd like to bring out one other small point, and that is that some of these agents are given for a specific reason. Whatever they're used for, the physician should explain to the patient exactly what the drug does and what side effects it has. The cross-mixing of drugs can have very serious consequences.

June Christmas: I agree. I think we do often have physicians prescribing without encouraging patients to explore the causes of stress and anxiety and pointing out that there are other ways of dealing with them.

Werkman: May I just make one point here? It has to do with deep values in the society. I don't think we can cast these values into a puritanical versus permissive dichotomy, with some people saying you should be able to bear whatever your suffering is and never mind the mechanisms or prescriptions that are available for helping people to cope.

Pellegrino: That kind of polarity has no place in the relationship between the physician and the patient. Before a physician prescribes medication he should make it clear to the patient the alternative possibilities and let the patient decide, based on that knowledge. If the patient wants medication that really isn't necessary, the physician must decide whether or not to prescribe it. That really has nothing to do with puritanical views.

Werkman: There are many different positions, but I've worked with physicians who felt very uncomfortable about these issues. There may be alternatives, but sometimes they're not very realistic, like taking long walks and cold showers and all the rest to control sexual desire.

Visotsky: There's another consideration involved here. We had a detoxification unit for patients who had been taking prescribed drugs and become psychologically or physiologically dependent on them. The point that I'm making is that many of these patients realized that the drug they were taking was affecting their lives, their health, and their relationships with their families, and literally wanted to get off the drug. They came in asking for alternatives. If they had been made aware of these alternatives before they were put on the drug, I think they would have chosen not to take it.

Willard Wirtz: I have a great deal of difficulty in formulating my question. For the last 15 minutes we've heard the experts here discuss carcinogens on a level totally beyond my comprehension. I'd like to get some idea of the scale of concern we ought to have about the things that have been discussed. I assume that the life expectancy of a 25-year-old today is longer than it used to be. Can anyone tell me, on a scale from one to ten, how great our concern should be about carcinogens?

King: I'll give you a personal opinion. Bruce Ames says that so much has happened in the chemical industry in the last 15 years that he thinks we're going to have an increase in cancer in the 80's and 90's. I disagree with that. Right now there are probably between 20 and 30 known compounds that are causing cancer in humans. If you take something like diethylstilbestrol, there are less than 300 cases of cancer from DES out of all the millions of doses that have been given in this country. Vinylchloride is another compound that has made the front page. It produces a peculiar type of cancer of the liver. There are less than 200 cases documented in the country. Arsenic is another example. There are less than 20 cases documented in the country of cancer produced by arsenic, yet every plant that has anything to do with arsenic is considered hazardous. My answer to your question is this. In those industrial situations where the toxicity is high—even for a small group of people—extreme precautionary measures should be taken. The same sort of things should be done in chemical plants. But to say that one chemical plant that is making a chromate is a hazard to the entire general population is crazy. The real hazard lies in the quantitative change that's taking place. Individual susceptibility of the cell to the great spectrum of 60 or 70 thousand chemicals that are now present in our environment is completely unknown.

Ernest Bartell: I think the eagerness of society to zero in on these rather small cases that we're talking about, and the eagerness of practitioners to prescribe drugs as opposed to other alternatives, really has to do with our preception of probabilities. The practitioner prescribing the drug knows well what some of the consequences are, but at least there's a probability of some kind of success. A quick fix of some kind is more certain than the alternatives. There's a bias toward making the choice whose chances of success seem greater. And that applies to a lot of things in our society. It also has to do with going after these carcinogens. We go after the ones where we can find something rather than trying to regulate the ones where there is still a lot of uncertainty.

Visotsky: Getting back to the question of alternatives to drugs. I think you really have to belong to what I call the "chicken soup" school of therapy. Maybe they won't work, but there's no harm in trying them.

Paul Bohannan: My concern is with the cultural dimensions of the drug problem. That we can maybe do something about. We can learn to think about it and talk about it.

Aldrich: I want to comment on another aspect of the drug problem. I think the ignorance of this age group about personal care and personal hygiene and that sort of thing is really rather large in comparison to what is known, and I think that those of us here in the medical profession can rightly be criticized for not launching a really major effort to educate people in elementary school and on up about personal hygiene.

There's a very interesting program of that kind going on in India, where, as you know, the child mortality is very high. What they do is teach personal hygiene of the simplest kind to elementary school children and then have those kids teach it to their younger brothers and sisters at home. This system works very well. We in this country could do a lot more than we're doing now.

Werkman: I'd like to follow up on that point. In this age group several of the most important health issues have to do with nutrition as well as ingestion of all kinds of drugs. Society in general has not come up with useful guidelines that would help young people make informed choices about their diets. What we have is a lot of advertising on TV and in the newspapers that pushes junk food. No wonder young people are bewildered about what they ought to eat.

Visotsky: That's the kind of anxiety that causes some people to take Valium.

15 Psychiatric Factors

Leonard Duhl

My remarks will be brief, but I hope they will be provocative. My thesis is that we are moving into a time when we must change the way we think about reality.

What we have called education is really de-education, because it stifles our natural curiosity and teaches us to conform to an established set of realities. It ignores the fact that there are other realities, and other ways of looking at what we call reality. Take, for example, John Schweppe's questions: Who are we? Why do we exist? What's the meaning of life? These are the big questions that somehow never get answered by science but get answered by religion and philosophy. My experience with students in the university, and interestingly enough with students as young as junior high age, is that when you start raising these questions they get turned on and are willing to talk about issues in a way that says: "Hey, those questions are interesting," while some of the scientific issues which get us into dead ends are issues that are bothersome to students. If you can show them how science and philosophy tie together they're really turned on. Their imagination begins to move in many, many different ways. So I think this raises questions for us about what education is all about.

Now, briefly, I would like to comment about health and medicine. We have talked a lot here about growth and development. I have a suspicion that what we are calling health is the optimum balance between one's internal and external environment. When a crisis or a difficulty arises, we sometimes break down and go into dis-ease. What tends to happen is that we develop a series of symptoms. The physician goes in and treats the symptoms. In modern medicine we do magnificently in treating the symptom, but we do very poorly at returning that person to the normal growth and development cycle — what I call healing. Treatment and healing, which used to go hand in hand, are now almost separate. In the process of getting more and more scientific, the physician has given up that healing quality which has in part been the laying on of hands. In part it's been the reconnecting of the individual to all the significant others. This is, in fact, the role the shaman plays—the role of doing both the treatment and the healing by reconnecting people. So networking and reconnecting people to the environment become part of the concern with health.

At the present time, there are a tremendous number of people emerging who are called alternative healers. Many of them, I have to say, are quite fraudulent. On the other hand, they are performing functions that the medical profession is not performing. This is a more general question I'm raising for all of us — that if our institutions do not deal adequately with the problems that face us, alternatives will emerge. Confronted by these alternatives emerging in medicine, the profession can do one of two things. It can absorb or co-opt them, which is what I guess will happen. Bioethics is just one of the opening doors. Acupuncture and biofeedback are other ways of co-opting the alternatives. I believe that a lot of these unorthodox healing methods will get absorbed by modern medicine. On the other hand, there are some things that modern medicine cannot do. Some of us work in cultures where there are not just white Anglo-Saxons, but also Chicanos, American Indians, and people from the Pacific. What we suddenly find is that we cannot treat these people primarily by Western medicine—that we have to find some kind of blending between the two.

We're living in a period where we have to search for many different ways of looking at things.

The Public Health Service, to my amazement, is supporting medicine men on the Navajo reservation. There are programs jumping up in many different places around the world in which the two kinds of medicine are being put together. So, just to sum up, the challenge that I think is now before us is that we're living in a transition period. We're living in a period where we have to search for many different ways of looking at things. But how can we live with the ambiguity of this period? As Harold Visotsky asked earlier, do we have to take Valium to deal with that ambiguity and anxiety? Or can we openly deal with it and realize that some of the problems do not have answers given our current reality? And that we have to start searching in new ways? Ways that may be very foreign to our education?

My experience with all kinds of alternative health methods during the past eight years has taught me that I have to accept the fact that people do believe in them, and that some of them are effective. It's like the argument I used to have with my father, who says: "Prove it to me and then I'll believe it." I've had to say that you've got to go back to the old religious notion: "You'd better believe it first, and then I can prove it to you."

I found that as I moved into alternative reality I had to go into it with an open mind — to suspend judgment first and then try to understand and appreciate it. But what does this have to do with adolescents? I think they are probably becoming better prepared for this world than any of us. And I must say that I am disappointed that there aren't any of them here in this room. It's very hard for us to talk about them without their being present.

Discussion

Auditor: I'd like to raise a point. I think Freud mentioned that the musings of the educated conscious mind do not fully explain reality. I was curious as to what you might say to that.

Duhl: Freud was talking about the unconscious, but I'm talking about cultures other than our own. An awful lot of us don't know how really civilized those cultures are until we get into them — until we actually wallow in them. The more I got into some of these other cultures the more I realized that some of them — for example, that of the Tibetan Lamas — have a view of the mind which is far more complicated than Freud ever dreamed. They took me into spaces that I haven't got the vaguest conception of yet.

Paul Bohannan: I would like to point out, from an anthropologist's point of view, what Len has been doing is participant observation — that is, going out to discover how other people in the world organize their reality and understand it. I'm going to say categorically that there is no better way to understand what our form of reality is than to go out and come to terms with another form of reality. It is the only cure for ethnocentrism. It is the only cure for thinking that you have all the right answers.

I think it's very important to pay a lot of attention to other cultures, because we all know that the multiplicity of cultures has gone into making this country what it is. And by going out and actually living in other cultures you begin to get another point of view. You really develop a stereoptic point of view that gives you an entirely different perspective on reality. The cultural dimension is not something that anthropologists just talk about. It's something without which we can't even understand our *own* little culture and our *own* little prejudices.

Robert Aldrich: I think that a great many people — and I'm one of them, or was until I began to see a little light at the end of the tunnel — who don't think of culture as being as firm and rigid and real as it is. I used to think that changing a culture was as easy as buying a new suit. I really did, and I know that's not so. Now I'm beginning to think that changing a culture is as difficult as changing a person's genetic makeup. I'd like to hear more discussion of this.

Duhl: It might sound crazy, but the current discussions about the problems of death are really quite interesting, because they are part of a concept central to all religions. It has to do with the question of change. You've got to die in order to move into another world. Every culture of the world that talks about death talks about it in that way. And it's no accident that death is a very important issue for adolescents. They're talking about it now like I've never heard it talked about before. They're talking about dying — giving up everything they have. Giving up what most of their searches have been about — all their religious quests. You have to give up everything that you hold dear to you in order to really be born again.

John Schweppe: Len, I understand certain elements of what you're saying and to me it's intriguing. I know that if you're going to understand something you have to understand what went before it, and I know it will take a long time to fill in all the gaps. But is what you're describing a state of mind or a new way of looking at reality?

Duhl: I guess I'm suggesting that science has to switch around, from linear thinking to something which is relatively holographic — which says in fact everything already exists. That we may not have to find it. That all we have to do is let it emerge. It's a very different notion of research, and a different notion of education. Instead of saying the child is an empty vessel that we have to fill up, it says maybe we have to invent ways for the child to emerge. That's a philosophy of education that hasn't done very well. It means in medicine and in health maybe instead of thinking that we can treat anybody, all we do is facilitate the self-healing of the person in tune with the universe around him. The native American healers say they don't treat anybody. They say all they do is transmit God's energy of the universe to heal. Now, that's a very different notion.

So you're perfectly right and I'm really asking for a radical shift in the way we conceptualize issues. But I still believe that science is relevant. I'm really asking for what Ed Pellegrino asked for earlier — a dialectic between these various points of view so we can start expanding the way we look at things. For example, I have a suspicion that really great creativity does not come out of rationality alone. It comes out of a blending of rational with nonrational thinking — a hunch, or a flash of intuition. So, John, what I'm really pushing for is a major shift in the way we think about reality.

Auditor: When you say you're pushing for a blend of rational and nonrational thinking, aren't you taking a step back?

Duhl: I don't think so.

Auditor: But putting the pieces together in a different way doesn't necessarily mean that it's rational. It could be a step back.

Duhl: I'll go so far as to say that it's the last gasp of the scientific model to claim that if we only keep doing something we will find the scientific answer. My point is that we have to do something else besides.

Bohannan: I agree. What you're suggesting is not as frightening as many who have not tried it seem to think. Finding this "something else" would establish a sort of dialectic—a way of looking at problems from the point of view of many cultures, instead of looking at them inside the environs of our own culture. Our problems, of course, are our problems. But, let's get a wider view of them. They're very much easier to solve that way.

Pellegrino: I think the key to what Paul said is that you really have to understand the place where you are standing in terms of the experience of other cultures. One of the problems of the age group we're discussing is that they have not yet established a place to stand on. That creates this tremendous confusion and shift in values.

Duhl: I agree. A lot of the young people we see as psychiatric patients are those who have gone out and experienced the alternative world without being grounded. Their problem, though, is that those of us who are their teachers do not help them go through the process, chiefly because we haven't gone through it ourselves.

A. J. Kelso: You don't have to go very far to make the contrast, either. You don't have to go to another home or another place. The value differences between us and the young people we're talking about are probably as great as the value differences between separate cultures. Try listening to your teen-age son or daughter.

Bohannan: I want to add something to what I said earlier. In order to understand what is happening right now we cannot stand where we were standing yesterday, because new realities are emerging all the time.

Aldrich: I want to raise a question that I raised last year at an international symposium on the future of children. I got thoroughly blasted, but I'll try it again, because we really shouldn't engage in a serious discussion about young people in our own country without looking at the youth population of the rest of the world. The numbers are staggering, and so are the differences in culture, in beliefs, and in concepts of reality.

One of the new realities, as we may learn first hand very soon, will come when the young people in the rest of the world decide to come and get some of the things they perceive as needs.

I would guess that if you look at the international cohort of 25-year-olds, the vast majority live in Third World countries. These are the rough dimensions of the world reality. The reason that these numbers are so staggering is that we in the United States are so concerned about the future of our own children that we fail to recognize that the resources needed to produce high-quality offspring in the rest of the world should be in the hands of the 15-to-25 or 15-to-35 age group right now. And they aren't.

It seems to me that we have to ask ourselves how to go about realistically sharing the world's resources so that we will have high quality future generations on this planet. That's a much bigger challenge than our problem in the United States.

16 Socio-Psychological Approaches

June Jackson Christmas

I shall discuss some approaches we use in New York City to help youth experience adolescence adaptively in regard to themselves, others and society. I also want to share some of my concerns and the challenges I believe they pose for programs and policy.

Adolescents experience stresses resulting from the normal developmental tasks. Their physiological drives are intense. They have to cope with bodily changes, sexual urges, feelings of anxiety, frustration, aggression, and self-doubt. They live in a nether world, as neither children nor adults. In early adolescence, impulsivity may be prominent. In later years, the outside world is a potent force. The struggle to free one's self from parents arouses guilt and causes conflict. Peer pressure can contribute to growth and also to anti-social experimentation. The search for identity is painful; the search for one's own values is uncertain.

Socio-psychological approaches which we have developed address the varied aspects of adolescent life. Their referents are intrapsychic, interpersonal and social. I have conceived of them as containing, in varying degrees, five elements: social, therapeutic, educational, vocational, and environmental. In each, there is a social component, for human relationships are essential to the implementation of all approaches.

Specific programs can be preventive, developmental, therapeutic or rehabilitative in nature. Conducted in the natural setting of community or in the mental health, youth, or health service system, many of these

169

programs are group-structured, taking advantage of the natural desire of adolescents for peer relationships.

Socio-psychological group approaches provide assistance to individuals in reaching one or more goals:

- To accomplish the developmental tasks of adolescence: psycho-sexual identity, independence; a value system; and vocational choice
- To develop effective coping mechanisms and adaptive alternative behaviors
- To decrease psycho-social, physical and psychopathological symptons, behaviors, discomfort, and dis-ease
- To reduce disabilities due to physical, mental, and social handicaps
- To gain increased ability to relate effectively to the familial and social environment
- To have a greater sense of self-esteem and awareness and to achieve a satisfactory level of personality integration

Now let me turn to the approaches themselves, emphasizing one or more examples of their usefulness.

In the *social* approaches, individual and group relationships are directed toward socialization, resocialization, and social learning so that youth can move in widening circles of social competence. These approaches may be carried out with varying degrees of structure. Some employ peer leaders; others rely on trained counselors, educators, or therapists. In peer group "rap sessions" the focus may be on problems of daily living. Other groups may deal with interpersonal relationships; decisions about sex, drugs, school; or habits and behavior. Topics for discussion and mutually arrived at solutions may arise from the group or be initiated by the leader. Sometimes the issues are hypothetical; often they hit close to home. Worries about the future, resentment over parental control, and difficulties in resisting pressure toward sexual activity (or guilt at not resisting) are typical. As members discuss ways of handling these situations, they use the group not for content but for the social learning that comes from problem solving, tolerance, choice, and support.

Therapeutic elements predominate in approaches based on psycho-dynamic principles, in which understanding of factors and influences beyond awareness is important. Group therapies involve the expression of inner feelings and the resolution of deep-seated conflicts through the use of group as a whole as well as through relationships with individual group members. This process can lead to improved self-awareness, impulse control, and clarity of identity. In group therapy, emotions, fantasies, problems, and reactions are expressed in an atmosphere of both acceptance and support and constructive response. Members come to see their own habitual ways of relating; by understanding their behavior within the group they gradually come to recognize "ghosts from the past" that distort today's reality and interfere with functioning. Appropriate for those with personality problems, neuroses, or behavioral disorders, group therapy is particularly useful in helping young people going through the normal processes of adolescence as all participants express feelings, experience self-discovery, and begin to care for themselves and others.

In the educational approach, content is more important than process, although both are used.

The third or *educational* element focuses on social learning also but in the context of problem solving and cognitive development. Avocational skills development, intellectual growth, and the acquisition of knowledge characterize this approach. Content is more important than process, although both are used. Interpersonal relationships are viewed more in the "here and now" than in the probing, therapeutic approach. This element is used in sex education, in teaching parenting skills, in education about responsible choices around drugs, drinking, and sexual activities. It is both the gaining of information which is important as well as the learning of its applicability.

The *vocational* element is present in approaches less related to emotions, relationships, or information, but more actional in process. For younger adolescents these approaches include group projects or tasks which teach planning, require collaboration and patience, and provide a sense of accomplishment or the opportunity to handle failure. For older youth, role playing helps in preparation for the unknown of tests, interviews, or the work world. Skills development leading to prevocational competence includes making choices and learning new social roles.

The *environmental* element relates to control over the social situation, decisions affecting one's life, and increased life options. Many youth have to deal with racism, discrimination, and poverty; if emotional disorders are also present, they are at greater disadvantage. The interventions in this mode aim to strengthen positive and decrease negative aspects of the social environment. Youth become more involved members of their community and the wider society, by social action and by links to community support systems which give them a sense of being valued, of belonging, or of having a shared purpose. Group forces are used to mobilize change potential in both individual and group as well.

Let me now turn to concerns and problems. Although I have emphasized groups of peers, using social interventions as an effective way to address some of the needs of adolescence, I must add that I do not consider these to be panaceas. Sometimes they are used as an approach which can best be described as "watered-down therapy" rather than to meet specific needs of certain youth. At times this is as a second-best substitute for recommended therapeutic programs or for more culturally appropriate therapies for ethnic minorities of color. This is shortchanging those youth, for all approaches need to be used selectively and appropriately for their intrinsic value. Complementary to other services or alone, they should not be used to explain away the lack of program innovation or to deal with a Proposition 13 mentality.

We need more multi-service centers, where adolescents can find help for a wide range of problems such as drugs, alcohol, jobs, or parents, without having to enter the health care system.

Using adolescents to help one another, both within the age group and with younger children, is a good idea. But we should not forget that young people can also help and be helped by older people, both in regard to tasks and in human relationships. In our society today, with its high mobility, small apartments, and numerous nursing homes, we lose sight of the fact that having grandparents around can be a wonderful way to teach the young how to relate to older people. I don't think we draw enough upon the kinds of people who are natural helpers — for example, the "surrogate parents" whom kids turn to for help when they run away from home. They have an enormous influence on young people, and they could be of great help to us if we learned how to use them effectively. Finally, I think we need more multi-service centers, where adolescents can find help for a wide range of problems, such as drugs, alcohol, jobs, or parents, without having to enter the health care system.

My time is running short, but let me finish by listing some of the problems that we face in trying to help adolescents:

- Adequate funding is extremely hard to come by, especially at a time when federal, state, and local budgets are tight.
- The scope and boundaries of the mental health field are in dispute. Is mental health really health, or is it education? Part of our problem here is a guild problem.
- We need to do more health promotion, to encourage constructive life styles and to do more in our schools to teach young people how to cope with stresses inherent in living.
- Some young people are more disadvantaged than others. We can't afford to short-change them in dealing with young people as a whole.
- How can we figure out what works and what doesn't work? Or, to put it another way, how can we evaluate experimental projects without running the risk that the evaluation plan itself will distort the experiment?
- There are many young people who went through periods of stress and came out knowing how to deal with stress. How can we build on their experiences in a constructive way to foster positive mental health?

Discussion

William Threlkeld: What about the peer groups you mentioned? Are they carefully structured peer groups, or are they the spontaneous peer groups that occur on the streets?

Christmas: We use various kinds. We have found that whether you have an informal group that you start on a street corner, or a formal group set up by an older teenager in our leadership program, or even one that just grows out of kids hanging around together, you can get the benefits of peer-group therapy. In the public schools we started out by talking about drugs, but we soon found that the kids also wanted to talk about other things, including sex and parents. The subjects ranged far and wide.

Sheila Kamerman: The thing that particularly fascinated me was that all of the therapeutic approaches you talked about were essentially universalistic approaches of the kind Al Kahn mentioned. Would you care to comment on that?

Christmas: You're right. I think that all young people need to know what makes them tick. That's one thing they all have in common. But they also want to know what makes *you* tick. What goes on in *you*. What makes you different from another person. That, too, needs to be part of the developmental approach.

Kamerman: What kinds of people make good peer-group discussion leaders?

Christmas: Well, from my experience I would say the best leaders are people who have the ability to empathize and not to be narrow and ethnocentric. People who care. People who can conceptualize. People who understand what kinds of interplay turn young people on.

Alfred Kahn: One of the obvious things to ask at this point is: Shouldn't the educational system be able to train such people, or is it something that you have to feel? If you have specialists, do you train them or do you untrain them and then try to educate them?

Christmas: I have helped train teachers in the art of leading peer-group discussions, and you're right. Some of them have had to unlearn some of the things they were taught before they could learn the right things to do.

John Demos: I think you've spread out a series of interesting and hopeful and quite possibly inspiring strategies for us. But I was wondering if you would comment a little more about the relationship of these particular strategies to the whole social environment in which the individuals involved find themselves.

Christmas: I certainly did not want to short-change the social environment. In answer to your question, some adolescents find the external world frustrating to the point where they feel like acting with violence against society, or against other people. But if you give them an opportunity to do something constructive with that anger—to turn it into positive social action or sublimate it—you have taught them how to deal with their feeelings and conflicts less destructively and with control.

17 School Counseling as an Intervention

Helen R. Washburn

I am a public school educator. I have been engaged in that endeavor for fifteen years. Twelve of those years have been spent as a school counselor working with students, parents, and colleagues in the schools of Boise, Idaho. This past year I have been on leave from my position in Boise to serve as President of the American School Counselor Association. This unique experience has offered me the opportunity to travel extensively and to be involved with education programs and educators throughout the country. When I discuss counseling as an intervention, I will be doing so primarily as the practice of counseling takes place in public schools. I have made what I hope is a clear distinction between education and public schooling. To attempt to discuss school counseling outside the context of the situations facing present-day schools would be meaningless. Therefore, many issues facing schools today have been included. In light of the age group that is the focus of this meeting and the fact that the public school, in most cases, is part of their experience, such a focus seems appropriate.

What actually occurs in schools in the name of school counseling depends on many factors, including:
- The values and priorities of the community.
- The attitudes of the school administrators.
- The training, skills, and values of the counselor and, yes, even to some extent, the needs of the students.

Before we proceed further, perhaps we need to explore the meaning of the term "counseling."

That may be easier said than done, however, for counseling is one of those words that everybody understands but no two people seem to understand in precisely the same way. The term was not invented by psychologists, educators, or social workers. It is part of our everyday language, and the activity it represents is a part of our everyday life. During this century there has been an ever-increasing amount of professional counseling carried on, and an increasing number of people who see themselves as counselors.

In schools there is considerable confusion as to what is counseling, what is guidance, and what is education and consequently a broader school responsibility. One can only understand present school counseling, I feel, by understanding from whence it came. I am, therefore, going to highlight some of the events that have led up to the model I advocate and which I feel makes sense in light of the complexities surrounding present-day schools.

School counseling is an American invention, and it emerged from the early vocational guidance efforts. The need to place people in the labor force of early twentieth-century industrial New England and provide some incidental "moral" guidance gave school counseling its start. Testing functions to assist with that placement process were added in the 1920's and 30's. Also added at that time was enforcement of compulsory attendance laws.

With Sputnik, the encouragement of the academically talented took on a high national priority with emphasis on college placement, and under the National Defense Education Act many classroom teachers went through a retreading process to become school counselors. They were trained in quickly assembled counselor-education programs which lacked commonalities as to purpose, philosophy, and development of relevant skills. Client-centered therapy utilizing a one-to-one approach was touted as the ideal. Those newly trained counselors re-entered their school settings to encounter administrators and teachers who saw them as having an abundance of free time to assume the paper work associated with college admissions and quasi-administrative tasks related to the running of the school. In 1959 James B. Conant published a book entitled *The American High School Today* in which he advocated large comprehensive high schools. He also advocated full-time school counselors for every 250 to 300 pupils. Conant's ideas had a tremendous impact on public education. The consolidation efforts caused a disruption of many natural communities. The large high schools resulted in depersonalization, which created new problems.

The Sixties brought the era of great social change. Increasingly, the school was viewed as the instrument to restructure society. The civil rights movement heightened concerns about the need for guidance efforts to reduce the effects of stereotyping and bias in student choice-making. The "do your own thing" credo and the rapid increase in drug use among the age group caused a rise in conflict between students, parents, and school staff. Group process as an alternative to one-to-one counseling gained credibility. Prevention and early intervention began to be discussed, and elementary school counseling came into vogue.

The Seventies brought increased federal legislation — busing; the Career Guidance and Counseling Act; Title IX, the anti-discrimination legislation; and probably most far reaching in its impact, Public Law 94-142, the legislation for mainstreaming special education students. The Seventies also brought forth the idea of doing prevention and early intervention through the teaching of psychological education. Parents and teachers were viewed as colleagues with counselors in this process. A large volume of curriculum materials on decision making, career education, substance-abuse prevention, and self-development became available for use by counselors and teachers. Incorporation of psychological education was questioned, however, as a back-to-basics philosophy also hit the country.

School counseling needs to move beyond the stage of being a service related only to the whims of the administration and a drop-in clientele.

Psychological education is educational jargon for presenting in a planned, sequential way the tools for assisting students to figure out the questions of:
- Who am I?
- Where am I going?
- How do I get there?

Over the years the role of the counselor moved from one of advisor to one of therapist to now, one of facilitator of problem solving and decision making and educator for personal development.

I will come back to more specifics on counseling as an intervention later, but let me say that there is tremendous diversity as to what counselors do in schools. There is a need countrywide to accurately identify student needs in individual schools, and to structure guidance and counseling programs which include services aimed at addressing those identified needs.

School counseling needs to move beyond the stage of being a service related only to the whims of the administration and a drop-in clientele.

To develop such counseling programs rather than offer a service represents a significant change. The determinants, however, of the priorities of a district—change or otherwise—are the attitudes, values, and economics of the local community. The American way is local control. Everyone thinks he or she is an expert on education by virtue of experience or because of financial investment. Will Rogers summed it up by saying: "Education ain't what it used to be and never was."

With counseling services, parents sometimes resist those aspects of programs that usurp prerogatives they feel are better left to the family. In some areas counselors are subject to sharp criticism, particularly where information relates to sex education and values clarification. In the very same community there may be strong need and support for such programs. In public education the role of the school, and consequently the services of the school, are in a continual state of flux. Students' needs are often secondary to other issues.

Communities exert their influence on guidance programs, but the real determinant of what happens at the school level is the school administrator. Guidance services for too long have been seriously affected by what was comfortable for and met the needs of the principal. In one school where I worked ten years ago as the girls' counselor, my chief value rested with my abilities to be a model of ladylike decorum for our young women and to keep students from throwing what seemed like the cook's favorite dish, peanut butter balls, at the acoustical tile ceiling of the cafeteria during the lunch period. There were times when those two expectations came into conflict. That administrator, up until last year, was still a principal. The only thing that had really changed was the reality of Title IX, which prevented separately identified girls' and boys' counselors.

A school administrator who is able to react to the needs of students as well as to utilize the unique skills, talents, and interests of the school staff can do much to assure the kinds of programs that really assist students to grow and develop in healthy ways.

Most administrators tend to make decisions themselves and few view themselves as group facilitators, yet research indicates that the principal's leadership and the climate of the school can be greatly improved by bringing staff members and students into collaboration in solving problems. Such a style also increases the likelihood of innovation and change.

Reward systems being what they are in schools, utilization of such a democratic leadership style in preference to an autocratic style is discouraged rather than encouraged.

Although most state departments of education offer models and guidelines for guidance and counseling programs, such programs are seldom put into practice without administrator and community support. Top-down change processes have never worked very well in education.

Few districts actively recruit school-level administrators who have skills at implementing innovative programs. The exceptional administrator is seldom given awards such as uncommitted time for research and reflection or positive recognition for achievements, let alone monetary awards. Instead, such administrators are assigned larger schools with more responsibilities and some are frequently treated as subversives.

Although most state departments of education offer models and guidelines for guidance and counseling programs, such programs are seldom put into practice without administrator approval and community support. Top-down change processes have never worked very well in education.

Counselors as persons generally are concerned people who work hard. Many feel that they are the "token caring person" and also a dumping ground for minutiae with little autonomy for changing the situation. Gilbert Wrenn, a leader in our profession, says "counselors in schools should be the ones to spot the person others don't see." Counselors would like to be that for students, yet the size of modern schools plus the complexities of today's schools prevent that kind of service. In a report I read recently a young black female said: "In my school I'm one of 3,000 . . . sometimes even I can't find me."

To make a point about student needs I'd like to share a story about a mother bird who was always building nests without bottoms. This became a community concern and some of the good citizens decided to pay her a call and find out what was going on. When asked about it, she replied, "I like building nests, but I hate kids."

One might think that we as a society hate kids when we look at our expectations of the schools and our lack of consideration of the needs of the adolescent age group. Worse yet, it appears some of the kids' needs are aggravated by or are a result of situations presented by the school. The stress of teachers over conflicting expectations — social change agent vs. back-to-basics — compounds the complexity of our dilemma.

People generally don't like force, yet currently social change is being implemented through economic and legal mandates, and as a result some unhealthy dynamics are occurring in schools. The learning climate is being affected and teachers' abilities to relate to students are restricted because of teachers' difficulties in coping. Students are reacting by missing more school, exhibiting violent behavior, and all those other myriad reasons for alarm that we read about in the newspapers.

There is a need for a more universal understanding and commitment to what services should be provided in the name of school counseling, and those services should more accurately reflect student needs.

Don't, please, misunderstand: I believe in the motives of the federal legislation. But from my perspective, the classroom teacher has been given a major role, maybe even *the* major role, in social change in this country and with virtually no preparation or psychological support for carrying out that role. If that function is indeed one which policy makers wish to continue, then counselors might well provide a valuable service by serving as a support system for teachers and as a convener of school staff discussion groups for the purpose of collective problem solving. For example, counselors could move into a role where they worked more with staffs and individual staff members to assist them in coping with the new dilemmas confronting the school and in developing skills for working with the increasingly diverse needs of students.

The implication for policy makers is that there is a need to:

- Study the impact of legislation on the people in the schools — staffs as well as students.
- Examine and alter the reward systems that are currently the norm in public education.

- Encourage smaller schools or school communities within larger schools so that trust, a sense of school community, and greater opportunities for responsibilities for students are established.
- And most importantly, continually strive toward community-level consensus on the purpose and desired outcomes of public education.

As I close, may I say that my perception is that many publics see the functions of school counselors as being a significant and positive intervention in assisting adolescents to gain understanding of themselves and their lives and to gain information and skills for making life choices. However:

- There is a need for a more universal understanding and commitment to what services should be provided in the name of school counseling, and those services should more accurately reflect student needs.
- Counselors should give greater importance to their educational or guidance functions and utilize group processes both in classrooms and in specially organized groups and in-service with teachers to carry out those shared functions.

There is a serious lack of coordination between in-school and out-of-school resources that serve youth, and between the schools and the private sector. This makes schools an artificially isolated experience for students, leads to frequent duplication of services, and wastes scarce resources.

- School counselors need to use maximally those others in schools — especially others involved in pupil personnel services — as well as specialists in the community, to provide the intensive therapeutic counseling for those students who need it.
- Finally, educators and policy makers must determine ways to improve schools as positive learning environments for people and then commit the resources to assure that growth occurs.

In summary, I offer the following conclusions and recommendations:

- Data collected in California indicated three areas in which students emphasized the need for information and guidance is greatest: planning for their futures, including work and further education; getting them to appropriate courses and seeing that they earn proper graduation credits; and understanding themselves and relating to others.

- The same study suggested that students in urban schools feel that guidance services are not accessible to them and do not assure them the confidentiality required to openly discuss their personal problems. They do not view guidance staff as the appropriate people with whom to discuss such problems.

- Sexist and racist stereotyping continue in career counseling, job placement, course assignment, and access to college admissions information.

- No effective vehicles seem to exist to communicate student needs to guidance staff or policy makers, and students are not included in designing or evaluating guidance services.

- There is a critical gap between the broad and idealistic rhetoric of guidance and what is actually delivered.

- Most guidance services operate without any program definition or clear priorities, and so are accountable to no one, do not deliver comprehensive services, misuse increasingly scarce counseling resources, and frustrate guidance personnel, students and administrators.

- There is a serious lack of coordination between in-school and out-of-school resources that serve youth, and between the schools and the private sector. This makes schools an artificially isolated experience for students, leads to frequent duplication of services, and wastes scarce resources.

- Innovative programs and counseling approaches do occur, but generally depend on the energy and inspiration of an individual counselor. Such programs tend to fade with that individual's energies. There are a few means of institutionalizing, sharing or evaluating innovative approaches.

- Ratios of students to staff in urban high schools range from 250:1 to 400:1. Almost all counselors complain that paperwork reduces direct counseling time. Counselors do not have classroom access or regularly scheduled times to meet with students. The lack of staff time for counseling is exacerbated in urban schools where government compensatory programs require extra paperwork, where counselors have larger student loads, and where students have fewer outside resources to deal with their problems. Counselors deal with crises and with the assertive students who demand their time, while most students receive little or no guidance.

My recommendations address the serious gaps between student needs and the guidance system's definition of services, between the potential of guidance counseling and its actual delivery.

- Develop comprehensive guidance programs that address student needs.
- Coordinate in-school with out-of-school resources.
- Involve students in defining, evaluating and delivering guidance services.
- Direct special resources to big city schools.
- Establish better means to share information about guidance programs and resources that address diverse needs in productive ways.
- Incorporate guidance into the curriculum.
- Assure confidentiality in the student-counselor relationship.
- Improve student-to-counselor ratios.

Discussion

Donald King: Helen, I thought Conant's whole premise was that if you had a big high school you had better art, music, better vocational education, better counseling, and so forth, and that the little country high school couldn't do as good a job in providing these things.

Washburn: That was his premise, but I think over the long haul we're finding that the small high school has some advantages over the large high school. Students in a small school have more of a feeling of belonging, more opportunities for relationships with faculty and students, and they have more freedom to grow in their own way. They have more opportunities for participation and leadership in school activities. These experiences are as important if not more so in preparing for life as the formal class experiences.

King: Wasn't Conant trying to give students of average ability a better chance to compete for admission to college?

John Darley: I think that's true, but it's also true that a lot of those students drop out of college after the first year or two.

John Schweppe: My question is how can we encourage young people to choose careers where the demand is great enough to insure them of a reasonable chance of getting a job? Or to put it another way, why don't colleges discourage students from preparing for careers that are already overcrowded? Couldn't career people from the community help with that problem?

Washburn: Let me start with your first question. I think there's a great need to identify the needs of individual students. Often times people outside of the schools have their own idea of what those needs are, and encouraging the building services around those ideas. If we did a better job of matching student needs with what is going on in the school or the school district, and making full use of the school's potential for meeting some of those needs, I think we might do a better job without spending a lot more money. We need to indentify some priorities within those needs, and if the money does not exist we should try to do the best job we can in some of the more important priority areas.

As to the question about utilizing community resources, I think school guidance counselors are in a unique position of being able to identify and coordinate those resources with the resources of the school. If they could feel better about that coordination role, and if that role could receive more importance in terms of the community, I think the schools could tap into all kinds of community resources that they are not currently utilizing.

I also want to say that school guidance counselors can't always get some of the data they need concerning student opinions and attitudes. It is increasingly hard when they have to have parental approval. Finally, I'd like to point out that schools tend to move from crisis to crisis, and guidance counselors don't have a heck of a lot of time to reflect on what's going on because they are also building up to the next crisis.

18 Counseling and Intervention in Colleges

John G. Darley

Instead of making a formal presentation, I would like to list a series of well-established facts about counseling at the college level, based on the findings of group or individual difference research and cohort studies. With our limited time for discussion, I hope we can then move directly to a consideration of their implications for action and policy change in the provision of counseling services for the age group we are concerned with at this seminar.

- **Counseling in the college years, as a form of intervention, is modestly effective. (See references 1 and 2.)**

The evidence for this comes chiefly from a 25-year follow-up study completed in 1965 by David P. Campbell, and from a more recent study done by Mary Lee Smith and G. V. Glass.

In Campbell's study, which was done at Minnesota, the 300 or more students who had been counseled, when matched with 300 or more who had not, showed superior performance in college grades, in degrees earned, and in life-accomplishments. But they also showed a greater amount of discontent and concern with their life. We're not entirely sure yet what this means. It may mean that a touch of "divine discontent" is a requisite for seeking appropriate professional help. And if so, that tells us something about the kinds of people who will profit from counseling and the kinds who will not seek it.

- **Educational environments can be measured and can differentially affect "staying power" and outcomes for various student ability levels and personality types. (See references 3 and 4.)**

185

These findings derive from the work of Henry Murray at Harvard, which has been carried forward by George Stern at Syracuse and Bob Pace at UCLA. Their studies show that if you put very bright kids— able to succeed at high levels of competition—into clearly differential environments, they may disappear from the system; they cannot be held in the system if it is intellectually impoverished or inappropriate to their needs.

- **In American higher education, a college can be found that will admit and may graduate any level of student ability. (See references 3 and 5.)**

In a study that we did some years ago, we were able to convert average scores of entering freshmen classes in a representative sample of colleges and universities on a single standard scholastic ability measure. The lowest average school represented a percentile of 1 on the norm, and the highest average represented a percentile of 94 on the same national norm. We pride ourselves on having an educational system of great diversity, but I think the system is not cost effective; no single master plan exists to guide its development and it has no "buying guide" for consumers.

- **The composition of a graduating college class would be significantly different in age, sex, achievement, and size if graduation were based on achievement rather than time spent. (See reference 5.)**

This is a classic study, by Learned and Wood. It has never received the attention it deserves.

- **Social class structures, sociometric variables, sex and ethnicity are all closely related to *but not inevitably causative of* academic performance, staying power, and ultimate station in life. (See references 6 and 7.)**

These data, from our colleagues in social and cultural anthropology, generally derive from studies of small communities: "Elmtown," "Middletown," and the "Yankee City" series. In such communities the intercorrelations of socioeconomic variables are quite high and determinative of one's fate.

- **But the correlational network between these variables is significantly lower in large urban areas than in small communities. (See reference 8.)**

The Hochbaum study, done in our social psychology program at Minnesota, dealt with these intercorrelations in the city of Minneapolis; they are significantly lower in an urban area and are therefore less determinative of an individual's fate. One major role for the counseling psychologist is to be an advocate for the individual by concerning himself with fostering the individual differences related to desirable outcomes and minimizing the effect of class, caste, and socioeconomic variables that too often block progress of subgroups in our society.

- **The global correlation between ratings of "goodness of family environment" and "adequacy of adjustment" of a college-aged sample is about .40; comparable values would be expected if the sample involved non-college adolescents. (See reference 9.)**

While this was a college-based study, it used a junior college sample where the I. Q. average of the sample was about 104, which is close to the I. Q. level of the typical high school graduate of most urban areas. If the sample were made up of non-college adolescents, I would predict that the correlation would be about the same.

- **Successful members of, and aspirants to, various families of occupations show characteristic patterns of interest, personality traits, values, and needs. (See references 1 and 10.)**
- **The relation between claimed and measured vocational interests, while positive, is too low to permit counseling based on claimed interests alone. (See reference 10.)**
- **Measured vocational interests are remarkably stable over time and within individuals. (See reference 10.)**

These three statements derive from our extensive research on the measurement of vocational interests.

Because of the long-standing friendship between the late E. K. Strong of Stanford University and the late Donald G. Paterson of Minnesota, our Center for Interest Measurement Research is the repository for all of Dr. Strong's original data on vocational interests.

Adequate counseling must take account of the individual's vocational interests, both claimed and measured, since they reflect basic motivational patterns.

- **Labor market requirements and vocational aspirations become more congruent and in balance in cross-sectional studies of age cohorts. (See reference 11.)**

If comparable findings emerge from our longitudinal studies, we will begin to understand how dreams are reconciled with labor force reality.

- **The origins of counseling (circa 1910 and thereafter) represent a confluence of four streams in American society and American social science: developments in psychometrics and statistics; problems and needs of an industrializing society; a strong strain**

of social idealism and reform; sociological forces, such as urban
development and population shifts. (See reference 12.)

● But the "intelligent layman" may be somewhat suspicious of the
findings and methods of the social and behavioral sciences; their
accepted "facts" are not immediately impressive and are sometimes threatening. (See reference 13.)

In this connection, consider the ambivalence and understandable
concern with which psychological testing is viewed today. Consider
also the issues of protection of human subjects and invasion of privacy
in both behavioral and medical research.

Berelson and Steiner conclude their excellent inventory with a comment regarding the human animal's unique ability to impose fantasy on
the reality of his world and the resultant problems of such an imposi-
tion. Some of the results of the social and behavioral sciences make
George Orwell's 1984 appear uncomfortably at hand: behavior modification, psychopharmacology, participating management, industrial/
organizational psychology — all of these may contain threats to the
layman's self-perception and self-esteem.

● Today's zeitgeist seems less favorable to the age range 14 to 24
than was true in more affluent times. (See references 14 and 15.)

Today's economic climate is far less favorable to this age range than
was true in more affluent times. Our attitude is: "Let's get these kids
out of the work force, and keep them out." Our society is not doing a
very good job for this age group, either in involving them, or being
advocates for them, or in understanding their concerns and fears and
needs.

● The problems of professionalization in American society may
have gotten somewhat out of hand. Who are the counselors in
the world of youth today? What standards prevail? What positions are taken by the American Personnel and Guidance Association, the American Psychiatric Association, the American
Psychological Association, the American Medical Association?
What impact do third-party payments have on the provision of
mental health care, broadly defined? (See reference 16.)

The quality of counseling services, both in high school and college,
ranges from atrocious to adequate, in my estimation. In the absence of
effective services to youth, new "professions" emerge; witness the
spate of chemical dependency counselors, walk-in counseling services,
and "hot-line" services. Each of these may seek recognition and vali-

dation by way of state legislative enactments or certification requirements.

If this summary statement of findings about college counseling sounds rather pessimistic, let me say that emeritus professors may be permitted to suffer a transient episode of pessimism. If the references I have cited deal with important facts in the provision of counseling services to youth, we must find ways of getting them into the minds and hearts of counselors on the firing line. Short courses, refresher courses, targeted publications widely and inexpensively distributed, closed-circuit television, mandatory recertification and relicensing requirements — these and other innovative educational devices must be used in upgrading the performance of counselors at all levels.

References

1. Campbell, David P. *The Results of Counseling: 25 Years Later.* W. B. Saunders Co., Philadelphia, 1965.
2. Smith, Mary Lee and Glass, G. V. "Meta-Analysis of Psychotherapy Outcome Studies." *American Psychologist,* Vol. 32, 1977.
3. Darley, John G. *Promise and Performance: A Study of Abilities and Achievements in Higher Education.* University of California, Center for the Study of Higher Education, Berkeley, California, 1962.
4. Pace, C. R. *The Demise of Diversity.* Carnegie Commission on Higher Education, Berkeley, California, 1974.
5. Learned, W. S. and Wood, B. D. *The Student and His Knowledge.* Carnegie Foundation for the Advancement of Teaching, New York, 1938.
6. Hollingshead, A. B. *Elmtown's Youth.* John Wiley and Sons, Inc., New York, Science Edition, 1961.
7. Warner, W. L., Meeker, M., and Fells, Kenneth. *Social Class in America.* Harper and Row, Inc., New York, 1960.
8. Hochbaum, G. et al. "Socioeconomic Variables in a Large City," *American Journal of Sociology,* Vol. LXI, No. 1, July, 1955.
9. Williams, Cornelia T. *These We Teach.* University of Minnesota Press, 1943.
10. Darley, John G. and Hagenah, Theda. *Vocational Interest Measurement: Theory and Practice.* University of Minnesota Press, 1955.
11. Gottfredson, Linda S. *Aspiration-Job Match: Age Trends.* Johns Hopkins University, Center for the Social Organization of Schools, Report #268, December, 1978.
12. Borow, Henry (ed.). *Man in a World at Work.* Houghton Mifflin Co., Boston, 1964.
13. Berelson, B. and Steiner, G. A. *Human Behavior: An Inventory of Scientific Findings.* Harcourt, Brace, and World, Inc., New York, 1964.
14. Pernell, Ruby; Whittacker, J. K.; Wald, Patricia M. *Proceedings of the First Konopka Lectureships.* Center for Youth Development and Research, University of Minnesota, 6 June 1978.
15. Pifer, Alan. *Perceptions of Childhood and Youth.* Carnegie Corporation of New York, 1978 Annual Report. pp. 3-11.
16. Moore, W. E. *The Professions: Roles and Rules.* Russell Sage Foundation, New York, 1970.

19 Shaping and Mentoring Influences in Adolescence

Sidney Werkman

A critical problem among the adolescents and young adults I work with is the lack of enough deeply committed, caring people in their lives—the lack of parents, teachers, bosses and other adults who really care about young people. This I find very disturbing. A society that doesn't cherish its young people is in deep trouble.

I will get back to that later on, but to set the stage for my remarks, let me make a few preliminary observations. The first has to do with the major concerns of young people. At the risk of pushing alliteration to its outer limits, let me list some of the ones I hear most frequently:

- Sex
- Sex roles
- Self
- Sense of self
- Separation
- Stable identity
- Some parental figures
- Sensible careers
- Super-solutions to their problems

As a further sort of preamble, I would urge you to ask yourself, as I constantly ask myself in working with young people: "What was I like when I was 15, or 17, or 21?" Because in order to understand adolescents, you have to be able to remember what you were like at their age, and yes, to some extent, be able to tolerate that!

191

Let us also remember that youth differences begin prenatally. It is not only whether a child is wanted but whether the parent wanted a child of that sex that matters. We increasingly recognize that what matters is not what parents do with children but what they *are* with children, and *who* they are with children.

A great deal of recent research suggests that parent-child bonding, and especially mother-infant bonding, is basically more important than we had believed to be the case in the past. The research data are coming in rapidly now on the first two years of life. They demonstrate that bonding is indeed crucial to later character development.

And, as those of us who work clinically with adolescents and young people recognize, what comes through as you work with them hour by hour is not whether parents spent time with the child but what the parents thought of the child, whether they wanted that boy or girl, whether the child was really in the way of parental ambitions, and how committed the parents were to their offspring.

I have seen many an adolescent who was a product of conflicted infancy. By that I mean a lack of mirroring, a lack of being able to idealize a parent or both parents and to be able to feel trusting, loving, loved and cared about, that the world was a safe and important place rather than a frightening, threatening place.

Children who have such conflicted early life experiences are very prone to a tendency toward terror and inhibition in adolescence or, if they were not soothed and able to internalize an acceptance of frustration, they have a legacy of fury, of lashing out and of mindless impulsivity. If such an adolescent is not remedially mentored, so that a group of inner goals and directions evolve, then adulthood and middle age become years of depression, of "Is this all there is?", of stagnation, of an inability to give of oneself or to give to the next generation.

I'd like to say a few words about gender differences among adolescents, with the *caveat* that the data in this field are indicative rather than definitive. We can't say that all adolescent males are this way and all adolescent females are that way. There is no question, however, that significant differences do exist.

Males, because of their superior three-dimensional vision, greater muscle mass, better gross motor coordination, and better ability to track moving objects, are better suited to stalking prey and recognizing predators.

Females, because of their better fine motor coordination, greater sensitivity to sound, greater skin sensitivity and better perceptual sensitivity, are better suited to caring for infants and young children, for assuring a kind of affiliative nurturant constancy in a community.

Such qualities would have significant survival value in an ideal hunting and gathering society in which sex roles were idyllically separated. We are not sure whether such a society ever existed and we certainly can't envision such a society in Denver or Detroit!

So what we find is that girls are different from boys, though not by much. These findings raise several questions. Is there a group of nurturant qualities and a group of aggressive qualities, and if so, are these qualities or patterns modifiable? Do we want to modify these patterns so that there is overlap between males and females or a cross-over between the two?

I will now suggest some potential directions for research, clinical work and planning for high school, college and work careers for youngsters based on some of the recognized differences between the males and females.

Parents should discuss goals with their children, in order to assure them a sex-free potential for developing their lives.

Though there are differences, one of our greatest needs is to foster an ethic that offers adolescents the opportunity to develop individual potentialities without restriction. Career and life paths should be chosen on the basis of ability and motivation, and not by sexually derived mandates. Preordained career paths or restrictions are not healthy. That is, a girl should not be hampered from becoming a mathematician, mother, physicist or soldier. A boy should not be forced to become a competitive salesman, an isolated entrepreneur, or a meaningless external achiever.

Young people get subtly directed toward one career or another. It is not that mathematics, physics or science are not available to girls, but that those very courses often conflict with opportunities for them to be "pom-pom" girls. It is not that boys are barred from child care programs, but that they get no support from school counselors to take such a direction.

There is an urgent need for a systematic study of high school courses and the informal direction and career counseling opportunities that help and hamper the autonomous development of students' abilities. Parents should discuss goals with their children, in order to assure them a sex-free potential for developing their lives.

At this point I'd like to express my strong support for an opportunity for free role experimentation in adolescence for males and females—a kind of psycho-social moratorium, an opportunity for exploration. What I have in mind is a modern equivalent of the Grand Tour—an opportunity for a person to find himself or herself, an opportunity for that very rare experience of solitude, for being alone and finding inner goals as opposed to societally directed goals. Time is needed to enjoy the creative experiment of adolescence. After all, adolescence is the touchstone experience of life for many people.

We all think back to what our ideals, our possibilities, our enthusiasms were at adolescence. Those ideals and enthusiasms must be nurtured and transformed in order that they not be frozen into a premature finality. You may remember what Scott Fitzgerald said about Tom Buchanan, the rich, handsome Yale halfback, in *The Great Gatsby.* Fitzgerald said of Tom Buchanan that he achieved such an acute, limited excellence at the age of 22 that everything afterward savored of anti-climax.

You can't be a halfback all your life, nor can you be a "pom-pom" girl. But it is important that even such goals be respected while other more durable ones are forged.

As we look at girl and boy roles in the schools, we especially need to develop structures for building self-confidence in women through awards and recognition for novel recreational and career paths. Some schools have already begun to offer athletic opportunities and leadership roles for girls. But in many schools, such opportunities are not available to both sexes. Some roles are reserved exclusively for boys and others for girls.

I think it is also important that we in some way correct the deficit image of both men and women that is constantly expressed through the media. That is: you'll be whole or good if you use this hair spray, this deodorant, this lipstick—that you're not good enough now but if you buy these things you will be able to function effectively in contrast to the sloppy way you present yourself now.

Not long ago, one executive among a group of American business-men who were going to China was asked: "What kind of markets can you find in China?" "Well," he replied, "there are 800 million pairs of armpits." This was a good example of the advertising ethic that says you're no good as you are, but with the purchase of the right deodorants you will be made acceptable. Such negative "image" appeals help to shatter the fragile developing sense of self in young people.

It is true that young people often invest in ideals and goals that don't last. You can't be a halfback all your life, nor can you be a "pom-pom" girl. But it is important that even such goals be respected while other more durable ones are forged. Young people are constantly engaged in a quest for goals, a quest for who they are and what turns them on. They need freedom to seek the unique meaning of their lives rather than be forced into molds made by their elders.

Now let me say a word about mentors. I think this is a critical issue, because boys typically have more mentors available to them than girls. A mentor in this frame of reference is someone like the old teacher in the Odyssey — a guide. Mentors are people who live exemplary lives and are willing to give of themselves to young people.

Because of the role-changing now going on, males may become more nurturant, and females more competitive.

Why is it that in the past boys have had more mentors available to them than girls? That is an important question, but even more important is that we set ourselves the task of developing mentors for girls. In any event, all young people — male and female — need the help of mentors in developing a variety of responses to others — assertiveness, indepen-dence, self-reliance, sensitivity and concern.

I might say a word about androgyny — a combination of traditional masculine and feminine qualities. Among today's young people, there seems to be a trend in that direction, but we don't know how strong it is. But what may be lost if we blur the sex roles in youth? When the society continues to value external achievement and economic success, both sexes will certainly direct themselves toward such goals, and nurturance may well be in very serious danger. It may be relegated to the poorest socioeconomic group in our society. Is it possible for the entire society to move toward more nurturing? Is it possible to teach nurturing and the valuing of nurturance to high school students?

Philip Slater, in his book *The Pursuit of Loneliness,* made several sharp distinctions between the United States and Western Europe. In Western Europe, he said, dependence has been a highly valued concept, whereas the United States is a country that values independence. Co-operation has been a hallmark of Western civilization. The United States is a competitive society. Community has been part of all Western societies. The United States has increasingly moved toward individu-alism. As a result, according to Slater, Americans have experienced a loss of meaning, a loss of relationship and a loss of the value of nurturing, both in adolescence and in later years.

One of the dilemmas now facing our society is how to save male lives without losing female ones. Let me explain. Males now have an excessive rate of cardiovascular disease, cirrhosis of the liver, cancer of the lung, and a lower life expectancy than females—largely because of the pressures generated by the demands of the roles they play in society. Because of the role-changing that is now going on, males may become more nurturant and females more competitive. No net gain would accrue if women's rates for diseases of life stress caught up with those of men. It is our job to help identify and direct needed changes in the understanding and shaping of the future for both female and male young people.

Discussion

Sheila Kamerman: I'd like to make a comment about nurturing—not only as a role but as a field, a profession. A matter of some concern, for example, is that as more and more women are working, particularly women with young children, and as these women have their children cared for outside of the home, there is a growing professionalization of child care. Will this result in the development of a new female industry and, if it does, will child care become a low-status economic and social role? These are questions that are worth pondering.

Auditor: Junior high is the level at which a girl usually decides not to go on to the next step in math, even though up to that point she may have excelled in it, because she thinks her career opportunities are very narrow and in the traditional fields in which math is not that important. I think that's changing now.

Another thing I want to mention is a study done this year by Phyllis Farmer at Temple about women going through middle age. She found out that no woman in her sample had had a mentor, but that all of them were mentoring. So it looks as though there's still hope.

Werkman: Yes, there is. For some reason or other, girls don't have to make the break with their mothers but boys do. Girls, in fortunate circumstances, have a ready-made model to follow.

Auditor: That's true. Girls are more likely than boys to be mentored by their mothers.

Paul Bohannan: I want to make a plea for social invention, because many of these nurturing and mentoring things can be done if we invent new ways of doing them, and I want to give one quick example.

I had the good fortune to spend over 30 months with a people in West Africa called the Tiv. These people have an age-grade system through which young men begin to move at 16 or 17. Forming their age-group has an important role to play in making them recognizable as entering adulthood. They are assisted by an age-set of men about 30, who help them to organize and who teach them systematically how to do all the things an age-set must do.

This is a good example of socially recognized and socially controlled mentoring, and there's no reason why we can't invent new ways of our own if we just put our thinking caps on. We don't need government action.

Werkman: I agree. What we need are some good ideas and some good publicity.

Merrell Clark: I've been impressed by the capacity of elders to mentor, and I think that the mentoring apparently does increase, at least in some areas, with age. And yet, we have in our society certain patterns that create distance between elders and the young. For example, there are no elders in the schools unless we inject them into the schools. We could use more elders in the workplace, too.

Werkman: The generation gap is obviously bridgeable in certain circumstances, but I think that the tastes of the young and the old in books, games, music and the like are so different that in these areas the gap probably will remain.

Duhl: I was just wondering whether the gap may not be the fact that in order to mentor somebody in their development, you have to have been there yourself. When the mentor-mentee gap is too great, you can't help the person develop. However, in any community with multi-generational interactions gaps are very useful, because of the continuous opportunity for interaction.

My Blackfoot Indian friend tells the story about how on certain issues, the elder women, the grandparents, who are no longer sexually active, can assist the young people, while their own parents, who are still involved in sexual activity, are still competing with their children. Different kinds of issues can be dealt with in different ways. What we really need is some sense of community, as a context within which mentoring takes place. I agree with Paul Bohannan's comment about the community — that to have all these interventions separate from community creates tremendous difficulty because they all become very artificial.

Bohannan: Could I say something about that? I think it's important to keep in mind that the community of the past was far different from the community of the present. We are in the center right now of a tremendous change in the nature of community and therefore when we say community I agree 175 percent, but I also think that we have to be clear about the kind of community we are talking about.

Duhl: Yes. I think that one of our dilemmas in talking about family or community is that we're stuck with old models. In fact, the critical problem may be that new notions of family, new notions of network, new notions of community, all seem to be emerging, but have not been formalized into the institutions that we've been talking about.

Werkman: This discussion has developed nicely the concepts of nurturing, mentoring, community and intergenerational potential. It has crystallized a vocabulary that will help in both definition and action.

20 Heterosexual and Homosexual Relationships

Philip Blumstein

I'm a social psychologist, and at this conference I'm a minority of one. That might become clear as I go along, but to give you an example of it I'll offer one brief, very brief reaction to what Sid Werkman just said and that is, from my professional prejudices or point of view, a few of the sex differences he described occur early enough in life to suggest that they might be due to biological factors.

I also want to point out that from day one minus nine months all sorts of non-biological factors enter in and, therefore, my principal prejudice — and I hope it's one that's shared — is that whatever those innate differences are, it is clear that they are readily modifiable through the social experience.

I don't know if any of you would disagree with that. In any event, any policy implications in my remarks should be focused on the modifiability of the differences, not the *status quo*. Since I'm not a policy person, I'll probably steer far away from that, and will be describing to you *status quo* kinds of things.

Second, I would say about myself that I am a scientist. No matter what your prejudices about the scientific status of sociology in 1979 might be, I believe in scientific models and, therefore, anything I say will either be based on fairly sound and replicable scientific research or if not will be based on observables. I think I heard a lot of things said in the last several days that are based on wishes rather than observables. That's all well and good, but I will not do that.

The other thing I want to say is that I will simplify my remarks, which means that for the sake of simplicity I will speak in over-generalized and often stereotypic ways. So, if I say something that you think needs qualification, please don't jump up and say, "Hey, wait a minute!" Please assume that I'm aware of it, but don't have the time to qualify my remarks.

In the time available to me, I'm going to talk about two things. The first is about the notion of sexual identity, if for no other reason than it has not been discussed here and if there's any time in the life cycle where it's an important consideration it's during the period that we're discussing at this conference.

Not only in adolescence but in early adulthood, sexual identity often gets short shrift. And based on research that I have done — and I've done it long enough to enjoy the luxury of being able to give a lot of thought to the topic — I have very strong feelings about it — feelings that might differ from the current popular wisdom on the subject.

My second topic — which is what I was asked to discuss and I'll devote most of my time to it — is couple relationships in general and couple relationships in the age group that we're dealing with at this conference in particular. But in doing that I must remind you that I'm talking about research that's still in progress. However, I've given a lot of thought to this subject — thought based on what I hope to find or might find rather than what I have found.

The focus of my remarks will be on sex differences. I think that was part of our mandate when we came here and, lest my friend, Paul Bohannan, go ahead and jump all over me, I'll have to say that my discussion will be limited to the Western contemporary situation — the one that prevails in the United States.

With those prefatory remarks out of the way, let me turn now to the question of sexual identity — and again, by way of preface, I want to make my own value orientation clear. I think of homosexuality in the same way as I think of Catholicism. That is, it's just a facet of American life and in a moral sense I find nothing to object to people who live as homosexuals.

The conventional wisdom and the professional wisdom on homosexuality and its development, particularly as it concerns us here is that sexual identity is fixed in early life and manifests itself in adolescence. This suggests that all we need to do is help people adjust and discover their "true" sexual selves.

After considerable thought and research, I reject that conventional wisdom. I also implicitly reject, in the absence of adequate evidence, the notion that hormonal and genetic factors play a major role in the development of sexual identity. My research has convinced me that sexual identity is subject to creation and recreation throughout the life cycle, and that *proximate* causes are much more important than has been previously considered to be the case.

In my opinion, therefore, we at this conference ought to try—from a human point of view—to come up with a program or with an attitude or policy which will help homosexuals at the age when the seed that was planted ten years before begins to germinate to deal with their experience, minimize their own unhappiness about it, and maximize their chances of having a good and happy and profitable life.

I should note again at this point that I am rejecting out of hand non-interpersonal or non-social causes of homosexuality in the human male or female. I do that simply because I remain agnostic on the question of prenatal factors—agnostic because I don't feel that any evidence we currently have demonstrates that there is any prenatal influence. However, if convincing evidence does emerge, I am prepared to reconsider my present position.

A few moments ago I said that events strongly influence the way people develop their sexuality. This is especially true among females. By way of example, let me describe a case that has happened to me time and time again in my research.

Sexuality is a much more malleable thing than we think it is.

I will meet a woman, a woman of forty, who after a divorce or maybe not a divorce—it doesn't matter—but after what most clinicians would consider a perfectly typical heterosexual adjustment will say for some reason she became rather close friends with another woman very much like herself and they discovered that they really had a lot to share — were very close—and all of a sudden it dawned on her that the intensity of this friendship really approaches the notion of romantic love that she was taught and preoccupied with as an adolescent.

At some point the relationship becomes sexualized and the woman will start to call herself a homosexual. Upon hearing this, I tell myself that what I need to do now is find out about the first five years of her life and find those seeds that should have told me that this is the life she's going to have. What I have momentarily forgotten is that sexuality is a much more malleable thing than we think it is, and the fact that we don't consider it malleable as professionals really reflects the culture's view that sexuality is fixed in early life.

I believe that perception of sexuality as being non-malleable is a self-fulfilling prophecy. Consequently, people see their sexual biographies in much more consistent and rigid terms than indeed they can be or need to be or empirically are.

But why are we so preoccupied with this non-malleable view of sexual identity that it permeates our perceptions and our research? I think part of the reason is that most of the research in this area has been done by males on males and that male sexuality is less malleable than female sexuality. In our culture, male sexuality is learned earlier than female sexuality. It is learned through intra-psychic processes, that is, males respond to their fantasy world through various stimuli and test various hypotheses from their reactions to those stimuli. They make random connections between those internal reactions and their physiological reactions, and impute a lot of meaning to those imagined connections. Consequently, males tend to come up with a self-definition based on unexamined experience.

Female sexuality, on the other hand, tends to be more reactive, less externally genital, of course, and certainly more tied to concrete experiences — experiences that tend to be interpersonal, emotional and romantic.

It's really a sloppy thing for scientists to say that happenstance or random events are the major causal factors in sexual identity, but alas, I believe it to be the case.

We are fond of saying that you can't really know certain things about a person until after that person has died. Well, I maintain that you can't really know a person's sexual identity until that person's dead.

That's an overstatement of the empirical malleability of sexuality, but I think it's not an overstatement of the potential malleability. I guess what it suggests for us in terms of policy is not only do we need to be prepared for people's sexuality to unfold during the critical period we're talking about, but we have to be prepared for it to change at later periods in the life cycle, and we have to be prepared to allow people to have a great deal of confusion about it. What eventually will unfold as people resolve their confusions will depend a great deal on happenstance.

If a woman happens to fall in love with another woman and says, "I must be a lesbian," then it's very possible to conclude that if it hadn't been for happenstance, she would have gone to her grave without ever having had that thought or that experience.

It's really a sloppy thing for scientists to say that happenstance or random events are the major causal factors in sexual identity, but alas, I believe it to be the case.

I guess that suggests also, if you're willing to buy my interpretation, that one of the things we need to do in our counseling and in our policy making is to try and persuade people who are confused about their sexuality to go to a clinician. Sexuality is not a hard and fast thing; your current feeling is not showing the true, essential you, because in fact there is no true, essential you. What it shows is what some of your options are. I don't practice therapy in my work, but people do come to me—people who have been married for 20 years—and ask: "Should I leave my wife (or husband) and try being homosexual?" And I tell them: "If you do feel you have homosexual urges or impulses or attractions, that does not imply that you must stop your heterosexual activities."

Gender is a much greater predictor of almost every realm of sexuality than is the distinction between homosexuals and heterosexuals.

So I think that's my message. But there's one more thing I want to say about sexual identity. It's a prejudice of mine, but I think it's also consistent with the data. That is that one of the best guides in understanding human sexuality in our culture is the concept of gender. This may seem very obvious, but if you want to understand, as I've sought to do, the nature of lesbianism and lesbian experience, the nature of male homosexuality and male homosexual experiences, the first thing you have to consider is the differences in gender. Lesbian courtship resembles the courtship desires of heterosexual females, and male homosexual courtship resembles the desires of heterosexual males. Gender is a much greater predictor of almost every realm of sexuality than is the distinction between homosexuals and heterosexuals. I wish

I had a dollar for every time a lesbian has told me she couldn't in any way fathom the sexuality of homosexual men.

I guess what I want to do now is to get off this topic and talk about couples. I'm not going to go into detail about what my study is about and how we're doing it except to say that we're studying four different kinds of couples and an enormous number of each of them, married heterosexual couples, cohabiting heterosexual couples, lesbian couples and male homosexual couples.

We chose to have those four for a number of reasons. First of all, we don't think it's unreasonable to say that the cohabitors, the lesbians and the homosexuals deserve in their own right, being citizens, taxpayers and real people, to be studied. I don't mean to do a song and dance about that.

But an important reason to study them is for the comparisons they provide. The first are very straightforward, practical comparisons. The cohabitors, the lesbians and the homosexual men don't have the force of law to legitimize their relationships so, therefore, if you're concerned with what the law and other institutional arrangements provide for a relationship then you can make a comparison between those three groups and the married couples.

Another reason why lesbian and male homosexual couples are of interest is because they have been facing problems, ever since God created gay couples, that are not very different from the problems that heterosexual couples in great numbers are only now beginning to face. So we come up with the irony that they are good models for heterosexual couples, namely, the dual career question and how to manage it, and the organizational problems that cannot be understated of two careers in one couple.

Second of all, male homosexual couples have been dealing with the issue of non-monogamy for a great period of time and my data suggests that it's the rule rather than the exception for them to have non-monogamous arrangements, and I think people are becoming more and more interested in that, how to work that arrangement into their relationships.

Lesbians in this culture don't seem to be any further along in this area or any different from most of their more avant-garde heterosexual counterparts—considering it, but can't seem to work it out.

But I guess the most important reason the comparisons should be made is that there has never been a set of institutional arrangements that support, encourage and almost force people who are homosexual into intimate relationships with each other. For them, there has been no social support or institutional basis for marriage—as we've traditionally known it — among homosexual couples. Consequently, homosexual couples are not talking about either getting married or staying in marriages because of the intrinsic gratifications or growth potential or whatever words you want to use, that heterosexual people are supposed to get from the partnership, so their motives are worth studying.

There's another kind of comparison that I want to discuss. It's a bit more theoretical, but I think it's worth spending some time on. And that is that there's a lot of concern, as a result of the changes from traditional forms of marriage in our culture to the current forms, with working out the way the husband's maleness and the wife's femaleness are going to be accommodated—in other words, working out the notion of gender in the relationship. There has been a tremendous social change away from traditional marriage.

Husband/wife roles need not be allocated strictly by gender. We can easily imagine a situation in which the "instrumental" roles are played by the wife, and the "expressive" roles by the husband.

Everyone is able to say what traditional marriage is. It's that awful thing that people used to have to suffer. Modern marriage can be defined as that wonderful state that we all dream about in our fantasies. People have seriously tried to understand the difference, and to understand what this incredible course of social change is going to mean for the institution of marriage. Some researchers are trying to find out by locating some traditionally married folk and some modern married folk and studying how they're different. Well, we have a different strategy, which I'll describe in a moment. But first let me ask what may seem like a very very silly question, and that is, what is a husband and what is a wife?

We all start out by saying the husband is a man and the wife is a woman. Let me define it in somewhat different terms. Husbands and wives are people in a marriage for whom there are different expectations, and those expectations have a certain consistency and coherence.

Traditionally, one of those persons was expected to work. Again, I'm being overly simple and I'm talking about the traditional model to which many people have never conformed. In any event, if the other partner also worked, he or she did so purely for financial reasons while the first person worked to gain the major psychological and life-orient-ing gratifications. A career was the person's world. Second, one of the persons had the responsibility for the daily maintenance of the house-hold—usually the wife, but if the work got too heavy or strenous then it was shunted over to the husband.

They had different forms of activity when it came to decisions. They had different amounts of power in making decisions in their relation-ship. One was legitimately seen as the person ultimately responsible for the decision making.

The wife was certainly more responsible for child-rearing activities, although there were certain chores for the husband. The most important thing is that they had different basic orientations and expectations of the relationship, which they brought with them. These orientations and expectations have been described in some unnecessarily technical terms by social scientists as "instrumental" and "expressive."

But I will describe them to you in simpler terms. Instrumentality —part of the husband's role—refers to the ability and disposition to be the boss, the leader, to require respect and subordination, to organize the couple's energy and to get things done.

The instrumental partner was the one who tended to mediate between the nuclear family and the outside world, went into the outside world and was expected to be able to handle it.

The expressive person was quite different. That was the person who had the ability and disposition to be mediator, conciliator, to deal with disputes, to resolve hostilities, to be affectionate, warm, emotional, the comforter, the consoler, and I would add, to be deferential.

I've described to you a set of expectations of behavior and role and I didn't say which were the husband's and which were the wife's, because I didn't want to impugn your intelligence. I assume you all figured it out. However, husband/wife roles need not be allocated strictly by gender. We can easily imagine a situation in which the "instrumental" roles are played by the wife, and the "expressive" roles by the husband. Or, in fact, some of the roles could be divided up, and some could go by the boards.

Now let me say something else about the traditional husband-wife relationship. First of all, for better or for worse, it's a very very efficient social invention. There are very few of us here who would say that in terms of general social institutions a division of labor based on spe-cialization is not a good thing. If you look at the history of social organization, you see that specialization has been enormously functional. Unfortunately, however, when one of the partners drops dead or is incapacitated, in any form of social organization, specialization can by pretty debilitating.

Secondly, the husband/wife roles in a traditional marriage are based on gender. We all know who is supposed to do what. We build this knowledge into our children, raising them adequately as male or female, and at age 20 we sort of add water and stir and presto chango we have a functioning marriage. This is a lot different from taking two unacculturated young people at age 20 and having them marry and sit down to figure out who is going to work and who is going to take out the garbage.

Unfortunately, as I said, the traditional role allocation is not the greatest arrangement when one of the persons is incapacitated. More unfortunately, because of social change and because of the change in consciousness, many of today's young people seem to think that this is a very unhappy state of affairs. They feel trapped in this kind of arrangement. So here we have a clever social invention flying in the face of people's emotions, desires, and aspirations.

Whenever you look at lesbian couples you can say: "What is female in this relationship is whatever they have too much of, and what is male is that which they lack."

We also have people in our study who are experimenting with departures from traditional marriage, but I believe it's terribly difficult for any heterosexual couple to put gender totally behind them. Not very many really do, but some claim they have done it. Unforunately, we don't have the scientific instrumentation to be able to say whether they did or not. In any event, we're stuck with people in our study who are departing — but only modestly — from what they have been socially programmed to be.

Again going back to my remarks about gender continuity, I believe that homosexuals tend to be, in many respects, socialized to the gender they are. Consequently, a homosexual couple is a better example of people having to cope with difficulty. In a lesbian couple, for example, we presumably have two people, by virtue of being women, with an overflow of expressiveness and an undersupply of instrumentality. So when you look at those couples in contrast to heterosexual couples, you can say: "What is female in this relationship is whatever they have too much of, and what is male is that which they lack."

Second, of course, there is experimentation with cohabitation—and I'm not talking about the college campus. I'm talking about the people who are trying it as an extended relationship. This may well reflect an attempt on the part of women to develop a greater instrumentality and on the part of the men to develop a greater amount of expressiveness. It may be that they are trying to avoid the kinds of differentiation that I have characterized as social invention.

Well, what do I think is happening and what do I think it means for policy? If we endorse this form of social change — if we believe that these new experimental life styles are more humane, more useful to people as personal life goals, then I think our policy will need to be geared toward what Sid Werkman was talking about, some kind of androgyny so that these people at least develop a repertoire of behavior, including some behavior that is appropriate to the other gender. Whether they choose to activate those behaviors is another matter.

My personal feeling is that instrumentality — that is, the ability to cope with the outside world, to get things done, to organize other people's activities — is a skill that can be learned even at an advanced age, and by women as well as by men.

To the extent that men do develop expressiveness, it is in two ways. One is in interaction with women (we like to say that women tame men by marrying them, and make them better persons for it) and the other is in rearing their children, which they do precious little of.

I think there is nothing like the real world to cause people to come to grips with life, to cope with their situation. Clearly, there are many people who don't. But I think relatively speaking they can. I feel very strongly that people have enormous reservoirs of strength within them and I think that many of the helping professions traditionally have ignored that fact. Women who have not been encouraged to show that strength can develop it. However, I think that in our educational system we ought to move towards recognition of the need for women to have that strength and not to stigmatize them for showing signs of it.

Expressiveness is another matter. It's the kind of thing that's hard to put your finger on and hard to measure. It's not part of our training of young people. However, there are a lot of expressive people around. Most of them are women, and they got that way somehow. We ought to be able to train people to be expressive, but we just don't know how to go about doing it. I think it's a great deal harder than teaching people to be instrumental, and I think it might be one of those things that's harder to pick up later in life.

Now that's just my hunch and my prejudice. I'll go one step further and say that to the extent that men do develop expressiveness, it is in two ways. One is in interaction with women (we like to say that women tame men by marrying them, and make them better persons for it), and the other is in rearing their children, which they do precious little of.

So I guess I see the great dilemma as developing expressiveness in men and I think it should be clear to you that one of the problems is that there has not been and continues not to be any reward for that in men. It's stigmatized. Just to illustrate that stigma, I would suggest that you try to imagine any product being sold to men by emphasizing male expressiveness. You know: "Hi, I take care of children and I smoke Camels."

Women have ambivalences about their husbands. They want them to be expressive, and they want them to be strong, and if that isn't a dilemma for males today, I don't know what is. You're damned if you do, and damned if you don't.

By the way, the single largest complaint of married women in our culture is that their husbands don't talk to them. What they mean is, what they're really saying is "Please, God, let my husband be more expressive." Unfortunately, it's not clear to me how men can learn expressiveness, given our institutions.

There's another aspect to this, but I don't have enough data on it, and that is that women have ambivalences about their husbands. They want them to be expressive and they want them to be strong, and if that isn't a dilemma for males today, I don't know what is. You're damned if you do and you're damned if you don't.

Many young couples are dealing with this dilemma in many different ways, and maybe our study will shed some light on how it's being resolved.

Discussion

Leonard Duhl: What is so striking in most of the comments you made is that you were talking about a two-party contractual relationship which is fantastically difficult for people to work out. It seems to me that what we really need in our society is a third pillar—a belief system which makes the two-party relationship—no matter what the obstacles are—much easier to deal with.

So to me the critical issue in some of the things you talked about, and some of the things Sid Werkman talked about, is how to develop belief systems that permit two parties to come together in a different form.

Blumstein: I agree with that entirely, but as a scientist I have a certain kind of conservatism. I think I agree with your long-term goal, but I'm not so sure that we can do much to modify all kinds of interpersonal contractual arrangements.

I might also add that social institutions are much more impervious to our meddling—whether positive or negative—than we think they are. So my feeling is that new institutions — and new contractual arrangements between the sexes — may emerge in spite of us rather than because of us.

John Cannon: I work as a minister in a community which is predominantly homosexual, and your remarks have been enormously illuminating to me. I keep thinking I know all I need to know, but it's obvious that I don't.

Could I ask you to speculate on how important monogamy is — among homosexual couples — in terms of personal growth and fulfillment, particularly at the later stages of the life cycle?

Blumstein: All continuing sexual relationships are contractual. Even though homosexual couples don't have a written contract, they do have a very clear understanding, and monogamy is a part of that understanding unless the parties agree in advance that it won't be.

Alfred Kahn: In some ways this discussion makes it very clear that there are two sub-groups within the 14-to-24 age group—adolescents and young people of marriageable age. My chief concern is with the latter. The vast majority are going to live in heterosexual relationships. They have to live in heterosexual relationships because the society needs children. And even though it may be desirable to shake up the gender attachment to certain kinds of roles within the family, we also need to make sure that we have adequate nurturing of children. At the policy level, the questions are: How do we make sure that society provides not only adequate *maternity* benefits but adequate *parental* benefits? How do you make it possible for both parents to visit a day care center, or a child care center, or to spend time with the child who is in the transition from being at home to being in a care center? How do you arrange for time to stay home if the child is sick? And how do you make sure that the father as well as the mother shares time in early parenting, because you want the child to have the different image that Sid Werkman was talking about?

These are not short-range questions, because we've also heard here that although we have a lot of unemployed teenagers at the moment, we're very quickly moving into a period of a labor shortage in the society. The phenomenon of both males and females entering into the labor market in large numbers is likely to be much accelerated in a period of labor shortage.

In any case, it seems to me that we need to find ways of gradually institutionalizing other kinds of supports for the nurturing role, that assume gender flexibility in dual career or one-career families. We also need better policy in the whole area of child care, maternity benefits, parental benefits, leave to take care of sick children, time for personal days, and so forth. This is not pie in the sky. Every one of these programs exists, and in countries that have a lower gross national product than we have.

In many ways, the Eastern European countries are more sexist than the West European countries or the U.S. However, if you look at the labor force distribution there, you have a very dramatic illustration of how women move into careers requiring a lot of the attributes that have been ascribed to men. Some quickly become "feminized" careers. The occupational distribution shows the enormous malleability of gender in professions, just as there is an enormous malleability in distribution of sex roles within families.

As people who want to intervene in this situation, we need to concern ourselves with policies that create social supports for families.

John Schweppe: I want to interject quite a different sort of question. We have to accept the fact that a certain number of people are going to turn out to be homosexual. What should our policy be toward these individuals?

Kahn: I don't think we ought to interfere in any way with people who want to have different life styles, but I'm very concerned that we not confuse that with what I regard as a much more fundamental issue —the shift of roles in families and the question of how we're going to make sure that both men and women play nurturing roles, and that somehow in dual-career families we make sure that we have both family and social support for the proper rearing of children.

21 Young People and Religion

John Cannon

Whether we choose to admit it or not, religion is a powerful force in American society. We have among us about 137 million people over the age of 15 who belong to religious denominations — 50 million Catholics, about 75 million Protestants, more than 2½ million Mormons, more than 4 million members of the Eastern Orthodox Church, more than 3,700,000 persons of the Jewish faith, and more than 2 million other religious adherents.

These statistics are not derived from the Bureau of the Census, but from the different religious organizations themselves, so they do indicate something more than nominal preferences.

Sixty-five to seventy percent of the American people participate with considerable regularity, and with a profession of satisfaction and confidence, in organized religious institutions.

Moreover, of the 75 million Protestants, 45 million are either biblical literalists or highly conservative in their values — an enormously important fact for social policy and planning. The political implications are obvious. And this number does not include about 30 million Lutherans, Presbyterians, Methodists, and Episcopalians. I would render an educated guess that of these, more than half are evangelically inclined today, and quite conservative in their approaches to social programming. So we have among us 90 million to 100 million conservative

213

Protestant evangelicals, and I do not need to spell out the social and political ramifications of this.

Some of you are familiar with the annual report I've written called "Religion in America," which is put out by the Princeton Religion Research Center. Quite aside from what you may think of Gallup's sampling methods, this is one of the few available sources of information about religious behavior in this country. Political prognosticators and political hopefuls are quite familiar with it. I've just finished the report for 1979-80 and I don't believe it has been released to the public yet. Without boring you with the detailed findings, let me just mention that according to the report, some 65 to 70 percent of the American people participate with considerable regularity, and with a profession of satisfaction and confidence, in organized religious institutions.

Curiously enough, the figures are not significantly lower for the age group 18 to 24, although there is a drop for this age group in the rating of the importance of religious beliefs, from a national figure of 84 percent to 72 percent for the age group 16 to 24. That figure increases to 78 percent in the age group 25 to 29.

Clearly, therefore, we should be cautious about drawing conclusions which minimize the role of organized religion in American society and, indeed, in the transition from childhood to adulthood. America, as the Princeton Religion Research Council reminds us, is still a nation of believers, whatever this may mean for behavior or misbehavior.

Academic observations about religious influence should be most cautious and most qualified. The fact is that there is precious little empirical research to support our theological and sociological conjectures about a post-Christian society. We have in this country religious pluralism, yes. And secularization, depending on what we mean by it, yes. But organized religion remains today the repository of powerful symbols of human aspirations and social transformation. Despite all the deadening weight of organization — much of it superfluous, archaic and dysfunctional—and despite all the religious illiteracy and superficiality, we cannot ignore the sheer presence for good or ill of a complex religious institutional system in our society. I emphasize this, because as I've already indicated, it has profound implications for social policy planners.

Now what about our age group in their relations to organized religion? Well, interestingly enough, Princeton's Religion Research Center this year undertook a special survey of the religious world of the American teenagers. It does not cover college students, but I think that we can extrapolate from the findings and to conclude that they are applicable.

There was an interesting paradox in these findings. Ninety-five percent of the teenagers — an extraordinarily high proportion — said they believe in God or a universal spirit. Seventy-one percent said that they are church or synagogue members, and there is no significant decline in teenage church attendance, although parental constraint may be an important factor in this.

However, at the same time, an overwhelming proportion of the teenaged Catholics (82 percent) and Protestants (74 percent) believe that a person can be a good Christian or Jew if he or she doesn't go to church or synagogue. A fascinating, fascinating finding.

Moreover, to a considerable extent, American teenagers are turned off by churches of organized religion, despite the fact that they profess to be highly religious in some key respects. In short, a large proportion of teenagers feel that organized religion or the church is failing to do the job it should and the reasons given, I submit, bear close attention.

Three of the major reasons were as follows:

- Churches are not reaching out to the people they should be serving.
- Churches are failing to deal with basics of faith, and are not appealing to youth on a deeper spiritual level. (I'm underlining that one.)
- There is little feeling of warmth in churches.

These responses bear close examination in light of challenges to growth during the age span we are discussing at this conference.

Before I proceed to that task, let me conclude these remarks about the religious character of youth with an observation about religious growth and maturing.

Obviously, what is important for our purposes is some perspective on why and how religion has failed. Our concern is for the process of encounter and crisis that is necessary for growth. Incidentally, biblical religion places a high value on crisis and encounter, not only as essential to growth but as revelatory. Indeed, my own view is that until we ourselves recover a sense of the sanctity and mystery of this particular transition process we will not be accepted as mentors by the young, particularly in those crises and encounters that we tend to be so troubled about.

The question so often begged by parents and clergy and religious educators alike — the crucial question, and I use the words of Tom O'Day, whom some of you know, is this: "Does religion enable people to grow in such encounters and crises or does it protect us from the challenge and maintain us in a stage of relative immaturity?" It is in light of this question—really a question of values—that I want to look at the phenomenon of turning off, turning on and tuning in.

It's obvious that religious maturity — again from a biblical point of view — requires that youth suffer an ordeal of sorts. They learn the principle of suffering in order to enter the world of adulthood, equipped not only for survival but for a degree of personal freedom and satisfaction and social responsibility.

The focus of youth is necessarily on the institutional corruption of primary religious experience, a corruption enhanced and exaggerated not infrequently by a certain sense of parental and clerical hypocrisy.

The first stage of this ordeal, and as a pastor I worry more when it doesn't happen than when it does — is turning off. We can relate this proverbial rebellion to the task of separating from parents and caretaking institutions.

This process requires disillusion, a sense that the institutional embodiment of the religious involvement is less than perfect if not a fabric of lies.

The focus of youth is necessarily on the institutional corruption of primary religious experience, a corruption enhanced and exaggerated not infrequently by a certain sense of parental and clerical hypocrisy, by the sense that parents and leaders who profess belief do not practice what they preach.

Disaffection and disaffiliation may be seen from this perspective and as necessary and transitional tasks—necessary not only for the expression of critical discriminating faculties but also for the search for primary religious experience, the "high," and not incidentally, for the perfect or ideal social order or social expression, if you will, of institutional and/or ideological embodiment — of that primary religious experience.

The quest for the perfect, for something or someone to believe in—an ultimate alliance as much as an ultimate allegiance—takes various forms, including a turning to the so-called rational and disembodied religious teachings of the East, the gravitation to charismatic embodiments of sublime truths, the embracing of Utopian ideologies and movements, and not the least the demand that organized religion cleanse itself.

The response of organized religion to the disaffection of youth more often than not is negative and defensive and not infrequently takes the form of youth ministries that are in fact calculated to re-enforce inappropriate dependencies.

Believe me, I know that well, and I simply ask you to believe it. The observation is in order here that American albeit a nation of believers is also a nation of spiritual illiterates.

There simply is no maturation from a religious point of view without radical exploration and experimentation, without a testing out of a vast array of realities and experiences.

The onset of adolescence invites organized religion to mediate the separation of youth from parents in a way that challenges levels of spiritual inquiry and religious learning previously neglected. This does indeed suggest that the clergy and religious educators should pay more attention to the communication of primary religious experience, including the teaching of classic Western spiritual disciplines and especially techniques of prayer and meditation.

Moreover, it suggests that the serious disciplines of biblical study and theological reflection be taught to the laity, and not reserved for seminary graduates.

The second stage of the ordeal is in a sense the other side of the coin, the more affirmative side of turning off, namely, turning on. I don't want to be altogether critical, but some of the cultural manifestations of turning on are indeed dangerous and do indeed result in significant derailment.

But there simply is no maturation from a religious point of view without radical exploration and experimentation, without a testing out of a vast array of realities and experiences. Incidentally, it is remarkable that there is no one here to discuss and interpret pop culture. I feel this even more than I feel the non-presence of youth of this age group. Pop culture is the prime mediator of the experience of youth, including forms of behavior that are not infrequently intentionally outrageous but also faithful, entertaining, sensitive, and sometimes edifying.

The volatility of adolescence, the highs and the lows, is all too familiar. I would like to have time to discuss this in depth from a theological point of view, since I'm convinced that a bona fide sense of transcendence, not to mention an expansion of consciousness, is dramatically and significantly experienced in these peaks and valleys that are all too frequently ridiculed by adults.

Let me make one final comment about turning on. The process is dynamic, and necessitates a rejection of the static, fixed truths, if you will, of the establishment. In religious terms, this means the codes and creeds are dismissed as irrelevant, which probably means inapplicable. There's a question in my mind as to whether today's generation of youth will ever accept those "establishment" truths in terms which previous generations have accepted them — as static guides to behavior and norms of conduct.

It's worth mentioning, incidentally, that youth will welcome adult mentors and guides in this process; indeed they will seek them out. What they will not accept is an authority figure representing parental or parietal interest in controlling or truncating experience.

I am terribly worried by the inclination of today's youth to "cop out," to eschew history as understood in its Western sense. This invites the state to command worship, either manifestly or latently.

The final stage of the ordeal is the one that interests and worries me the most, and that is the challenge of tuning in. There aren't many words that we could use to describe the dynamics of adolescent closure — if you wish to use that term — of the entry into adulthood. Clearly at this point our discussion becomes most value-laden.

We are speaking of the necessity of orienting the self to the society, a particularly difficult ordeal in an age when the self itself is not perceived as a fixed entity, but as a history and a process — when the society itself is changing at a constantly accelerating pace.

The problem is that youth must be poised for change. Protean man has arrived, and is here to stay, and if theological anthropology can't be comfortable with that fact, it's because it has not read the Bible. It has lost the sense that man is by definition a wanderer and a discoverer, both the subject and the agent of new creation.

Which brings me back to my subject. Somehow youth must discover and identify some sense, perhaps only a relative sense, of place and of purpose. No, not a life plan—God forbid!—but a sense of orientation to society that is compatible both with the gifts and the limitations or, if you will, the vulnerabilities of the individual.

Youth believes that organized religion is failing it in this regard, and in my opinion it is. But then again, I'm a religious eccentric, or eccentrically religious.

But I'm also inclined to think that religion is not being given a chance. Social planners have decided to organize religion in America. They have decided that it should be an adjunct to the family, and that the church should be primarily at the functional level a multi-service center. Incidentally, I run one, and to the degree that I run one, I'm not engaged in the far more difficult and important task of being a priest and a pastor, nor are thousands of other clergymen in similar situations.

I want to conclude by reading from an essay by Robert Bellah, who in my view is the most cogent and incisive observer of the American religious scene. He says much better than I could what I really want to say:

> It is in society that one's individual life history and history as collective experience intercept. History is the proving ground for both personal values and social values.
>
> Given structures, personal and social, may prove inadequate in the face of the contingencies and the catastrophes of history. In history reality is encountered as judgment and no finite structure is ever entirely adequate to that encounter.
>
> Since history reveals the inadequacies of every empirical society, it becomes clear that society much more than the individual is the final repository of transcendence. Rather, every society itself is forced to appeal to some higher jurisdiction to justify itself, not entirely on its adequate performance but through its commitment to unrealized goals or values.

I will end by saying that I am terribly worried by the inclination of today's youth to "cop out," to eschew history as understood in its Western sense. This invites the state to command worship, either man- ifestly or latently. The only antidote that I know is to cherish and to keep alive the biblical mythos that has shaped Western civilization and, in fact, informs even secular humanism.

With that, I would like to raise some questions: What do you see as the role of organized religion with respect to social policy and planning? How can the clergy, conventional clergy, be drawn into policy and planning, especially concerning this age group? What is the role of the clergy as mentors and guides concerning youth? And finally, what should be the role of organized religion in setting boundaries?

Discussion

Merrell Clark: I'd like to comment on something you said at the very beginning of your remarks. I am more impressed by the towering influence of organized religion in this country than I am by the loss of its efficacy. For example, as powerful as we believe prime-time television to be, there is nothing you can do in this country that will draw as many people out of their homes at one time than an 11 a.m. Sunday morning church service. More than 40 percent of the American people are in one place at that one time.

Religion is the only institution we have that provides a touchstone for the entire life span. It brings all generations together, and relates them to one another in ways that are not possible in any other kind of institutional setting. It's the only kind of institution given absolute freedom by our Constitution to be what it will be. It's the central repository of our value systems, and it's the best forecaster of our future institutions.

Just as in the eighteenth century we saw the emergence of schools and colleges and universities, today we see in the churches the emergence of self-help groups, of child care centers, of many kinds of human organizations which will ensure over the decades ahead that we will meet the needs of our human family and our society. Too often we look to public finances, public institutions and public policy to provide for those human needs.

Cannon: That's one of the central dilemmas of our time, and I'm of two minds about it. There are enormous pressures on churches to provide those services, and I think they do it fairly well. However, we pay an enormous price for that. Our priests and ministers have become spiritually illiterate. They may be good therapists, but as spiritual guides—gurus—forget it.

Robert Aldrich: John, what do you see as the role of women in organized religion?

Cannon: I am an eccentric religious, and some of the things I say get me into a lot trouble, but let me say quite candidly that the church is sexist. It is re-enforcing the oppression of women. It's still limiting

and defining their roles in sexist terms, and I think this is going to be
the next big revolution. It's going to be internal. It's not something that
can come from outside. It has to come from within the church. I think
it's the enterprise that will engage us most seriously over the next ten
years.

Willard Wirtz: John, I'd like to try to answer your question about
what turns people off about organized religion. What I can't take is
that song "Onward Christian Soldiers." Maybe it's because I hate war.
I also believe in contraception, and I don't see why the established
churches should oppose it, especially at a time when overpopulation is
such a big problem.

Cannon: The Unitarian Universalists have done a wonderful job in
revising some of the traditional songs, and particularly "Onward Christian Soldiers." They have also developed and have used for about a
decade now what I think is still the best model for education in sexuality
both for parents and children. It's still about ten years ahead of its time,
and I commend it to you. I've been through it as a father, and really it's
incredible.

Sidney Werkman: There are many things that I'd like to comment
on, but our time is limited so I'll just bring up one. In much of religion
the teachings about sexuality do not square with the realities of adolescence. There is no way for young people to deal with these teachings
except through hypocrisy. Premarital chastity, for example, is a very
tough thing for young people to accept.

Cannon: It would be hard to verify this, but I think in fact that
churches do as much as any social agency in helping young people to
deal with their sexuality. I can recall my own adolescence in a small,
puritanical town, where we stopped Sunday School altogether. We
would eat coffee and doughnuts and talk about sex. It was enormously
liberating and that's happening all over the country. Also, pastoral
practice is quite different from traditional teaching and I think kids
catch on to that fairly soon. Am I right, Ernie?

Ernest Bartell: Oh, yes. I belong to the church that's generally
regarded as the model of the great monolith and all of that, and yet,
having grown up in it, I never was overwhelmed by that monolithic
aspect, because there was always the pastoral dimension, and some of
your experiences were the same as those I went through as a teenager.
Now maybe being partly Italian to begin with, I've got the Italian
viewpoint. Italians know how to handle the church very well.

Seriously, though, I've always seen the church as an institution that's conservative in the very best sense of the word. Human institutions all have their failings, but nonetheless, it's hard to live without an institution that preserves a kind of continuity with the past. The value of the church as an institution has been that it provides a framework within which you can rebel and still not have to drop off the deep end, and go absolutely crazy in the process. I don't want to give that up too easily.

I've certainly seen colleagues who have been badly treated in the church by excesses of conformity, but I've also seen a great deal of ferment. The Second Vatican Council that Pope John called seemed to the outside world to be quite earth-shaking—a new breath of fresh air. In fact, what it really did was tend to canonize a lot of the discussion that had been going on up to that point—ferment that was finally going to have to be dealt with institutionally and dealt with in the Council.

Harold Visotsky: I'm concerned about the stage at which young people really want a dialogue and whether that dialogue can be open enough in organized religion so that they don't have to give up and drop out. If there are inconsistencies between what the pastor says and what is taught in the classroom, young people pick them up very quickly. They may discard the value system along with the dichotomies. That's what worries me about the institution of the church. It is those inconsistencies that bring about massive rejection.

Cannon: To date it hasn't been as massive as you might suppose, but I agree with you that it will be. There's no doubt in my mind but that there will be massive repudiation of the established organized religions in this country, unless there is radical reform much sooner than I expect there will be.

The thing that worries me even more, however, is the enormous power—the massive success, if you will—of the fringe evangelical and revivalist groups, which appeal to the desire for action. Some of them can be dangerous.

22 Cults

Harold Visotsky

A phenomenon of the last decade has been the surfacing of the dozens of Eastern religious cults and movements attracting thousands of American youths searching for truth, structure, relationships, brotherhood, and authority.

This surge of cults in the United States began in the late 60's and became highly visible as a social phenomenon in the early 70's. In 1978, cults came to our critical attention as a result of the mass suicides and murders at the People's Temple commune in Jonestown, Guyana.

It is estimated that two to three million young adults in the United States have had varying contacts with groups and cults. These contacts have resulted in their leaving home, dropping out of school, leaving jobs, and breaking their continuing relationships with their family, spouses and even children. A good article by Margaret Thaler Singer, *(Psychology Today,* January 1979), lists a variety of reasons for these phenomena. Dr. Singer, who is a member of the Department of Psychology at the University of California Berkeley, has studied, with her colleagues, the young people and their characteristics.

The term cult has been applied to various groups who are off the beaten track of traditional religions. Many of these groups have made exploratory excursions into non-Western philosophical practices. There are intense relationships between group members and those managing the group, and a particularly intense relationship between the followers and a powerful idea or leader. The well-known groups include the Unification Church of the Rev. Sun Myung Moon, the Children of God, the Krishna Consciousness Movement, the Divine Light Mission, the Church of Scientology, and the People's Temple. Lesser-known cults include The Way International, The Love Family, and there is even one called the Bo and Peep cult.

In interviewing cult members and ex-cult members, one is struck by a number of similarities, including:

- The pre-cult personality problem and family relationships of the members
- The methodology of the group's recruitment
- The indoctrination procedures

There are a variety of reasons why young Americans are turning to Oriental and cult-like religions. The participants tend to be young, in their late teens or early twenties, but there are some cults that are now working rather intensively with early high school youngsters. It seems, from a sociological point of view, that the Eastern religious movements are made up almost exclusively of white, educated, middle- and upper-class young people. Most have at least begun college, although some have dropped out after a year or two. The men and women seem to join in equal numbers, but the men control the leadership structure. The young people come from all religious denominations with relatively more from liberal Protestant and Reformed Jewish backgrounds than the proportion of these groups in the general population would suggest.

Most of the young people who join cults are looking for warmth, affection, and close ties.

The cults supply ready-made friendships and ready-made decisions about careers, dating, sex and marriage, and they outline a clear "meaning of life." In return they demand total obedience to cult commands. There is constant exhortation and training to arrive at exalted spiritual states, altered consciousness states, and automatic submission to directives. There are long hours of prayer, chanting or meditation, and lengthy repetitive lectures, day and night. The exclusion of family and other outside contacts, rigorous moral judgment of the unconverted outside world, restriction of sexual behavior are all geared to increasing the follower's commitment to the goals of the group, and in most cases, to its powerful leader. Leaving a restricted community such as this can pose problems, particularly when one is aware that the reasons for joining the cult in the first place may have been due to a variety of emotional problems.

Dr. Marc Galanter, of the Albert Einstein College in New York, and several colleagues reported that some 39 percent of one cult's membership said they had serious emotional problems before their conversion, and 23 percent cited a serious drug problem in their past. Given this pattern, it seems that most of the members of these movements are looking for friendship. The reply most often heard from those actually living in religious communes told the story of loneliness, isolation, and the search for supportive community. They are looking for warmth, affection, and close ties. They did not find these at work, or at school, in churches, or even in their homes. They do seem to find them, at least for a while, in the communities and in the communes.

Another reason, cited by Harvey Cox, a member of the Harvard School of Divinity, is the search for a kind of immediacy. Cult members are looking for ways to experience life directly without intervention by ideas and concepts. Some of the young people move and drift from movement to movement, looking for an encounter with a meaningful life experience—perhaps an encounter with God or simply with "life" or "nature." The Neo-Oriental movements all include instruction in some form of spiritual discipline. Members are taught the primary techniques of prayer or chanting, contemplation and meditation. The teachers rely not only on words, but also on techniques. Ideas are kept at a minimum; repetition and direct exposure to repetition seem to be the common pattern.

Some are rejecting what they consider to be the effete, corrupt, or worn-out religious traditions of the West.

A third reason why young people join cults is that many of them are looking for authority. They have turned to the East to find truth, to be able to believe a message or a teaching that they can trust. They join these groups as refugees from uncertainty and doubt. Their trust in the leader, or guru, seems to be infinite. The quest for authority results from a wide range of factors documented by dozens of sociologists. It may stem from a disillusionment with conventional moral codes, the erosion of traditional authority, or the emerging of "over-choice." This is a term which Alvin Toffler used in his book *Future Shock*; it referred to the fact that people are in a condition of choice fatigue. Too many choices, alternatives, complicated arrangements seem to confuse and overwhelm them. They hunger for authority that will simplify, straighten out, assume something, or somebody who will make their choices fewer and less arduous.

A fourth reason that a small number of people seem to turn to these religions is because it is "natural." They are talking about "natural" in the sense of pureness, simplicity, and honesty. It is a deliberate response, rejecting what they consider to be the effete, corrupt, or worn-out religious traditions of the West. They see in Eastern spirituality a kind of unspoiled purity. In contrast to the Western faith, the East seems artless, simple, and fresh. Many of these young people can tell why they turn away from Western religion more clearly than why they turn toward Eastern religion. They look upon Western religion as corrupt. "It is nothing but technology, power and rationalization," they say. "It is corrupted to its core by power and money; it has no contact with nature, feeling or spontaneity. Western religion has invalidated itself; now only the East is possible. The Oriental peoples have never been ruined by machines and science, and they have kept close to their ancestors' simplicity." The mysterious Orient is what they are seeking. Those who yearn for what they call an Oriental approach are opting for an archaic rather than a historical way of life.

The cultists claim that they have found peace, happiness, love, and mental stability in their groups.

This inner seeking is reflected in a less malignant way by what we can call the decade of narcissism, which many of us are involved in — a turning inward, a search for health through biorhythm systems, through jogging, through some type of regaining a pattern which will keep us intact in this fragmenting, stressful world.

Now, I would like to turn for a moment to the process of indoctrination. What is there about this indoctrination pattern that can explain the series of events that have occurred within cults and have at times horrified our country? What can account for the murder of a U.S. Congressman and three newsmen, to be followed by the mass suicide-murder of more than 900 followers of Jim Jones's People's Temple in the jungle of Guyana? What explanation is there for the families who are horrified by the changes they have witnessed in their children — changes such as leaving home, shaving their heads, changing their names, signing over their money and possessions to a group leader, denouncing their families, suing their families, and attacking their families as agents of Satan's world? The cultists claim that they have found peace, happiness, love, and mental stability in their groups, have freely chosen their forms of worship, and are entitled to the full protection of the law under the First Amendment of our Constitution.

The techniques utilized by cults are very similar overall, but each one seems to have worked out its particular style. All of the groups appear to have worked out ways of gaining access to what they would consider to be susceptible individuals. Those who are recruited seem to be divided into two rather distinct groups. The first is composed of seekers. These seem to be schizophrenics with chronic symptoms or borderline personalities. It is quite clear that the existence of emotional or personality problems is one reason for becoming involved in the cults. Most mental health professionals consider only this reason at the present. This is an oversight, resulting perhaps from a superficial look at the groups. These inductees involve themselves in order to feel better because they are uncomfortable with the outside world and themselves. Such motivated conversions are restitutive in that the seekers are trying to restore themselves to some symbolism of comfort in a fresh or false reality. We also see this attempt at restituting in the so-called secondary symptoms of schizophrenia and other forms of mental illness as the effort of troubled or damaged individuals to put together a new, simplified mental world and style of reasoning. They do this to compensate for their terrible perception of the world and their personal vulnerability. In some studies, approximately 58 percent of the inductees seem to be found in such groups. This, to me, seems a little bit high, but it may well be within the ball park.

The purpose of indoctrination is to bring to complete subjugation the mind of the individual who is a candidate.

The remaining inductees were not ill or damaged in the sense of a mental illness; they were apparently normal developing young people, going through the usual crisis of development on the way to becoming adults. They joined for a number of reasons. Many of them were facing, as I said earlier, the pains of separation from their families, the normal depressions, the complexity of outer reality, which is a part of early college life, and they were seeking some form of adaptive response to their internal pain.

The purpose of indoctrination is to bring to complete subjugation the mind of the individual who is a candidate. The first event is the gaining of access to these potential converts. This is raised to a high art by all of the successful cults. Some have printed manuals describing where to approach prospects, exactly what type of initial pressure to put on each of them, and what the odds are that they will acquire a certain number of converts from a given amount of pressure, well applied. The general openness of manners of this age group adds to the ease of access. Once such a prospect has agreed to investigate the rather simple propositions expressed by representatives of the cult, he or she is brought into the next and highly sophisticated activities of the conversion process.

From the first, intense group pressures — lectures, lies, and other interpersonal pressures unexpected by the individual — are brought to bear. Singing, chanting, a constant barrage of the kinds of rhetoric which catch the young idealistic minds are constantly played. So intense is this that individuals who are under such pressure and are susceptible, tend to enter a state of narrowed attention, especially as they are more and more deprived of their ordinary frames of reference and of sleep. This condition must be described as a trance or a trancelike state. From that time there is a relative or complete loss of one's own mind and will, which are then placed into the hands of the group or the individuals who have the direct contact with the individual inductee. Once in this state of passive narrowed attention, willingness to be influenced is achieved; the true work of conversion or thought reform begins in earnest. This is always a program of unbelievable intensity.

The victims are induced rapidly to give up all familiar and past loved objects—parents, siblings, home, city. They are physically and emotionally moved to as foreign an environment as is possible to imagine.

During this, the cults step up their ideological reform pressures by showering attention and affection on inductees — sometimes called "love bombing" — by a change of diet, and by the introduction of elements of guilt and terror. The question of supernatural pressures that one must face in the future are brought out more and more explicitly and concretely. Many promises are made of redemption or safety, in the certainty that the world will soon end, at which time there will be enormous rewards to believers, terrible punishments to nonbelievers. The threats may be implicit, but are sometimes increasingly physical and explicit.

Preaching is constant from all sides, supervision is absolute, and privacy of body or mind may not be allowed for days or weeks into the future. All relationships with other people are organized and stereo-typed, and no chance is given for idiosyncratic expression. The victims are induced rapidly to give up all familiar and past loved objects — parents, siblings, home, city. They are physically and emotionally moved to as foreign an environment as is possible to imagine. Thus, it becomes increasingly hard for them to reconstruct in imagination what they have experienced some time in the past. Reality becomes the present, and it includes elements of the supernatural, magical, terrifying thoughts which have been expressed constantly all around. There is no base for reality testing. I could go on, but the process is intense and repetitive. I have a collection of speeches that are read over and over again by the Rev. Moon groups.

There's a tremendous commitment to a leader who's all-knowing, all-powerful, and never wrong.

There are number of reactions that are important in terms of those individuals who have left the cult and have a particular spectrum of problems and difficulties. Many of them are of an emotional nature. It is reported that it takes anywhere from six to eighteen months to get their lives functioning again at levels commensurate with their histories and talents. Symptoms vary, but among them is depression. A second is loneliness. A third is indecisiveness. A fourth is slipping into altered-consciousness states, then a blurring of mental acuity. A fifth is un-critical passivity. A sixth is fear of the cult. A seventh is what we call the "fishbowl effect," that is, the constant watchfulness of family or friends who are on the alert for any signs of difficulties with real life that will send the person back. The eighth is the agony of explaining why one joined the cult. The ninth is guilt, and the tenth is perplexity about altruism. After these come problems with money and the feeling of being "elite" no more.

It seems to me that the uncritical willingness, even eagerness, of young people to join cults should be a matter of great concern to all of us, not only because it's so pervasive among the age group we are discussing, but also because these cults represent a "solution" to many of the problems that beset young people, including some of the problems we have been talking about at this conference.

Discussion

John Schweppe: Are the techniques used by these cults somewhat akin to those that were used in the Nazi youth movement and in Maoist China?

Visotsky: In some respects yes, but not in others. What you see in the cults is more like the organized brainwashing that went on in the Korean War.

I would say that the techniques used by the cults have at least three major elements. One has to do with fatigue — that is, keeping people awake for long periods of time in order to get an altered state of consciousness. The second has to do with repetition of doctrine, over and over and over again, whether it be through chanting or through lectures. And the third has to do with altered nutrition. The nutrition is seriously altered to weaken the resistance of young people.

The relationship to the techniques you asked about is that in the cults there's a tremendous commitment to a leader who's all-knowing, all-powerful and never wrong.

Leonard Duhl: What is the dividing line between cults and other religious or self-help groups? I've run into some that I wouldn't call cults.

Visotsky: You have to take a look at the prime motive. Most of the cults are more interested in power and money than in helping people. Many of them start out by saying, "We want to help you," but there seems to be a drift away from that the longer the cult operates. A prime example is Synanon, which started out as a program to help drug abusers. Many programs in the United States modeled themselves after Synanon. It really was a high-class, well-thought-out, succcessful program. Then it changed in a way that was quite frightening. It turned violent. It hired lawyers to sue anybody who criticized it. It has managed to drive several small corporations out of business, and it has taken over real estate and property by literally driving out of the community elements that it felt would be inconsistent with its primary goals. Now it spends very little time on helping people who have problems with drugs or alcohol.

There's also another distinction, and it lies in what happens to the young people who join cults. I'm referring now to the sudden and drastic alteration of their value systems, including abandonment of academic careers and career goals and other changes that are relatively sudden and catastrophic.

I mean sudden, like in three to six weeks. There is a loss of cognitive flexibility and adaptability. When you ask them questions, they give you stereotyped answers.

There's also a great deal of regression of behavior, sometimes to almost a childlike level. Many of the youngsters really do very well soliciting contributions at the airport. They approach people very well. Then you see them sitting on the curb waiting for the mini-bus to pick them up because they have their hundred dollars, and they're sucking their thumb. They're looking into space. They seem to have regressed.

Duhl: I was struck by what you said about the eagerness of some young people to join cults. Maybe there's a lesson in that for us. Maybe we need some new pattern of governance that will permit us to be both individualistic and to belong, and to be able to move back and forth. There are very few organizations that I know of that could serve as models. Maybe we can create one that will give us a sense of wholeness and at the same time permit a certain amount of individuality.

John Cannon: It's obvious to me that in some sense the attraction of youth to cults is a judgment on organized religious institutions. Do you see any way in which the liberal, mainline religious groups from which many of these people are drawn can relate to their teenagers differently, so as to meet the need that's there?

Visotsky: Yes. Let me begin my answer with an analogy. Don King talked about the mutation of certain microbial organisms through the pervasive use of Tetracycline and other types of chemicals. Cults are highly mutational, too. We can fight them in a number of different ways, but I don't think that's the answer.

The answer is a competing system which is relevant to youth and relevant to our own value system. What I'm worried about is that we're not designing that competing system. All we're trying to do is shut the cults off. We're not going to shut them off, ever. They have an immense capacity to be adaptive. Their lawyers can tie up our lawyers. They do it every day. We're bogging down in the courts. We can't even get them on the docket. They have had more deprogrammers brought up on kidnapping charges than we have had *them* brought up on kidnapping charges. So the point I'm trying to make is that we'd better come up with a competing system, a system of values, a system that the kids can relate to.

Cannon: May I interject an observation on that point? In the late 60's, the so-called liberal, mainline religions as well as the liberal Jewish groups began what has been ever since an accelerated withdrawal from campuses throughout the country of what had been traditional campus ministries, because of financial constraints and because of the fellow-traveling style of campus ministries during the 60's. That has been a matter of great concern to me, because I agree with you.

Duhl: Our problem is that most of our alternatives are very narrow alternatives. The religious organizations are fairly narrow. We haven't had any holistic organizations. Our real social innovation can come if there are broadly based psychological, social and spiritually oriented alternative institutions. I also believe that there should be freedom of exit and entry, so that you can belong if you want to, but you can also move to something else without anybody going out after you.

There is one other point I want to make, and that is that young people today are searching for a kind of spirituality that goes beyond what the churches are talking about, and they think that cults are the answer.

Visotsky: I don't disagree with you, but there's a joker in the deck. It's part of the pathology. Some of the kids who join cults are borderline schizophrenics who are looking for a refuge. They do a lot better in cults than the ones we've turned out of institutions into a hostile community that rips them off. So when you talk about holistic groups, don't forget that cults are holistic. They find a place for the borderline schizophrenics. They find a job for them. The job is meaningful. They provide a sense of security. That's the joker. I sometimes think I wouldn't have any objection to Reverend Moon if he'd take all of my sick kids, instead of taking the healthy ones.

Sheila Kamerman: I have two questions. First, is there any indication that kids who are part of this cult movement will outgrow it? That is, does age make a difference? And second, since kids who join cults are a very small group, how much study has there been of those kids who don't join cults, and find alternatives? What have been their solutions?

Visotsky: Some youngsters do outgrow their attraction to cults, and a lot more are dropped by the cults themselves because they really become too much of a burden. The cults, like other groups in our society, do their own share of dumping. If they get a very sick youngster, they'll dump him off at one of the state hospitals or see that he gets outside of the group and doesn't return.

Many of the other youngsters with problems—those who don't join cults—form groups of their own or join groups like Alcoholics Anonymous. We had a bull session at Northwestern designed for a small group of people to answer questions about life styles. We started with fifteen kids and it went up to 400. They were asking such questions as, "If you try heroin, will you get hooked?" "If you join a cult, who should you call to get you out?" Kids like that really need a validator, someone to validate what they're going to do. They don't need a director.

23 Juvenile Crime

William E. Threlkeld

I will discuss some trends and patterns of juvenile crime as well as some problems faced by young people in the age group we are concerned about here. The figures and trends I will cite are from Denver, but through my discussions and association with juvenile officers of major jurisdictions, I have discovered that these trends and figures will not differ much from those in other cities across the country.

I agree we need to approach the subject of this age group in a broad context. I also know we need to deal with the major underlying issues rather than isolated immediate problems. However, I am in the business of dealing with delinquency problems, so I will discuss the source of juvenile crime as well as some of the things that might affect it.

The most disturbing trend we have seen among young people is that toward violence. Crimes of violence, while not increasing in number (except for homicide), are increasing in severity.

First, understand that all young people do not involve themselves in crime or delinquent behavior. Of the 45 million persons in this age group, only about 2.5 million will come to the attention of law enforcement as delinquents. Nonetheless, that is a great number and creates large problems for society. More than 50 percent of the Class I felonies — murder, manslaughter, aggravated assault, larceny, and auto theft — are committed by juveniles under the age of 18 years. Probably another 30 percent of these felonies are committed by people 18 through

235

25 years of age. These figures indicate that the young people we have been talking about are the criminals of our society.

Let us now turn to trends in juvenile crime. We must first understand that we are now only dealing with those youths up to the age of 18. The statistics in the accompanying table do not exclude youths between 10 and 14, but those from 14 through 18 are, by far, the largest number of problem youths.

Denver Juvenile Arrest Trend
1969 through 1978

	1969	1978	Percent Change
Murder	5	10	100.0%
Manslaughter	1	3	200.0%
Rape	18	19	5.6%
Robbery	108	134	24.1%
Aggravated Assault	201	139	-30.8%
Burglary	901	1,103	22.4%
Larceny	1,784	2,144	20.2%
Auto Theft	869	662	-23.8%
Other Assaults	388	494	27.3%
Arson	62	42	-32.3%
Forgery	10	22	120.0%
Fraud	32	8	-75.0%
Stolen Property	11	25	127.3%
Vandalism	522	650	24.5%
Concealed Weapons	144	134	-6.9%
Prostitution	11	78	609.1%
Sex Offenses	51	40	-21.6%
Narcotic Law Violations	308	280	-9.1%
Liquor Law Violations	66	117	77.3%
Intoxication	223	0	-100.0%
Disorderly Conduct	377	309	-18.0%
Vagrancy	18	9	-50.0%
Driving While Intoxicated	10	105	950.0%
Loitering and Curfew Violations	784	433	-44.8%
Runaways	1,401	1,086	-22.5%
Traffic Violations	189	86	-54.5%
All Other Offenses	1,876	1,344	-28.1%
TOTAL	10,370	9,476	-8.6%

The most disturbing trend we have recognized is that toward violence. Crimes of violence, while not increasing in number (except for homicide), are increasing in severity. In the past, a youth might involve himself in armed robbery; now it escalates to aggravated robbery and aggravated assault, with the senseless shooting or beating of the victim more and more being a fact of the crimes which are committed by youths. Many people who have done in-depth psychological studies of these young criminals attribute this alarming trend to TV violence.

More and more we see a trend toward more than one youth in the family being involved in delinquency. In the past, generally only one child would become involved in delinquency; now we see two or more children in the same family involved.

Juvenile prostitution has increased dramatically. No programs exist to help these young girls, and no answers seem to be forthcoming from the people who have studied the problem.

Our attention is also drawn to a disturbing rate of gang activity, not as we knew it in the 40's and 50's, but, nevertheless, gang or group involvement in delinquent acts. This involvement differs from our past experience in that there seems to be no true dedication to "protection of turf," nor do there seem to be defined leaders and subleaders in the group. Instead, the leader seems to be that individual most accomplished in the delinquent act the group undertakes, and will change with the involvement in different types of crime. As an example, the youngster who seems to know most about burglary will lead the group in a burglary, while one who is most accomplished in theft will be the leader if that is their undertaking.

In the past ten years, juvenile prostitution has increased dramatically. Arrests for this act doubled in 1978 over 1977. Based on figures to date, they will undoubtedly double again in 1979. This is particularly a problem that society has not addressed. No programs exist to help these young girls and no answers seem to be forthcoming from the people who have studied this problem. Of all crimes that involve both adults and youths this is, undoubtedly, the most heinous in that the pimp coerces, convinces or cons the juvenile into a life of prostitution. As Dr. Visotsky pointed out, the recruiters for cults are very accomplished in their business; the pimp is, unquestionably, the master of the trade. He can recognize a confused, troubled girl and, within a matter of a few minutes, will have started her on a way of life which will destroy her self-respect and set her as an outcast to the rest of society.

There has been a sharp increase in the use of alcohol, but a decrease in the use of hard drugs.

Denver has had a decrease in juvenile arrests over the past ten years. This does not indicate a decrease in juvenile delinquency, it indicates an increase in juvenile understanding. We now do things for children other than arrest them. We have developed programs to cope with youths and provide alternate solutions other than detention.

Sexual abuse is especially troubling.

Another quite different trend, which is truly frightening, and is certainly escalating, is that of child abuse. While better reporting of such incidents is much of the reason for the increase, the number of reported cases continues to rise and can really be explained only by the fact that people are more violent toward each other, and toward children. Child abuse and neglect involve the age group of 14 through 24, both as victim and as perpetrator, to a far greater degree than any other age. While physical abuse is not so great at the ages above 15, sexual abuse certainly is.

A large percentage of the abusers of small children fall into the new-parent range below the age of 25. Sexual abuse is especially troubling, because most girls think that because it occurs at home it is normal. When the young girl discovers it is not normal she very often does not know how to handle the problem, believing it is her fault — and to report it will cause trouble not only for her, but for the entire family. The result is a young girl very mixed up, often troubled and unable to cope for the rest of her life.

Child abuse has a self-repeating aspect to its nature. As Dr. Henry Kemp of the National Center for Child Abuse has stated: "If you were abused, you will abuse." This is true to an alarming degree. It appears to be a vicious chain that drastically needs to be broken.

All of these problems of delinquency and abuse are a tremendous financial drain on society. The cost of supporting the juvenile justice system is huge and uses a great many of society's resources that are needed elsewhere. Very few new approaches have been tried and, of those that have been tried, only a few have met with any success. I would like to offer one possible program which just might help. I do not expect the results to be immediate, but I sincerely believe they would be well worth the expenditures which, taken in the scope of the cost of today's existing problems, would be very, very small. I submit this idea to you because most of you are in the business of education and it is from there, I think, the help must come.

Let me explain first why I believe education is the answer. In the early 70's, when the drug scene seemed close to overwhelming us, drug education was started. Enforcement did not lower the indiscriminate use of drugs; facts and information did. While drug education still is not as widely and extensively presented as it should be, it has still shown remarkable success. When young people came to understand drugs and their danger, there was a rapid decline in their use. I believe this is also possible with delinquency and abuse.

The schools simply have not kept pace with the needs of society, and particularly the needs of youth.

My proposal is simply this: Through the use of structured, mandatory classes in the subjects of delinquency and abuse, I feel young people can better understand the impact of these incidents on society, not just the cost and the suffering to the victim but the cost to society and the cost to the perpetrator in terms of the loss of self-respect and self-esteem. The courses I have in mind would explore and discuss crime and delinquent behavior honestly and openly, not necessarily to elicit more reporting, but to promote a better understanding of them and, hopefully, teach our young people that they are not normal and not acceptable in the kind of society they want.

I, frankly, see this sort of education being best done, at the present, by those school counselors whom we have previously talked about. For the most part, they have an understanding of the problem and would only need to perfect their delivery. Sex education should be included in these classes.

In my opinion, these classes should be required and should start at the junior high school level and continue through high school. Without being overly critical, I would like to point out that the schools simply have not kept pace with the needs of society and, particularly, the needs of youth. I feel that this type of curriculum would support those important family values we all agree are essential to society and would help prepare young people for a better life.

Discussion

Paul Bohannan: I agree with what Bill says about the need for better education, but I don't think it belongs in the schools. They're not equipped to do the job. What we really need is more backup institutions to do the jobs that are not done by the family or the schools or the other institutions which sometimes can't do what society says they're supposed to do.

Leonard Duhl: I'm very concerned. We're beginning to talk about human services organizations again, forgetting that the big problem is in society itself. So the real issue is something way beyond what any of us are talking about. How do you redesign society so that it offers a meaningful role to these people?

Threlkeld: Do you want to redesign it or do you want to return it to what it used to be? We did not have these kinds of problems when we had communities — when everybody in the community knew little Johnnie and when he stepped out of line any one of a number of people would be glad to help him back.

Duhl: I don't think we can go back to the old communities, but the question is can we move to a new kind of community in the next few years which will take care of all these kinds of functions and make a place for people to really be part of the whole?

June Christmas: That sounds like Utopia.

Sheila Kamerman: This is a multifaceted situation that we're discussing and, obviously, different types of support systems, different types of interventions, and different types of policies have to be developed in response to different kinds of needs. But let's not forget the fact that for the kids in this age group who are under 18, the school is the most important non-family institution they are exposed to. So, if we're talking about developing support services for young people within a formal institution, the school is a critical place to begin. That doesn't negate the fact that other types of approaches have to be developed for older youth.

The second comment that I would like to make is that we have to recognize that one of the problems in getting some of these programs into the schools is political. One thing that has not been adequately addressed in this conference is how to implement any of our recommendations. Indeed, implementation is a critical aspect for all policy and program development.

The issue is not just to come up with some nice-sounding plan, policy, or program, but how to get political support for having these implemented. This is a very, very difficult problem and it seems harder right now than it has ever been before. I hope at some point between now and the time we close that we address the issue of constituency-building around some of the kinds of policies that we think are significant — and the whole question of political feasibility at a time when we seem to be moving in a very conservative direction.

A. J. Kelso: I'd like to raise two questions that occurred to me while listening to this discussion. The first is have we already extended beyond our biological capability of adapting to the problems that we're addressing? Have we already placed so many stresses on individuals and institutions that any policy we might come up with would take us beyond our biological capacity to adapt to it? Is what we're looking at the kind of stress, or the kind of result of stress, that the human being is no longer able to address in an effective personal way? If we are, then we're addressing something that ultimately means extinction of life as we've known it. If we've already gone beyond our biological capability, we're just wasting our time.

Another question which it seems to me worth raising at the moment is not what policies are going to be adopted, what solutions need to be found, but what is it that you and I can do as individual human beings in our own spheres of influence that might address these issues in some effective way? It's very easy for us to make the problems so remote that there isn't a damned thing we can do about them as individuals. I'm interested in what we can do as individuals in our own spheres of influence, in our own lives. And I don't mean offering courses in sex education. What I mean is what does Jack Kelso have to do to make a difference in helping to ameliorate some of the problems we're talking about?

24 Careers and Vocations

Willard Wirtz

I am going to speak from notes, partly because I've reached that stage in life when I find it wiser to speak even to my wife only from notes for fear of saying too much or too little, but mostly because what I have to say here is closely linked with things that have already been said, and I'd like to note those interlinkages.

I should warn you that I cannot speak with the same certainty Phil Blumstein expressed. Phil prefaced his extraordinarily illuminating remarks by asserting a belief in science, in observables, not in wishes. I can't proceed on the same firm basis.

Indeed, I'm prompted to think of those three haunting questions of T. S. Eliot's: "Where is the wisdom we have lost in information? Where is the understanding we have lost in knowledge? Where is the life we have lost in living?" I can't find the answers to those questions if I proceed only from observables.

With those two warnings out of the way, let me turn to my assigned topic. In discussing careers and vocations, I'll be talking particularly in the next few minutes about the 16-to-21-year age group, not only because that's the group for which figures are most readily available, but also because for my purpose it is the group that merits particular attention.

Four things need to be noted about their vocational activities. The first is a sharp increase in the participation rate of the 16-to-21-year old group so far as vocational activity is concerned. We hear a great deal about youth unemployment figures going up. It's equally true, and no paradox, that the employment figures are also going up. There is a sharp increase, as Arthur Norton noted, in the extent to which kids in school are also working, and in the extent to which young women are entering the labor force.

Second, there is a very marked shifting of youth from jobs in the production sector to those in the private service sector, which also has been remarked on, and from jobs in large establishments to jobs in smaller establishments. Moreover, among large employers in the manufacturing sector — that group upon which we relied very heavily for youth employment in the 60's — the general practice has developed of no longer hiring males under the age of 21, which is a very sharp shift.

Third, fewer and fewer of youth jobs are now on career ladders. More and more of what the 16-to-21-year olds are doing would be described in common jargon as "dead-end" jobs. Another related characteristic is that there's a very high turnover as they move rapidly and frequently from one of these jobs to another.

Fourth, so far as the 21-to-24 year age group is concerned — that's the older part of the age group we're discussing — there is a parallel development in the increasing percentage of jobs that don't make the maximum use of the individual's education and talents. This is called "underemployment."

Now I've given you no statistics, nor do I mean to, partly to emphasize what is in my judgment the virtual uselessness of the available employment and unemployment statistics as far as youth are concerned.

There were six years during which I bore personal responsibility for the release of the unemployment figures the first Friday of every month at precisely 10 a.m. Throughout that whole period I would be constantly reminded of the line in "The Man from La Mancha," just after Don Quixote is finished singing about the impossible dream: "Facts are enemies to the truth."

When you pick up any set of figures based on the current national reports of youth employment and youth unemployment, keep in mind that they have been developed as part of an economic indicator; the definitions have little relationship to the truth as far as youth are concerned, and are almost useless as far as working out an administrative program.

With those four specifics, I turn to one other aspect of the employment/unemployment situation which has been referred to earlier in this conference — the oversupply situation resulting from the baby boom. Sheila Kamerman and Al Kahn have suggested that the converse of this is a new labor shortage or undersupply situation starting in about 1985 and going on up to 1990.

Al made my position very difficult by referring to the "false alarm" of the early 1960's as a lesson prompting humility among those of us of apocalyptic tendencies who see a more serious problem ahead. I take a different view, and I hope we'll have an opportunity to talk about it later.

The prospect I see now for the 1980's is one of an aggravation of what we have called the "oversupply" situation. I think that situation may well get worse rather than better.

It seems to me that the bearers of good news from the demographers forget the fact that their reports are based only on a head count; they take insufficient account of several other developments. One is the increasing competition these kids are going to have from women, from seniors, and from adults in general at a time when the unemployment rate nationally is 6 to 7 percent. They're going to feel that crunch increasingly, I believe.

There's a particular temptation when we address the problems of this age group to do our thinking in mollycoddle terms; we would be much better off to think about them in broader terms — as societal concerns.

I have left out a more arguable point — the effects of machines rapidly getting as much education as these kids have. The other thing that's left out in the head count, it seems to me, is the realization of the fact that what we're talking about here is about the education of the people involved as well as their numbers.

The demand for developed talent has been going up in this society, and it keeps pushing up the supply curve. I venture the though that the developed talent supply curve has already crossed the developed talent demand curve.

The subject, unfortunately, cannot be talked about publicly in the political forum because what I'm saying in an oversimplified way is that as things go nowadays, there is not enough for young people to do and there's not going to be, given the present situation.

And so, in general, I suggest — as a basis for proceeding to a consideration of programmatics—a picture of the career and vocational aspects of 14-to-24 year olds' activities that includes some structural elements which warrant some different manipulative pathologies, but second, which involve a much deeper question, putting it oversimply, of whether there's going to be enough for this age group to do. And, of course, that question requires the consideration of much more radical surgery.

Now about the program possibilities. My discussion of them will be divided into two parts. The first will assume the present economic realities; the second will take on what has been called "the Utopian purpose" of redesigning the society.

Even accepting the overall context of present realities, there's unquestionably a good deal that can be done to provide young people with what Sid Werkman called "sensible careers." That's *our* concern as well as theirs. There's a particular temptation when we address the problems of this age group to do our thinking in mollycoddle terms; we would be better off to think about them in broader terms—as societal concerns.

It seems increasingly clear that in our search for new ways of dealing with the career and vocational problems of youth we are going to have to look more to the local community rather than the federal government. I think we've probably reached very close to the limit of federal support.

What would have been my essential point about careers for young people has already emerged as the unifying theme and certainly the key good sense of this seminar, which is that the first essential step in moving toward new programs is the establishment of procedures and structures through which specific right answers are most likely to evolve.

The establishment of those procedures is itself an action program, and a tough and demanding one. Several of the speakers at this conference have emphasized that. I'd like to add three thoughts of my own about this process. First, it seems increasingly clear that we're going to have to look more to the local community for the development of the new processes and structures which we need. I don't think it's going to come at the federal level alone. I don't think it's going to come at the state level, and Bob Aldrich properly reminded us that even a large city presents problems that are very hard to meet.

I think we've probably reached very close to the limit of federal support as far as this particular area is concerned. I believe it's more than a matter of succumbing to the Proposition 13 philosophy. I think it's primarily a matter of there not being a breathless immediacy about it, of there being no sufficiently noisy crisis involved in what we're talking about here to expect very much more as far as the federal government is concerned. There are just too many other demands.

I think the principal reason we have to look to the community for the processes and structures we need is that it is only at the local level that people in general can acutally participate in the handling of their own affairs.

At the state and federal levels, they are merely represented. We are at the point in the history of government in this country where a vast majority are fed up with representative government. People have come to identify it with corruption of one kind or another. The only antidote I know to that is to give them an opportunity to participate, and it is only at the local community level that this can be done. It is particularly only at the local community level that the members of this particular age group can participate in the handling of their own affairs. Right now they can do nothing but march on Washington.

My guess is that this conference can be replicated effectively much more readily at the local community level than at the federal government level and that we can expect more from the interplay of the disciplines locally than we can federally.

Now I come to the point Jack Kelso raised when he asked whether we'd gone beyond the biological capacity of individuals, and the traditional institutions, the family, the school and the church, to handle these problems. My answer, of course, is a resounding no, at least in the case of individuals and very much in the case of the institutions. I don't think the problem is overload. I think it's under-use, as far as the individual is concerned.

Then Jack asked what we, as individuals, could do about some of the problems we have been discussing. I would like to answer that in what I hope are not maudlin terms, although they involve the recollection of a conversation I had some fifteen years ago with my father, then not at all well. We were watching on a Sunday afternoon on television the Mormon Tabernacle Choir. When the program was over, he said, "You know, that means quite a lot to me." I asked why. "Well," he said, "if one member of that choir stopped singing, I wouldn't notice the difference and I don't think anybody else would. Similarly, too, if two or three dropped out. If a quarter of them stopped singing, then I'd notice the difference in the quality, and if a half did, I certainly would. Then, eventually, only one would be singing and that would be all the difference in the world—between a choir and a solo. And finally, if the last person stopped, the loss in numerical terms would be no greater than that which occurred when the first person stopped, but it would dramatize the fact that the individual does make a difference, even though it can't always be noted."

The second point I will try to add involves what I think is the importance of developing new forms of effective collaboration between established institutions, which can be done at the local level but not possibly at the federal level. The Department of Labor and the Department of HEW are only four blocks apart, but you have to go around the world to get from one to the other.

The programs developed for 14-to-24 year olds would almost of necessity, if they are to be effective, have to be broadened out into programs meeting the parallel needs of other age groups.

Len Duhl has noted that networks do not make communities. I agree thoroughly and completely. I guess I would urge very strongly that the implementation of the sense of community in an institutionalized society depends on the development of new working relationships between these various traditional institutions. I think it's appropriate to note that this problem of inter-institutional collaboration is particularly important as far as this age group we're talking about is concerned, and indeed, I suspect that this is equally true of the transitions from one age group to another. We're terribly bad at handling the transitions, because so many of them fall in "no-person's land," between the turfs of established institutions.

I also want to suggest that the problems developed for 14-to-24-year olds would almost of necessity, if they are to be effective, have to be broadened out into programs meeting the parallel needs of other age groups. I would make this argument first in terms of equity. For example, I think the reaction of most people my age is that if you talk all that much about the problems of the youth maybe you better start talking about my problems at the same time. They are at least comparable problems. But it's also a matter of pragmatic politics. I think what we're talking about depends on the development of a coalition constituency which we don't have as of right now. For example, I see the decline in educational appropriations as relating very directly to the sharp decline in the number of people who have children of school age. As that number of taxpayers and voters goes down, there's less support for education. Maybe the obvious answer is the development of a concept of continuing education so that more people in the community have a self-interest, if you will, in the process rather than only a derived interest, and they don't need to think necessarily of somebody else's children.

The most commonly suggested way of dealing with youth unemployment would be to exempt young people from the minimum wage law. I might be in favor of that if someone could devise an exemption that did not have the consequence of taking Peter's job and giving it to Paul's son or daughter.

Before leaving the subject of structures, I ought to mention the National Manpower Institute, which is engaged in the development of community education work councils. This is an attempt to implement some of these things that I've been talking about here. It has been a remarkably successful enterprise. It grew out of a previous catastrophe in which we tried to develop a citizen-involvement network. That was too broad. Instead of that we've tried to take a narrower focus — the relationship of education and work—and have found there a degree of local community interest and inter-institutional collaboration which I can only commend very strongly.

I come now to specific program possibilities. My time is running out, so I've got to go through them almost on a checklist basis.

First, the commonest suggestion is to establish an exemption to the minimum wage law. In the country at large, if a referendum were held, there would be an overwhelming majority who would say that the way to create more jobs for young people is by exempting them from the minimum wage law. I might be in favor of that if someone could devise an exemption which did not have the consequence of taking Peter's job and giving it to Paul's son or daughter. The point is that the exemption would help with the youth unemployment problem by worsening the adult unemployment problem.

Second, Helen Washburn and Jack Darley talked about counseling and guidance. We ought to talk a lot more about that, especially about the possibility of an expanded counseling and guidance function that would be community based instead of school based.

There's a great shortage of adequately trained counselors and guidance people in the schools in this country, but there is a surplus of such people in every community. The organization of a community-wide system of counseling and guidance opens up the prospect of organizing an opportunity inventory for every community, a placement function, and most particularly, a follow-up function which starts when the individual gets a job. Failure to follow up is one of the worst weaknesses in many guidance and counseling systems.

It seems to me that we can draw on the tenets and history of democracy to make as good an argument to a child about the importance of work as the Chinese do.

My next suggestion is one that I would call, for the lack of a shorter title, better mobilization of the available private service sector opportunities. Here I'm including such things as taking care of somebody's yard, mowing lawns, or working at MacDonald's. We call those dead-end jobs, but I would like to see that expression stricken from the vocabulary. Almost everything I did between the ages of 15 through 25 was a dead-end job, and some of the things I've done since were dead-end jobs. The attitude that has grown up about dead-end jobs seems to me absurd.

This brings me to the important matter of changing the attitudes of youth toward work. John Schweppe brought that subject up in an interesting way when he referred to the China experience. I was there a year ago. What they have done in the development of work attitudes leaves only the question of whether we couldn't do the same thing here, without the enslavement of the mind which extends to political indoctrination. Those kids really want to work, and I refuse to think that there's something about Communists that gives them a monopoly on the development of the attitude toward work that those kids have. To put it more affirmatively, it seems to me that we can draw on the tenets and the history of democracy to make as good an argument to a child about the importance of work as they do with communism.

There was an earlier reference to a youth service program. I would call it a *voluntary* youth service program, because you just can't sell the "mandatory" idea at this point. I'd also leave out the word "national," because it seems to me that any such program, to avoid the dangers which both June and Len have pointed out, must be community based and must involve a large degree of local discretion. If it is unidirectional, or if it becomes an elite corps of one kind or another — if either of those things happens it will not have been worth it. I would also emphasize that it's service we're talking about, and not employment in the traditional sense, because there is a great deal to be done in the communities in the name of what we now call public service and it is not going to be done by traditional employment.

A one-with-one guidance and counseling program would not be too costly, and it would meet the adolescent's greatest concern: "Doesn't anybody give a damn about me?"

My fifth programmatic suggestion is that we must pay attention to June's concern about not cheating the hard-core disadvantaged group. It's not just a matter of not cheating them, but also a matter of trying to minimize the horrible cost to the community of not doing enough about that group. So we've got to have some kind of program that meets the special needs of the disadvantaged. Our discussions of mentorship may contain the elements of an answer.

Incidentally, I think mentorship is a bad word. People don't understand what it means. I'd call it a "one with one" program. I think the validity of this approach comes from doing some arithmetic. There are about 45 million young people in the age group we're discussing. Only about eight or nine percent are what you'd call hard core, out of school, out of work, in trouble. That's a manageable figure. John Schweppe asked very properly about the cost. I would put it at $2,500 a head, to do something about them on an essentially mentorship, one-with-one basis. That kind of approach would be feasible, and it would meet Sid Werkman's point that the adolescent's greatest concern is, "Doesn't anybody give a damn about me?"

I'd start with the names and addresses of the hard-core group in any particular community and then work out a federal program administered at the local level. I'd divide that $2,500 in half. I think we could assume that the federal government — the national taxpayers — should put up about $1,250 a year. The other half would be reimbursement for in-kind services to the public and private sectors of local communities. That's about $2 billion a year in federal funds, which I would simply transfer from the present programs designed to meet the needs of this age group and pay it out to the local communities.

If there were more time I'd tell you about the success of our Wheeling, West Virginia, Community Work Council, working along precisely those same lines. About once a month, a group of 12 to 15 representatives of various community organizations sit down and start with the names and addresses of the hard-core disadvantaged youths and they simply divide up the responsibility for helping them. They always find some institution in the community which is willing to take on the responsibility for helping each kid.

My sixth programmatic suggestion is also something that we're already working on. For the moment we're calling it a "community-based lifelong communications learning curriculum." Let me simply note the elements that are involved in that awful mouthful of words. It's community based rather than school based. Communications rather than writing; and learning rather than education. It's an attempt to respond to the very strong feeling in the community today — I think well founded—that what these kids need most is the ability to communicate, to write if you will, and if we're talking about basics that's the one on which we ought to concentrate.

But you don't learn writing just in school. You also learn it in the home, and in the community as a whole. It's a lifelong process.

The sooner we start using the word "learning" in place of "education," the better off we're going to be, because education is an institutionalized term. It vests the school with some responsibilities which only the community as a whole can exercise. If you once start trying to work out the role of the family, the school, the employer and so forth in developing the learning of writing, you will be far ahead.

Let me mention two little specifics of the Wheeling program. One would be a requirement that no examination involve more than 50 percent objective answers. Why? How can these kids be expected to learn how to write when they're going to be graded and credentialed on the basis of which box they put an X into? That's self-defeating, because it doesn't require any writing! A second requirement would be that every applicant for any job has to sit down on the spot and write one paragraph of prose, or if it's a particularly important job, four lines of poetry.

Finally, I'd like to say a few words about another programmatic possibility that has already been mentioned—the idea of redesigning our society. Not because of time pressures, but because of the limitations of my own understanding and knowledge, I'll get through this very briefly by asserting three propositions, which you may accept or not.

The first is that this is critically yet only superficially a matter of developing a new economics. A model is already available to us in Chapter V of E. F. Schumacher's book, unfortunately entitled *Small Is Beautiful,* which it isn't in this country. The economics he talks about puts people in first place instead of someplace else down the line. It's an economics which says that our needs are limitless, that our human resources are limitless, and the two have got to be matched up.

I am totally persuaded that the only effective instrument for basic change in this country is education . . . an education which is as fully committed to the teaching of values as it is to the teaching of facts.

It seems to me the economists commit a fraud on the public by leading us to assume that economics is necessarily confined to the parameters and perimeters of the grossest national product in history. I know nothing which confines economics as an abstraction to this particular framework. I know nothing which justifies an index that excludes all the services women perform at home. I think, and with all humility—a humility born of ignorance—that any given economics is only a form of rationalizing whatever happens to be the current political orthodoxy.

So my second proposition would be that what we need is a new politics. This need is more basic than the need for a new economics. The new politics would be based on the value judgments which have been repeatedly expressed throughout this seminar. I reject the notion that such a politics is impractical and impossible. I think of Antigone's rebuke to her sister's timidity: "Until we have tried and failed, we haven't failed."

I think the new politics will have to evolve at the grass roots level, because that's where people can accomplish something. I have difficulty with the fact that as things now stand, the judiciary has become the principal agency of change as far as the government is concerned. If you look back over the major social changes of the last twenty years, you find that the judiciary has been the principal institutional instrument for change. Now that's a kind of horrifying thing, because I think the reason for it is that the judiciary is furthest insulated from public opinion, and if you reach that conclusion your faith in democratic tenets is in serious trouble.

Somehow or other, we've got to change people's minds rather than rely on remote institutions like the judiciary. Most of us have grown up thinking that education's role is to conserve, and that the role of politics is to change, but I am totally persuaded, after having spent my life about half and half between those two disciplines, that the only effective instrument for basic change in this country is education.

It would be presumptuous of me in this group to try to identify the proper elements of a new education. At the risk of oversimplification, let me say merely that what I have in mind is an education as fully committed to the teaching of values as to the teaching of facts.

If these propositions seem to depend too much on faith and to have too little support in either experience or reason, I can only say, in conclusion, two things.

First, I think youth's vocational and career prospects, and whether they will contribute to a sense of worth rather than to an infecting questioning of life's integrity, probably depends on reordering the current irrational realities.

Finally, I recognize that as I get older it seems more intellectually satisfying, when reason and experience don't carry me as far as I think I have to go, to proceed the rest of the way on faith — if only in the attainable integrity of the human opportunity.

Discussion

Robert Aldrich: For many years I have been of the opinion that the study of the human life span ought to be a life science, and after hearing what you just expressed I'm more and more of that opinion. It could well be a part of the new education you mentioned. I agree with you that it very definitely has to be offered at the grass roots level—in the elementary schools, and in the family. We have to address the parents, the children, and the grandparents — different generations — with the same topics, at the same time.

Wirtz: Thank you. Let me say that twenty years ago I would have scoffed at the notion that you could rely on what was going to happen at the local community level. I was completely convinced that it takes the leadership of the national government to do something of this sort. I've changed my mind totally and completely.

Aldrich: I think many of us are beginning to recognize the hazards of federal grants. Having served, as you have, but not for as many years, as head of a federal agency, I also have developed some second thoughts about them. The more I look at them now, the more dangerous I think they are.

I think that the opportunities have never been better for private resources, whether they're large or small, to seek out points of leverage in a community that has thought through what it wants to try to do and needs $10,000 or perhaps more to move from just a concept to some pilot work. Hundreds and hundreds of foundations, to say nothing of the industrial and manufacturing companies who have money to give to this sort of thing at the grass roots level, would be willing to help.

Wirtz: That goes a little too far for me, Bob, but only a little bit. We're basically in agreement, but when you referred to a reliance on corporate sources, I think we've got to make a judgment as to whether we think the expenses of these things should be borne by consumers or by taxpayers, because when we talk about shifting the burden to private employers, it gives me pause. We're already at a serious disadvantage in some international competitive markets, and I think we should go easy about transferring the cost of social programs from the taxpayer to the consumer.

As far as the federal government is concerned, it would be very easy for me to suggest cutting back on federal grants except for the fact that there are disparities between the situations in various communities which we've got to level out. So it isn't that I would cut down the level of federal expenditures. I wouldn't, and I guess if I could do something about it, I would increase it some. It is rather that I would carry the process of decentralization further than it has been carried even in the last ten years.

Decentralization makes no sense in my book if it means only transferring the responsibility and authority from federal politicians and bureaucrats — those being two groups of which I'm a card-carrying member—to local bureaucrats and politicians. The point of decentralization is that it can get us back into the area where people in large numbers *participate*. Therefore, it seems to me the answer is to transfer these funds to local communities with no more strings attached than necessary to meet the minimal demands of fiscal responsibility.

Aldrich: That's precisely where I stand. I won't continue this any further, except to say that for about seven or eight years I have been working with political scientists in helping to educate small, rural towns on how to recruit and hold a competent medical and nursing staff. It has been very successful, and so far we have not had to ask for a single outside grant. The interesting thing is that the money is in the community. The lead time is about 18 months until you really see results.

June Christmas: I was very much touched and stimulated by many things that you said, Willard, but a couple of questions occurred to me. Not questions, really, but troublesome thoughts.

The first has to do with your reference to China. One of the things that I was impressed with in both China and Cuba was that what made people willing to do the kind of work that even for their countries would be considered menial work and seem to enjoy it, was a sense that they belonged to the society and that they were a part of it, and that what they did in some way would affect what happened in that society.

Even in a place like Cuba there was a very strong feeling among people on the local level that they mattered — that their vote, their work, and their participation was part of a common goal, a national goal.

I don't think that we have that in our country. We don't have generally shared values that are important enough to make most people really feel that what they do is going to be valued ultimately. I see that as a real problem. Maybe ultimately, as John Cannon said when he spoke of people searching for something to believe in, maybe there will be some way to get there, but I don't see us getting there yet.

My second troublesome thought is that having experimented as a local bureaucrat with ways of enabling people to be involved in our city-wide and neighborhood mental hygiene committees and councils, what we found in some instances was that at the local level people are not always interested in the fine, humanistic ideals you spoke about.

They act out of self-interest. For example, they will decide not to allow a halfway house in their neighborhood, or not to have open hiring in their particular agency. If we had let those grass-roots decisions stand, we would have moved the community away from humanistic goals. How do you deal with this?

Wirtz: I have two answers. One, I think the experience you mentioned is partly a reflection of the fact that in certain situations the bad people are more active than the good people. That's an oversimplification. My second answer is that I think you could get a working nucleus of good people. These are soft answers, but I feel a little better about them for having participated in the attempts through the community education work councils to develop local leadership. I think we've got to find some way to develop that leadership at the local community level.

Sidney Werkman: Have you discovered any inventions or transmittable ideas from your community education work councils that perhaps other communities could profit from?

Wirtz: Yes, we have. Through the support of the federal government and some private foundations, we have made arrangements to permit not only the development of a consortium of these communities but also the development now of quite a literature about their experience. The extent of the interest manifested by other communities has got to be significant. We find it everywhere.

Sheila Kamerman: At the risk of running counter to the tide, and in the face of your experience with the community education work councils, I'm going to disagree with you about community-level participation. The reality is that most people don't really want to participate. For example, parent participation in child-care programs is very poor. Some parents don't have time to participate, and others don't want to.

Harold Visotsky: There's also another problem. The federal government tends to co-opt local programs that prove successful. Look at what happened to Saul Alinsky's Woodlawn project, for example.

25 Meeting the Educational Needs of Young People

Alvin C. Eurich

I can think of no better way of beginning than by raising once more a question that has been raised many times in the past, is still being asked, and will continue to be asked in the future: "What *is* education?"

Many thoughtful answers to this question have been expressed, going all the way back to Plato. We have heard a few at this conference. But there have also been many pompous answers, long on bombast but short on substance. It was in response to some of these answers that Gertrude Stein gave us a few of her own thoughts about education, in a column published by the *New York Herald Tribune* on March 16, 1935. Let me quote a few excerpts:

> Education is thought about and as it is thought about it is being done it is being done in the way it is thought about, which is not true of almost anything. Almost anything is not done in the way it is thought about but education is it is done in the way it is thought about and that is the reason so much of it is done in New England and Switzerland. There is an extraordinary amount of it done in New England and Switzerland.
>
> In New England they have done it they do do it they will do it and they do it in every way in which education can be thought about.
>
> I find education everywhere and in New England it is everywhere, it is thought about everywhere in America everywhere but only in New England is it done as much as it is thought about. And that is saying a very great deal. They do it so much in New England that they even do it more than it is thought about.

Later on in the same column, Gertrude Stein said she thought education does not make very much difference, but I disagree with her on that point.

More seriously, Edward Gibbon wrote in his autobiography, "Every man who rises above the common level has received two educations: the first from his teachers; the second, more personal and important, from himself."

And, about a century later, Cardinal Newman said about the same thing: that, no matter how much a faculty can do for its students, it is nothing compared with what the student can do for himself.

There's a great difference, as Ed Pellegrino pointed out, between schooling and education, and I want to develop that distinction a little further and really show the complexity of the problem of educating a person.

We generally assume that what we teach is retained, but, actually, only a small part is retained in the long run.

Suppose we represent a school by a circle that encompasses everything that goes on in the school—all subjects taught, the relationships between teachers and students, the library, laboratories, and so on. One student really experiences just a very little bit of what goes on in school. But then when you think broadly of education, which is essentially what is taught, the individual student covers very little of what is actually taught. And when we go on to learning, very little of that which is taught is actually learned by the student. Finally, we go to the end result, which is what means the most, and that is what's retained from this process in skills, knowledge, ideas and so on. We generally assume that what we teach is retained, but, actually, only a small part is retained in the long run.

Some years ago when I was teaching educational psychology at the University of Minnesota, I had classes of several hundred students and I was curious as to how much they retained. After spending several years developing a comprehensive examination, one year I gave that examination at the beginning of the course and again at the end. I also gave it to a group of the same students the following September and again in June. All the students made some progress during the course. Then there was some loss during the period between June and September, but, by the following June, the retention was only about ten percent. We've all experienced that ourselves in courses that we've taken. As we look back to some courses, we don't even remember the name of the teacher.

The second point I want to make is that the program offered by the school is only a small part of the total educational process. Take, for example, television. We don't know the extent to which television is contributing to education today. We do have some figures on the amount of time spent with television, and they show that the average high school graduate has spent 15,000 hours looking at television. And by the time he's finished high school, he's spent only 11,000 hours in school. In other words, he has spent more time looking at television than the total time he has spent in school. We don't know what the effect of television has been, but obviously it has become a very major part of the education of students.

In addition, of course, we must consider the educational influence of the family. Now with that complexity, we can recognize how difficult it is to determine where to put our finger on the educational process in order to have some effect on the development of the individual. I don't mean to be discouraging, but we need to look at the whole process when we're talking about education.

All the experimental work that has been done on the student-teacher ratio over the past 75 years adds up to one conclusion — that the ratio makes no difference in the learning.

In the process of formal education, we perpetuate a number of myths, some of which have been with us for centuries. Take, for example, the teacher-student ratio. This has been all-important for many many years in our budgets for colleges and universities and for our local school systems.

A rule of thumb is that, in elementary schools, we should have one teacher for 30 pupils; in high school, one teacher for 25 students; in colleges and universities, one teacher for not more than 13 students. I think we ought to just reverse this. We ought to have a smaller ratio for elementary school children and a very large one for university students.

European universities have paid no attention to student-teacher ratios. In fact, in most of the German universities there's only one professor for every department and he has assistants. It doesn't make any difference how many students he has, he is *the professor*. We've ignored that European tradition.

When we look at the experimental work that has been done on student-teacher ratios over the past 75 years, it all adds up to one conclusion—that the ratio makes no difference in the learning. But, of course, we pay no attention to that at all. In education we can carry on experiment after experiment, but if we don't believe the results, we don't put them into practice.

The American ideal for many years has been "a student on one end of a pine log and Mark Hopkins on the other." But that is only part of what President Garfield said. A fuller statement reads: "If I could be taken back into my boyhood days and had all the libraries and apparatus of the university, with ordinary routine professors offered me on the one hand, and on the other a great luminous rich-souled man, such as Dr. Hopkins was twenty years ago, in a tent in the woods alone, I would say, 'Give me Dr. Hopkins for my college course rather than any university with routine professors.' " We, as educators, forgot what Garfield said about "routine professors" and just summarized it in terms of a student on one end of a log and Mark Hopkins on the other.

The late Charles Johnson, who was President of Fisk University for many years, summed it up in another way. He said: "Small classes make it possible to transmit mediocrity in an intimate environment."

Now, we might ask, where did this idea of a fixed teacher-student ratio come from? I was curious about that for many years but couldn't find any reference to it. One day I was lecturing at the University of Michigan about quality in education and, after my lecture, one individual in the group asked this question: "But you haven't said anything about the student-teacher ratio in relation to quality in education." I said, "No, I have not deliberately because I haven't found any evidence that indicates that the teacher-student ratio makes any difference whatsoever." Then I said I was curious as to where the teacher-student ratio idea came from. Well, one of the deans at the University of Michigan spoke up and said: "I'm not wholly sure of this but I think it comes from the Talmud." I was so curious about it that, after I left the room, I called the great Talmudic scholar, Dr. Samuel Belkin, President of Yeshiva University, and said: "I want to ask you a question that I've been looking for an answer to for years. Is it true that the teacher-student ratio is recorded in the Talmud?" He said, "Oh, yes." "Well," I said, "can you give me the reference?" "I'll do better than that," he replied. "Where are you?" When I told him I was in Ann Arbor, he said, "I'll send you a telegram with the exact quotation."

A few hours later I received a telegram with the exact quotation from the Talmud, the authoritative body of Jewish tradition: "Twenty-five students are to be enrolled in one class. If there are twenty-five to forty, an assistant must be obtained. Above forty, two teachers are to be engaged."

That particular rule was recorded before we had books to read, before we had any other means of communication other than the teacher speaking to a group of students. And yet, with the vast range of communication opportunities we have available today, we have rigidly adhered to that formula, thinking there's no other way to teach a class except in a small group.

Another myth about education is that the number of years spent in school determines a student's educational attainment. We all know how ridiculous that is, but we still cling to it.

We've made major changes in our educational system over the years, but I'm not sure all of them represent progress or improvement. That's still an open question.

A third myth, which I mentioned earlier, is our assumption that what we teach is learned, whether it's subject matter or a skill. We overlook the fact that many students learn very little in listening to a teacher, whereas other students learn a great deal. Very few superior teachers are effective with every student in every class, and very few of the poorest teachers are not effective with any. So the class becomes a situation where, even though there may be twenty-five or a hundred students, it's still a person-to-person relationship, which we generally ignore.

A fourth myth is that students don't know what is taught until it is taught. I think if we developed the habit of giving examinations at the beginning of a course, we would find that many students had already acquired the content in other ways.

We've made major changes in our educational system over the years, but I'm not sure that all of them represent progress or improvement. That's still an open question.

First, we have greatly expanded the educational opportunities for youth, essentially in the past century. A hundred years ago, attending secondary school was regarded as a privilege. Most of our secondary school students went to private schools instead of public schools. In fact, in the state of New York, it took one hundred years from the time the first public high school was established until every community in the state had one or more. That was a relatively slow transition. We had private academies that eventually went out of existence.

With the establishment of public high schools, education became a right, and then in the early part of this century it was made compulsory. So we went through three stages—a privilege, a right, and then compulsory education up to the age of sixteen.

In higher education, we've greatly extended the opportunities. A prime example is the City University of New York. For many years, City University accepted only about the top 7 percent of high school graduates. From that it went to open admission. That changed the University considerably. We still don't know the effect of open admission upon the learning of students.

From 1963 to 1968, President Johnson persuaded Congress to appropriate more money for education in a five-year period than had been appropriated by the federal government in our entire history prior to 1963.

In addition to making higher education available to more students —and really as part of that process—we have appropriated enormous sums of money for education in recent years. Let's take what has happened with appropriations from the federal government.

I happened to serve on President Truman's Commission on Higher Education and later on President Kennedy's Task Force on Education. In both cases we appealed for more federal funds for education. We didn't have much success. But then President Johnson came along with his Great Society, and, without any commission or task force, he persuaded Congress to appropriate more money for education in a five-year period, from 1963 to 1968, than had been appropriated by the federal government in our entire history prior to 1963. The decade of the 1960's was a period of building, a period of expansion. As a result, we have a very elaborate school system—great college and university buildings, high school buildings, elementary school buildings. Now we're confronted with the problem of closing many of these buildings. We're in a new period where we're closing hundreds of elementary schools and high schools, and colleges are going bankrupt because they can't be sustained on the basis of their income.

Another major change that has taken place in education is that we have greatly expanded our in-school and out-of-school learning resources through technological inventions of one kind or another, including radio, tapes, television, and now computers.

We have also developed new types of institutions, such as the Open University in England, which, although opposed by all the regular universities, was established by a separate act of Parliament. It offers instruction through television, radio, written materials, send-outs, and small conference groups, and it is now the largest university in England. In the United States, we have something like it in the University of Mid-America, which covers about 11 or 12 states in the Midwest, with headquarters at the University of Nebraska. It now has approximately 10,000 students learning by television and getting their credits by examinations from the old established universities.

In New York, we have the Empire State College which is similar to the University of Mid-America. At the community college level, we have Miami-Dade Community College in Florida, which has an enrollment of approximately 50,000 students, more than half of whom are older students taking work over television or in classes set up in apartment houses in Miami.

We've also changed teacher education. We've changed our teachers' colleges into general purpose institutions. I had an interesting personal experience in connection with that. A little over a quarter of a century ago, I was urging that teachers' colleges be changed or be abolished because we ought to be much more concerned with the substance of teaching rather than spending so much time on methods of teaching. I was burned in effigy on a teachers' college campus in Kansas for making those statements. Today we have very few teachers' colleges left. I don't know whether they've actually changed their program, or whether they just changed their name.

Well, these are some of the major changes that have taken place in education. But how did these changes come about? Essentially, they did not come about through the work of educators. They came about wholly because of other pressures. All we have to do is to look at our system of higher education today and see how it developed.

First of all, the liberal arts colleges of Colonial days were established to provide liberal or general education for ministers, doctors, and lawyers — a complete departure from European education. In European universities, students upon entering went on with professional work. The universities assumed that all liberal or general education was completed by the end of the secondary education. But we, in contrast, established the liberal arts college, and it's still unique in the world today.

But we were an agricultural economy and liberal arts colleges were not serving our agricultural economy. What did we do? We established new institutions, namely the land grant colleges. They were looked down upon by liberal arts colleges because they were degrading higher education. In fact, most of them were called "cow colleges" by the people in liberal arts institutions. But they were meeting an important need.

Neither the liberal arts colleges nor the land grant colleges were preparing teachers, and superintendents of schools became concerned about this. They first established county normal courses, and then some of these became two-year normal schools. Then these became four-year teachers' colleges, growing up wholly outside of the rest of our educational system, which was I think a very unfortunate development.

The whole system of higher education in this country has grown by establishing new institutions to meet new needs, rather than by adapting existing institutions.

With rapid advances in science and technology came the need for better teaching and research in science. Not much was being done in our regular institutions, other than the experimental work in agriculture by the land grant colleges.

The first scientific laboratory was established at Yale by Professor Silliman. He was laughed at by the other professors. Imagine bringing playthings into the classroom! Laboratory equipment indeed! It was only after we established institutions like MIT, for example, that science became respectable in all institutions, although it had been growing up gradually. But it eventually gained a very great deal of respect and was incorporated in our college and university curricula.

In more recent years, we reached the point where most of our youth were finishing high school and there were no jobs for them, so junior colleges—or community colleges—were established in order to provide additional education for these young people.

My point is that our whole system of higher education in this country has grown by establishing new institutions to meet new needs, rather than by adapting existing institutions.

The courts and legislative bodies have brought about other major changes in our educational system. Although parents have always exercised an influence over school affairs, parental pressures have become increasingly important in recent years, in community after community throughout the country.

Another outside force, a more recent one, has been the unions and professional organizations of one kind or another. They have been particularly important in setting pay scales, fringe benefits, teaching hours and so on.

Schools and colleges have not adapted to the individual differences among their students, nor have they adjusted their programs to the tremendous knowledge explosion.

Before closing, I would like to mention some of the major problems confronting education today.

- First is the financial problem. I've talked to many college presidents throughout the country in the last year, and frequently when talking with them have raised this question: "What is your major problem today?" The universal answer is retrenchment, reallocation of limited resources, establishing priorities, and eliminating parts of the program. This is a very, very difficult period.

- The second major problem is that schools and colleges have not adapted to individual differences among their students. Psychologists have emphasized individual differences for the last fifty years or more, but we've hardly made a dent in the way we handle these differences in the classroom.

- Third, we have not adjusted college and university programs to the tremendous knowledge explosion. Confronted with this knowledge explosion, which is probably the greatest revolution that we've had, we've merely added courses rather than rethinking our whole educational system in terms of what the new knowledge means for the education of students.

- Next, schools and colleges are not using effectively the modern means of communication. And yet, as we look at it, all of education is communication. It doesn't matter where you put your finger on it, whether it's in the classroom between the teacher and student, it's communication; whether it's the textbook material, it's communication; whether it's in administration between the teacher and the administrator, it's communication. Whether it's a matter of getting support for the school, it's still communication. Yet, in our educational system, we have virtually ignored the new and effective means of communication now available to us. That's one reason why television gets more attention, more hours, from our youth than the schools.
- Fifth, our current system of preparing and upgrading teachers is inadequate to meet current needs.
- Sixth, we've not adequately adjusted to integration.
- Seventh, we do not recognize that schools now provide only a small part of a student's total education, so we don't coordinate what we do in school with what's going on outside of school. We don't look at it in terms of the whole, the total process in relation to the individual.
- And, eighth, although there is a definite trend toward more and more career education, we are not coordinating education with the real world of work.

This leads me to my final observation. We need to adjust our entire education system to meet the needs of each individual in our day.

Thomas Carlyle expressed exceptionally well the ultimate aim of all education when he wrote: "Let each become all he is capable of being."

We don't know fully how to achieve this. There are no rules of thumb. What Justice Cardozo said about the legal process applies equally to education. When he was a young lawyer, Cardozo wrote that he thought the practice of law was a matter of learning the rules and applying them. The older and more experienced he became, he observed, the more he discovered that the legal process is a creative one, in which principles that have served their day expire and new principles are born.

(Editor's note: Mr. Eurich's remarks were discussed along with the remarks that follow.)

26 Ferment and Experimentation in Education

Rev. Ernest J. Bartell, C.S.C.

The big question Al Eurich left us to wrestle with is whether our institutions of higher education can meet the challenges of the future by internal evolution, or whether we need a whole new system.

What I will try to do is point out some of the opportunities that face our colleges and universities today in coping with what they see as a problem of institutional survival; then I will try to pay some attention to some of the constraints and the limitations that hinder their activity, and then let's see if the potential looks good or not.

I think there is no question that whether we have reform or revolution we're still going to have institutional education in this country. Several of the speakers at this conference have emphasized that education is finally the answer. People who have been deprived of educational opportunities have come to that conclusion very quickly, with a lot less deliberation than we have engaged in.

We all know that one of the major problems colleges and universities will face in the decade ahead is declining enrollment. I would like to make a comment or two about the possibility of compensating for that with additional participation in higher education from the age cohort we are discussing. Obviously, we don't have 100 percent participation now. People think that there is perhaps an opportunity for much greater participation by the minority cohort during the 80's.

Unfortunately, the current statistics suggest that the participation rate is not increasing among blacks and Hispanics. Despite all of the effort we've made, we're not experiencing an increase in the percentages of blacks and Hispanics going on to college. So I'm not sure at the moment that this is a very optimistic solution.

Many argue that the adult participation in postsecondary education has just about reached its peak, so those institutions that think this is the answer are likely to be disillusioned.

I do think that what we're really going to see is a great deal of competition among institutions for the available supply of students. This competition is going to force institutions into more differentiation of identity if they're going to survive. That differentiation will be up to them, of course, and it can take place in a variety of ways, including differentiation of clients served. As Al Eurich pointed out, we haven't done a good job yet of developing educational programs for nontraditional students.

Institutional definition based upon career orientation versus liberal education is futile.

Some institutions will, I think, clearly make it their business to develop new forms of higher education. Parsons and Platt did a study of the post-1960 period in higher education—after all the experimental programs of the late 60's — and they found that by and large the majority of our colleges were still dominated heavily by traditional academic models.

I define the traditional academic model as one dominated by the disciplines and the professions that educate the faculty for our institutions, as opposed to models that are created to meet the needs of the particular student served. Even in the colleges, not to mention the universities, this model continues to be structured along disciplinary lines, with all the values of the individual disciplines being particularly dominant in translating the work of the faculty into education for the students.

When I say that competition is going to demand a new definition of institutional mission, I don't mean the catalog definition. Many institutions in recent years have come out with new statements of purpose and mission. When you read them and try to identify in the grossest terms what type of institution is behind them—is it a four-year college, is it a research university, is it public, is it private, is it small, is it large? — you can't tell the difference. Somehow they all merge in glowing rhetorical statements about doing everything for everybody, with little about educational realities of the institution.

I think myself that institutional definition based upon career orientation versus liberal education is futile. Vocationalism has a certain immediacy, I suppose, in response to contemporary economic concerns and the current groping of education administrators to attract students, but as Al has pointed out, the development of higher education in this country has always had a career orientation. The original liberal arts institutions were established to train people for a career, and so were the land grant colleges, the teachers' colleges, and institutes of technology. Even today, most undergraduate colleges are really prep schools for graduate schools, and the principal purpose of graduate schools is to prepare people for academic careers.

One task often not accomplished effectively in general education courses is development of multicultural awareness.

We are seeing, however, an increase in concern about core curriculum and general education, as a reaction to the vocationalism of the last few years. (You've probably seen Ernie Boyer's book on the core curriculum, and you've probably heard about Henry Rosovsky's great effort at Harvard to develop a new core curriculum.) Despite differences, there is an emerging consensus, I think, that somehow the core curriculum or general educational component of a student's education must integrate useful verbal, analytic and methodological skills with fundamental personal, social and moral issues of human nature, history and culture.

One task often not accomplished effectively in general education courses is development of multicultural awareness. Traditional classroom approaches may simply be too dull. Fortunately, the legitimate arrival of experiential learning has made it possible for students to take time out, as part of their structured educational program, to experience another culture, domestic or foreign, at first hand.

Experiential learning goes beyond conventional undergraduate programs that combine internship experience and classroom instruction. It means finding a way to incorporate a student's life experience into his formal education. There have been some attempts, especially for older students, to evaluate and academically accredit life and work experience in our credential-conscious society. However, I believe it's more significant to find a way to build that experience into a formal education program. This requires some kind of mentoring link between formal education and the outer world of experience. People who can build these bridges will help both the institution and the students. Peer-group mentoring is a very real possibility here.

One of the challenges still remaining for institutions that want to identify with new clienteles is how to design general education for people who come from backgrounds different from those of the traditional college-bound population.

There's a growing concern, in addition, for more effective integration of values and beliefs into general education. Part of the difficulty stems from the value-free training that faculties themselves have had in their own disciplines. Nevertheless, I think there's an increasing awareness that colleges and universities must deal with both value education and commitment, perhaps because of the failure of other institutions in society to do it.

One of the challenges still remaining for institutions that want to identify with new clienteles is how to design general education for people who come from backgrounds different from those of the traditional college-bound population. This problem, for example, extends to minorities, to adult women coming in and out of the work force, and to an increasing number of part-time students. It's an interesting demographic fact that more than 50 percent of the enrollment in higher education now is part time. How to design coherent general education for people who can devote only a limited amount of time to course content is very difficult. However, there are some good examples.

Some of the programs in the Southwest for first generation Mexican-American students have been very successful. They bring together faculty and counseling staff and financial aid people—all the auxiliary service people—and they use a team approach in dealing with the needs of the individual student. This simple method of pulling together the resources of rather small institutions has reduced dropout rates by as much as seventy percent. Pressures against perseverance abound in that constituency. There is often not much family support for young people leaving home and going off to college, because of the economic necessity for their labor in agriculture. In addition, the student is confronted at college with an unfamiliar Anglo environment, so the high dropout rate is understandable.

The legitimacy of affective learning or noncognitive learning is another area of inquiry and experimentation for developmental learning. Simply, the intellect does not operate in a vacuum. Obviously, it operates in a biological, emotional and social system that must be taken into account.

In some institutions — traditional as well as nontraditional — there has been increasing collaboration between the faculty and the so-called auxiliary service people. Traditional faculties do not always look upon people who come up through the auxiliary service ranks as peers and professionals, but in some successful experiments they are collaborating in the teaching of courses, often in subjects that relate to the student's own personal life.

Unfortunately, innovative collaborative activities are often carried on outside the mainstream of the university.

We talked earlier about sex education for teenagers, but the subject is obviously not exhausted in high school. When I was a student at Notre Dame, Father Hesburgh, who was President even back then, addressed something called "The Marriage Institute." In the course of the address he recited for us a long list of characteristics that we should look for in a life partner. The list was great, but perhaps that's why I never got married. I never found a woman who met those standards. Today's students rightly expect to approach a subject like marriage from a variety of viewpoints, and to receive the assistance of experts, mentors and role models in and out of the classroom.

Moreover, collaboration extends beyond campus segments. In serving new clienteles, there is increased collaboration between educational institutions and other agencies. FIPSE is currently sponsoring a program to induce colleges and universities to work with prime sponsors in the community to provide educational opportunities for CETA clients that go beyond short-term manpower training. We're trying to get colleges and universities to look at CETA youth not just as a source of subsidized labor on campus — which colleges have already discovered, in their cost-cutting efforts — but as clients for their service. Unfortunately, I have to say that very often innovative collaborative activities are carried on outside the mainstream of the university. Unfortunately, they either take place in the continuing education division or in other community-relations segments in higher education that don't always have much credibility and clout among the dominant personnel on the campus, both faculty and administration, who determine effective policy for the institution. Continuing education in the past was often seen as a form of advertising or good will in the community. It became a money-maker when it was discovered that faculty are willing to moonlight for a few extra pennies to offer whatever educational smorgasbord will sell. Continuing education today finally is evolving into a more integrated, coherent type of program for different groups of adults and community people, offered in collaboration with community institutions.

At FIPSE we have also found ourselves helping museums, libraries, and other community-based organizations to do educational jobs for postsecondary clients. For example, we are trying to get the traditional museums that normally serve a very limited segment of society — relatively highly educated people who troop their kids through the museums on weekends instead of taking them to ball games — we're trying to get these museums to use the artifacts of the museum as learning resources for the community. We're doing that through a voucher system whereby community-based groups are given grants in the form of vouchers which they can give to a museum that develops a program that responds to their needs. It has worked quite successfully with a coalition of museums in the New York area.

I don't want to duplicate what Al Eurich said about the use of the new technologies in higher education, but I do have a few caveats.

First, to the extent that the new technologies are successful, they will assist individualized learning. One of the problems at the postsecondary level up until now has been that both the hardware and the software were relatively centralized and not easily adaptable for individual student use. That has been particularly true of cable TV and computer centers. They have not been conducive, either educationally or economically, to the development of a diversity of programs to meet diverse needs of different students.

But the newer technologies, including mini-computers, video disks, videotapes and so on, are much more flexible and much less costly.

They might finally make possible the technological breakthrough in postsecondary education that we've all been waiting for over the past twenty years.

What I fear is that some of the novelty of the techology may increase the attractiveness of information-transmitting type activities to the detriment of the more fundamental learning activities — especially those that happen when good teacher meets good student.

One area of experimentation which has not been uniformly successful but which must continue to be addressed is the area of assessment — assessment of programs, and assessment of students' performance. Unfortunately, it is not easy to assess the outcomes of educational programs, especially those that are non-quantitative, using the quantitative assessment techniques that are now being developed. As more and more life and work experience is incorporated into educational programs, the task of assessing individual student performance, of measuring individual competence, especially in arts and humanities, becomes increasingly complex. The costs of such assessment can partially offset the savings typically associated with credit for nontraditional educational experience.

The introduction of experimental programs for the disadvantaged learner, even in institutions on the fringe of conventional higher education, is having an important impact on the traditional mainstream learning.

Now let me say a word or two about credentialing. In the early days of FIPSE, the hope was that support for the right projects could downgrade the importance of the credential in our society, and induce people to look more productively at the underlying quality of education.

I'd have to say that the possibilities of one small organization doing that are about zero. We live in a credential-conscious society, and some of the experiments in the colleges and universities to substitute portfolios of student work for grading as a more comprehensive statement of the student's development have not proved to be operationally acceptable. I'm afraid that present methods of credentialing are here to stay until we can find a better and more efficient way to do it.

June Christmas has told us, and properly so, that in any program for addressing the problems of youth we mustn't cheat the disadvantaged.

I would like to go a step further. There is growing evidence that the introduction of experimental programs for the disadvantaged learner, even in institutions on the fringe of conventional higher education, is having an important impact on the traditional mainstream learning. I think we owe a debt of gratitude to the minorities and to women for improvements in education now reaching the mainstream.

One example has been the creation of alternative colleges within large systems. Some far-reaching experimental ideas are incorporated into these models, such as the downgrading of competition as a motivator in student performance by creating an alternative kind of learning community, a task that can demand new roles for the faculty. In the program up at Stony Brook run by Pat Hill, faculty members enroll as students in interdisciplinary courses taught by their peers.

Some of these experiments have a great deal of promise. On the other hand, I cannot say that under the current financial situation there is likely to be much proliferation of totally separate experimental colleges within large systems, so the problem still remains one of creating more individualized learning opportunities for students in the very large institutions, because the vast majority of students are now enrolled in large institutions.

On some campuses, winning an award as the "best teacher of the year" is a good way to lose points in the competition for tenure.

There are additional institutional constraints that are likely to retard educational improvement. One concerns faculty. Part of the problem is the familiar role of tenure. However, a more fundamental issue concerns the preparation of faculties for academic careers, taking us back to the goals of graduate schools in developing and cloning their own faculties, in preparing people to be professionals in disciplines without much emphasis on the teaching roles that they're going to have to play. Most of the students coming out of the graduate schools during the current job shortage are likely to be placed, if at all, in institutions where the teaching role is the predominant one. Increasingly, with the survival of the institutions so dependent on undergraduate enrollment, the teaching role is becoming dominant.

And yet, there's very little emphasis on teaching at the graduate school level, and often inadequate recognition of it in the process of hiring and promotion. Young faculty members thus begin with mixed signals. Some are anxious and eager to fill the teaching void in their training, but they also know that the reward structure within the institutions of education is tilted in favor of research rather than teaching. In the quiet closets of rank and tenure committees, promotion and tenure are based on professional performance, with a conservative inclination to favor the person who has had a few articles published, even if those articles will never be remembered by anyone. On some campuses, winning an award as the "best teacher of the year" is a good way to lose points in the competition for tenure.

I think the loss of individual cultural roots in homogenized preparation for competition in the academic meritocracy can be a real hindrance to an educational mission that requires empathy and sympathy with the needs of nontraditional students, but this problem may only reflect the biases of status in the larger society.

The governance of traditional institutions is often not responsive to contemporary needs. The increased democratization within traditional institutions, with additional layers of collegial bodies, has not necessarily resulted in more effective education. To jaded administrators faculty senates have often evolved into little more than therapeutic sandboxes for vocalizing dissatisfaction and for killing educational initiatives that might threaten established interests on campus.

The roles of faculty and administration during a period of financial stringency can become increasingly antagonistic. The increased application of labor law in higher education doesn't always help. Finally, the craft guild mentality of the professions and disciplines in clinging to their own prerogatives can produce institutional paralysis in educational innovation.

Ultimately, the fear is that constraints such as these, plus the now familiar financial problems of higher education, may lead to conservatism throughout the entire system of higher education — to an inertia that centers on institutional survival rather than student learning. We can only hope that good people will not be deterred from creative responses to educational needs because of hardening of institutional arteries.

Discussion

Leonard Duhl: I'm really very grateful for what I think was a magnificent diagnosis by both of you, because the problems you described are very similar to the problems we see in health and in a lot of other areas.

What I'm concerned about is what Bill Wirtz talked about yesterday, and that is that despite our diagnoses and despite the fact that we talk about the complexity of the problem and all the issues, we haven't really designed the social architecture that will enable us to cope with change.

Even in education, social processes and social architecture are not considered very relevant subjects to teach. So that the very people who are talking about the processes of change really know very little about those processes.

The amount of money that we can get for either research or teaching on change processes and institutional processes and organizational processes and all the things Bill brought together yesterday, is very minimal.

A lot of the faculty members working in the universities trying to change governance know nothing about governance. They're still stuck in their subject matter, and they're still stuck with holding on to their bureaucracies and their prerogatives. When you start talking about social change, they think you're some kind of a nut. They say that that isn't even a relevant subject to discuss.

This is also true of almost every field. In the field of health — and there are a lot of us here who have had this experience — the minute you talk about social change you're considered second rate because you're an administrator or you're a process person or a policy person. You're not dealing with real things.

I think from what we have been saying here that the tools for building the kind of social architecture Bill mentioned are available to us. Our problem is, how do we put those tools together? We've had the dialogue here and we've had the diagnosis, but we haven't gone that next step of how do you turn what we're talking about into real action. I think that's where we're always stuck.

Alfred Kahn: I want to add another element to the complexity facing adequate reform, but first I want to express my appreciation for the two presentations. I certainly was inspired by them. The other dimension I wanted to add is that in our society education is a passport to status. It's a rather important passport, and the fact that it is a passport complicates the concern with content and with access.

The new Christopher Jencks *et al.* volume, *Who Gets Ahead?*, not yet published, but being reported in the magazines, suggests that education is a different kind of passport for blacks and for whites. It does not take them to the same countries, so to speak, but education is an important passport for both, and it is a passport whether or not there is much retention of content. Going to school and learning have independent impacts.

The mere fact that education is a passport complicates the effort to reform it. I don't think that is adequately understood by people who are concerned with educational reform.

Robert Aldrich: I'd like to sound once more a theme that was sounded earlier. I think there are some built-in hazards in the whole federal granting system, even though it was quite properly pointed out that they're necessary.

It seems to me that one of the things that has replaced the classical role of the professor, which was to teach and to do scholarly work, has been a preoccupation with grantsmanship.

If you do a time-motion analysis of how faculty members spend their time and for what objectives, you're going to find that getting grants is a major concern, and that this has more to do with their prestige among their peers and also the administration than any other single thing they do, except maybe win the Nobel Prize.

This has not received the emphasis that I think it deserves. I don't think that grantsmanship is something to be thrown out, but it needs to be looked at and analyzed to see what it is doing to higher education.

I just wanted to say that again because, having been on both sides of that particular fence myself, I have developed some rather strong feelings about devising a more sophisticated way of using grants to prosper education, to prosper research, and also to coordinate these activities in terms of what is best for the students and the faculty.

Paul Bohannan: I would like to underscore what Bob just said. I teach a graduate course in which the assignment is a first draft of a research application that can be sent to any granting agency.

For ten years at Northwestern University I tried to get that course through the curriculum committee, and I was told it had no academic merit. It was a required course for people in the anthropology department there, and we taught it under one of those blanket numbers called "problems in anthropology."

The other observation I would like to make has to do with the whole matter of social structures. During the past twenty years, in the interest of consolidation, the number of school districts in this country was reduced from about 120,000 to about 16,000. This had two significant results. One was to throw the whole relationship between the community and the school into a cocked hat, and the other — even more significant — was to create with the bigness a complexity of social structures in the schools that the students have to deal with. It's this complexity, as much as anything else, that's the problem in our schools, because it takes so much time for the student to learn the system and how to manipulate it. So we have to keep in mind the fact that the social structure of the schools has an important bearing on the whole matter.

That brings me to the last point I want to make. School at all age grades is the place where one's lifetime networks are laid down, and I think it's terribly important that these networks be maintained, because the networks may be the most important thing you get out of the school. Where you go to school depends on who, twenty years later, you can call up on the telephone and say: "Look, I've got a problem and I'd like your help with it."

John Schweppe: I'd like to cite two examples, out of my own experience, of how difficult it is to get new ideas accepted in higher education. One relates to an effort I made two years ago to try to get the life cycle into the medical school curriculum. Outside the department of psychiatry and one or two others, there was little if any interest. The chairmen of the departments were quite reluctant to do anything other than cover the life cycle in maybe three or four lectures in the introductory course to medicine, which was given as a general course that included a potpourri of everything.

Second, I've tried to pick out ten of what I consider the most pressing problems in the medical field — problems like cardiovascular disease, pulmonary diseases, osteoarthritis, and other diseases that have a very high incidence, and use the team approach in teaching them. The comments that came back from the dean and other people were to the effect that this would disrupt the departmental design, and they can't be taught in an interdisciplinary fashion.

I hope that sometime in the future a student entering college can cut across the boundaries of the traditional disciplines and, with the help of his advisors, work on ten or twelve of the most pressing problems that his advisors consider appropriate for him to study. In other words, the structuring of his education would be according to what he individually wants and needs, rather than what is prescribed by the college.

I also want to say that I think the computer is just great for teaching certain types of sequences, but it falls flatter than hell when it comes to inspiring an individual. One of the most important things I experienced in college was the human element in teaching. I still remember the great teachers I had.

Alvin Eurich: I frequently think that the best teachers on college and university campuses are the athletic coach, the music teacher, and the art teacher, because they all start from where the student is at the time. They know where he is and they try to head him in the right direction. Too many other teachers only cover material.

Sidney Werkman: I think one of the greatest dangers of the various kinds of structural changes that are going on in higher education is that they will destroy the excitement and thrill of learning.

John Cannon: What has happened to the involvement of students in decision making and particularly in the shaping of the curriculum?

Bartell: It's a salt and pepper situation. In the late 60's there was a lot of pressure by students—and it didn't always have much to do with the love of learning — to get into the governance of colleges and universities. They won token representation on every committee in the institution, sometimes from boards of trustees all the way to determining the menu at the departmental Christmas party. But their interest and that of their successors in such matters didn't last. In fact, on many campuses now it is difficult finding students to fill those slots. In the heavier slots the students often found themselves overwhelmed by their fellow committee members from faculty and administration. I've been on search committees for deans and presidents that included student representatives, and I've not seen a student participate very constructively, because it's just something they're not prepared for.

On the other hand, students are still very active in curricular affairs, often forcing development of better advisory systems. In fact, some institutions have few rigid curricular requirements; they depend instead on an advisor and a student working together to design his or her curriculum. Brown University is a prime example.

Merrell Clark: I'm intrigued by the extent to which some of our major proprietary schools are providing a broad range of liberal arts and sciences in addition to specialized training. What is the future of these education-for-profit institutions?

Bartell: Back in the early 70's, there was a very deliberate attempt to legitimize proprietary educational institutions. The thought was that they would provide good competition for the rest. There were some good ones, and some bad ones, but there hasn't been as much development of proprietary education for the public as might have been expected. Some of them got out of the business. They found it didn't pay. Also, the consumer movement in education is growing, so there's going to be more public sophistication in the educational market.

Eurich: On the question of private and public funds for higher education, it interests me to see how little support comes from philanthropy. For last year, the total expenditure for higher education in the United States was $48.5 billion. Of that, 63 percent came from government, 35 percent came from direct charges, and only 3 percent from philanthropy. So philanthropy really plays a very small part in the support of higher education.

27 Third World Perspective

Aklilu Habte

My remarks will be very brief, but I hope they will add a new dimension to our discussion, because most of the world's young people live outside the United States. In spite of this fact, very little attention is paid in the United States to the international dimensions of education. Very little is taught here about other peoples and other cultures. That is indeed unfortunate.

In a world that seems to be in a constant state of crisis, I cannot stress too strongly the importance of improving international understanding, not only at the government level but, most importantly, at the person-to-person level. Mutual understanding and trust are more indispensable in today's world than at any other period of human history.

The importance of a people-to-people exchange, particularly one which involves living and working and travelling in another culture for three months to a year, cannot be overestimated.

The crisis seems to arise because of man, and the solution also rests with man. The approach cannot be mechanical, because people are not mechanical. It is therefore essential that we spend time and money to learn about each other's history, culture, languages, and development problems. In this respect the Peace Corps program is worth remembering. It has provided an opportunity for thousands of U.S. citizens to learn about other cultures and work with other people. I would like to see it expanded. Two or more years of service in the Peace Corps, in another culture, would provide rich and valuable experiences to the volunteers and would be of great service to this country when the volunteers returned and participated in their own nation-building process.

There are other endeavors and activities which encourage young people to travel and live in other countries, and these too should be expanded. However, this kind of experience should not be restricted to a small segment of the population, or to any specific, exclusivist age group. The importance of a people-to-people exchange, particularly one which involves living and working and travelling in another culture for three months to a year, cannot be overestimated.

For young people of this country, as well as older ones, it would be helpful to understand the broad characteristics of their counterparts in the Third World countries. For example:

- There are far more young people 14 to 24 years of age in developing countries, as a proportion of the total population, than in this country. Third World countries in general have a younger population than developed countries, and the implications of that are highly significant.
- People in developing countries do not have anywhere near as much to eat as their counterparts in the developed countries. It is worth remembering the shocking truth that nearly two-thirds of all the world's people live on the equivalent of thirty cents a day.
- Young people in developing countries have much less opportunity — and in many cases no opportunity at all — to go to school. As a result, millions of them are still illiterate and destined to remain so for years to come. The disparity between the sexes is even more worrisome: there are two or three times more boys in school than girls.
- Even the limited number of children who are fortunate enough to go to school suffer from the handicaps of irrelevant education, inferior physical facilities, untrained teachers, very few or no teaching materials, and a style of teaching that encourages nothing but memorization and repetition.
- A disproportionate number of children who begin their schooling in grade one end up outside of the classroom. This is doubly tragic, because it is wasteful of the resources that society has invested in their education and wasteful of their own human potential. For example, of every 100 students enrolled in grade one in 17 low-income countries in 1965, only 33 reached grade five. In 1970 the figure was 36. The situation is a little bit better if we include middle-income

countries. Of 100 students enrolled in grade one in 1965, only 38 reached, but did not complete, grade five. In 1970 the figure was 46. The situation is even worse in the higher grades.

- The millions of unschooled children of today are the millions of adult illiterates of tomorrow, and their lifestyle is not likely to be very happy.
- Millions of young people in the Third World countries — both schooled and unschooled—are without jobs. The mismatch between what takes place in the classroom and the job market needs no elaboration. The underutilization of human resources—the one thing Third World countries have in abundance — is becoming a vicious cycle in development efforts. We are all familiar with the unemployment situation in Third World countries from our own experiences and the analyses made by the International Labor Organization and other agencies.

The problems of youth in Third World countries clearly dwarf those of youth in the United States — in magnitude, complexity, and urgency.

- The health conditions of youth in the Third World countries are dramatically and substantially inferior to those of young people in the developed countries. Life expectancy is low, and the death rate among children under five years of age is very high. In several of the poorer countries, nearly 50 per cent of the children die during the first year of life. And for those who survive beyond that age, their life expectancy is considerably less than that of their counterparts in developed countries. It is estimated that life expectancy at birth in sub-Saharan Africa is only 43 years, in Asia 53 years, and in Latin America about 59 years. By contrast, it is estimated that life expectancy in Western Europe and North America is about 72 years.

From the illustrations I have cited, and from others that abound in the statistics gathered by world organizations, it is obvious that the problems of youth in Third World countries clearly dwarf those of youth in the United States — in magnitude, complexity, and urgency. I emphasize this as a matter of deep concern not only to those of us gathered around this table but to all of the people in the United States. Just as it is somewhat inequitable for a tiny minority in any country to enjoy wealth and abundance in the midst of the poverty, ignorance, ill-health and illiteracy that are characteristic of the majority, this also applies to nations.

Let me turn now to the role of the family in Third World countries. There, as in the industrialized countries, the family is a very important unit in society. It is a matter of regret that the role of the family is waning in importance in industrialized societies, and I hope that this does not happen in the Third World countries.

I believe that the family should be the dominant unit in any society. In several Third World countries the head of the family, usually the father, plays a central role in the affairs of the family even long after the sons or daughters no longer live with their parents. There are, for instance, certain areas in which I as a grown-up do not question the wisdom and authority of my father.

Grandparents are taken care of by their sons and daughters as a matter of routine and duty. In this respect I find nursing homes horrible in developed societies. The previous practice of keeping parents at home or as close to home as possible seems to me to be more humane.

The sociopolitical atmosphere in Third World countries seems to contribute to unsettling the minds of many young people. It is important to understand some of the factors that contribute to or influence this state of affairs. For example:

- Several Third World countries are still in the process of establishing and experimenting with the style of government and political order they want to live under.
- In several countries this is constrained by the colonial heritage, which did not take into account traditional values and forms of living and associating together.
- The big-power rivalry — Western style democracy versus Moscow style socialism—keeps on waxing and waning in popularity, extending its influence in international meetings and other arenas.
- The emergence of dictators, power-mongers, self-serving cliques and the military—all controlling the mass media and so controlling the life of the people — these and other situations contribute to the unsettled situation and the frustrations of young minds.
- Such political situations, coupled with the educational, cultural, and development problems I have mentioned, often erupt into serious and ugly confrontations between the young and the current authorities and establishments.

Let me conclude by stressing that the problems of youth in Third World countries are tremendous. Imaginative and innovative use of youth at the secondary and tertiary level for the benefit of the less fortunate mass of people—for example, in literacy campaigns as was done in Cuba, in road construction, in consciousness-raising among the peasants—these are areas that need further exploration.

But in the last analysis, development efforts must aim at self-reliance and the equitable distribution of resources at the national and international level. In this sense the New International Economic Order assumes importance and significance. If we are interested in world harmony, in peace and justice to each and all, some restructuring and redistribution of the world's resources becomes a necessity towards which all of us should strive.

Discussion

Paul Bohannan: I would like to say thank you for dragging us into the real world, because what you described *is* the real world. The real world is out there in the developing countries, whether we like it or not. That's where most of the world's people live. Therefore, it seems to me that anybody interested in studying our own culture must be cross-cultural. We must not only be cross-cultural in our analysis, but we must all get cross-cultural experiences. They're really not difficult to come by, and they are not frightening.

There's one more thing I would like to add. When we create our theories and our analyses, we must not collect from the English language and our own culture the categories of comparison we use to look at other countries in the world.

It is very very easy to take problems from our own country, from our own disciplines, and then go out and gather data about them from other countries. This is like putting parakeets into pigeonholes. They don't fit.

What we really have to do when we go to the Communist world, to the Third World, to any other world we can find, is to see how people *there* organize the processes of being human. We can then use those sets of experiences as core data to help us understand *our own* view of our culture.

If my discipline does nothing else, it says that in order to understand yourself you have to understand what you are not. Therefore, we've got to begin with this understanding of what we are not on a worldwide scale. Only if we do that we can begin to understand the whole process of human development.

Leonard Duhl: We have another interesting problem, too. We tend to project the future out of our own past. We tend to think that the best way to look at the future is to look at those parts of our own culture and other cultures that we can impinge upon. We forget that our own culture, and other cultures, are constantly changing. Our future, and the future of the rest of the world, may well lie in the sharing of resources and the sharing of people.

This is why Paul's point is tremendously important for us. Most of the people I know who are projecting the future and social plans for the world are projecting out of the present. They're projecting out of the current dialogue and the current concepts, ignoring the fact that there are other conceptual models and other realities.

Interestingly enough, the age group we're talking about here may be much better prepared to understand other cultures and other realities than we are, partly because they're wandering around the world very freely. Also, they are rejecting some of the realities that Americans have been emphasizing for a long time and they're leaving themselves open to the idea that other realities exist. The Peace Corps was a good experience, but we haven't had enough experiences like it.

Alfred Kahn: I would like to introduce some political realities. As someone pointed out quite correctly the other day, if you're enamored of grass roots participation, you have to realize that grass roots communities are very often the most parochial in the world.

I think you also ought to recognize that we're in a period of great nationalism, chauvinism, and racialism throughout the world and that the world political realities are not at all consonant with the values we have talked about here.

The Western welfare states are moving to policies of non-mercantilism that are becoming dominant in their economies. They are moving away from the assumption of interdependence. The African and Asian nations, in their concern with resources, are becoming racially aggressive and chauvinistic. The European welfare states are, within themselves, nationalistic. Within the major countries of the West, there are regionalisms which are tearing them apart.

So I'd like to make clear that although the values and the goals we have been talking about are imbued with the hope for a new future that is most attractive, I think that the realities make me say: "Don't expect that to be the picture tomorrow."

June Christmas: As I listened to Aklilu speak about interdependence, I realized how little in this country we really believe in an interdependent world. We even ignore the statistics about the number of people there are in the rest of the world. I'm reminded that when I came back from Sweden a few years ago, I mentioned that I had been to Parliament and had heard a very exciting debate on the New International Economic Order. I mentioned this to an economist teaching at Columbia University, a well-known person, and I said I wanted to learn more about this New International Economic Order. He said, "What do you mean? What is that?" This person was an advisor to state and federal governments, and he knew very little about what was going on in the rest of the world. That makes you a little pessimistic.

Aklilu Habte: That raises a very interesting question. What do the nations of the world do to make their people aware of others? I'm afraid that the United States doesn't do all it should.

John Cannon: How receptive would Third World countries be at this point to significant entry of our youth, whether through the Peace Corps or some type of exchange?

Duhl: Having spent a lot of time at international meetings, I think I can comment on that. Most of what goes on there is conversations like the one we're having around this table. They're very Western and they're very high-level intellectual discussions. But the kids who travel on their own, meeting people and living on a relatively low subsistence level, may do more good than those of us who go back and forth to all the international meetings.

Cannon: I certainly agree with that, Len, but are there significant numbers of our young people who are getting this kind of global experience?

Duhl: The numbers are small, but compared to 20 years ago, they're gigantic.

Habte: In answer to your question, John, I would like to say that most Third World countries would be very receptive to American youth. I visited Botswana and Kenya not long ago, and I found a large number of American youths in both countries. They were very well received.

Jack Darley: As you know, I'm quite interested in the Peace Corps. Can you tell us a little bit about the impact of the Peace Corps on Ethiopia, where there were a large number of volunteers back in 1963?

Habte: Yes. They contributed enormously to the tremendous expansion of secondary education in rural areas, especially at the junior high school level. In a matter of five years, junior high school enrollment increased from less than 8 or 10 thousand students to more than 80,000 students.

Darley: Does this mean that they created an educational opportunity that had not existed?

Habte: Absolutely.

Darley: How could the Peace Corps, having only 300 or so volunteers there for a period of two years, do that?

Habte: Simply because they were willing to go to areas where other volunteers wouldn't go.

Darley: So that the kids from the United States were willing to move into areas where your own Indian colleagues and your own Ethiopian colleagues would not go?

Habte: That's right.

Darley: Well that's something we can be proud of. Even though there was chaos wherever we went, something did happen.

Sheila Kamerman: I want to go back to the earlier comments about the need to internationalize education. It's something that I support very strongly, but I think that we shouldn't forget how much of it is already occurring in the area of informal education in our society. First of all, the number of young people who actually travel may be relatively small, but they underscore the fact that this is an experience that was limited to a very small elite group before World War II. Don't underestimate the significance of what's happening now among young people in terms of the availability of airplane travel and the low fares for kids. It's a very important element in internationalizing education.

A second element is the growing importance of television and the availability on television of all sorts of special programs dealing with other cultures and other countries.

The third element is, quite frankly, the professional international experience, and by all means don't underestimate that. The amount of learning which has occurred in a variety of professional disciplines as a consequence of exposure to other countries and other individuals working in the field in other countries and other cultures has been enormous. I can see the results in my own field.

Finally, there's the pop culture which several people mentioned. I don't think there are any of us who have not experienced the homogeneity of American pop culture throughout the world. I've heard

American rock music every place from Peking to Hong Kong, through-out Eastern Europe and Latin America — almost *ad nauseam,* I must admit. But the fact is, that very often I have been approached by young people all over the world, wherever I have been, with a lot of curiousity about experience in other kinds of countries. So my guess is that despite the inadequacies of our formal educational system, far more is already occurring outside that system than we recognize, and we shouldn't underestimate it.

Habte: I agree, but there is not much systematic education about other peoples and other cultures.

Robert Aldrich: We probably can learn more from the Third World countries than they can from us.

Habte: That may be true, but many of the Third World people expect leadership from the United States.

28 Priorities for the Future

John Scanlon

One of the major assignments the conference participants received on the opening day was to come up with recommendations for research and action programs that might be useful in helping society do a better job of meeting the needs of young people, and in helping young people themselves cope more effectively with the stresses and strains of growing up.

The conference participants took this assignment to heart. During the five days they spent together, they came up with scores of ideas, suggestions, and recommendations. Some were made as part of their prepared remarks, and others grew out of the discussion. Some were expressed as hopes and wishes rather than as formal recommendations, and a few of these were Utopian in nature. Others were not explicit, but were inherent in remarks made about the shortcomings of society and social institutions.

Many of the recommendations were not new, but were regarded as of sufficient importance to merit reiteration. A few of them were contradictory, because on some issues there was no unanimity. But all of them were thoughtful, and to the point. Most of them have been reported in earlier sections of this report, in the context within which they were made, but for the sake of convenience they are summarized here, according to the areas of concern to which they were addressed.

Education
The conference participants were almost unanimous in their belief that in the long run, more and better education is the best approach to the task of helping young people.

The consensus was that parents, educators, clergymen and policy makers need to be educated in how to deal with the needs, hopes, aspirations and concerns of youth, and that young people themselves need a better education in how to cope with the realities of life.

The importance of education was articulated in various ways. John Cannon expressed it this way: "Lord God, give us a better understanding of life." Aklilu Habte said that peace in the world depends to a large extent on helping the peoples of the earth come to a better understanding of each other. Edmund Pellegrino maintained that the greatest contribution our educational institutions can make to the welfare of young people is to help them meet the challenge of fashioning a personal identity.

There was also general agreement that our formal educational system is not as effective as it ought to be. The need for more and better education for young people, for parents, for policy makers and even for educators themselves, was mentioned time and time again. It cropped up in the discussion of drugs, health, sexuality, values, belief systems, cults, nurturing, mentoring, careers and vocations, guidance and counseling, religion, education, and international relations.

Edmund Pellegrino charged that one of the greatest weaknesses of our educational system is that it fails to teach a set of values. Several of the other participants agreed with him on this point.

June Christmas indicted the schools for their failure to prepare young people for the realities of life. She said the schools are not helping young people to choose a vocation, and they are not helping them to learn the skills they will need when they enter the work force.

William Threlkeld stressed the need for high school courses about the damage that youthful crime does not only to society but also to the offender. Helen Washburn said that the reward system in education is "completely out of whack"—that creative teachers and administrators do not get adequate recognition. Ernest Bartell and Robert Aldrich agreed. Bartell pointed out that research is more highly valued than teaching at the college level, and Aldrich maintained that too many college professors are preoccupied with "grantsmanship."

Mass education in school and college also came under attack on the ground that students get "lost" in large institutions and soon develop the feeling, as one participant put it, that "no one gives a damn" about them.

Willard Wirtz recommended more stress on continuing education, not only for its own sake but also as a means of involving more people in the solution of educational problems.

John Schweppe argued for more emphasis in medical education on the study of the human life cycle, the characteristics of each stage, the transitions between stages, and the relationships between the biomedical and social factors that influence behavior at each stage.

Aklilu Habte emphasized the necessity of teaching about other peoples and other cultures.

Paul Bohannan pointed to the need for what he called "backup" institutions that would perform the tasks that are assigned to parents or schools when the parents or the schools cannot or will not carry out those tasks. June Christmas agreed, citing as one example the need for multiservice centers where young people can find help for a wide range of problems like drugs, alcohol, jobs, and troublesome parents.

Alvin Eurich and other speakers pointed out that formal education is only a small part of the total education young people receive, and that more attention needs to be paid to the total educational environment. As an example, he cited statistics to show that by the time a student has finished high school he or she has spent more time watching television than in listening to teachers in the classroom. Other speakers stressed the importance of mentoring. Mr. Eurich and Ernest Bartell made several specific recommendations for improving formal education.

William Threlkeld argued that "education is so important that we can't let the educators make all the decisions about what is to be taught," and Leonard Duhl maintained that "we are entering a period where we need to develop an entirely new way of looking at reality." He said that young people are constantly searching for alternative ways of looking at life, and that their quest should be encouraged.

Historical Perspective

John Demos said there is a growing belief among scholars that there is a life-cycle stage called "youth" between adolescence and maturity, and that this stage deserves a lot more attention than it has received.

Leonard Duhl cited the need for more research on the adolescent experience of blacks, Chicanos, American Indians, and other minorities in American society. June Christmas agreed, saying there was a need for more information about the historical effects of slavery on present-day black adolescents.

Demographic Data

June Christmas pointed out that aggregate statistics often conceal more than they reveal, particularly those aggregates that lump white and blacks together. "We need to tease out the ethnicity," she said, "because some differences between whites and blacks are highly significant."

Donald King suggested a longitudinal study of the 1.1 million unmarried people who are living together to find out whether the divorce rate among those who eventually married was lower or higher than that for young people who had not lived together before marriage.

Youth as Individuals and as Family Members

Sheila Kamerman maintained that research scholars and policy makers tend to look at young people only as individuals, and not as members of families. "We need to raise some different kinds of questions about youth than we have in the past," she said. She went on to list some of these questions, including the question of what happens to the level of family income when young people leave the family they grew up in and strike out on their own, and the question of what impact the growing trend toward both parents working has on young people. She also pointed out the need for better data on how well the family is performing its role as the principal agency for socializing youth.

Robert Aldrich said there was a need for more information about adolescents growing up in single-parent families.

June Christmas emphasized the need for helping young people to establish better relationships not only to one another but also with their parents and other adults.

Guidance and Counseling

Most of the participants agreed that schools and colleges do not do a very good job with guidance and counseling. Helen Washburn and Jack Darley made a number of specific recommendations for improving upon what is being done.

Willard Wirtz said that guidance and counseling should be done on a community-wide basis rather than being limited to the schools, and he went on to say that each community should develop an "opportunity inventory" for young people.

Several of the participants recommended that the traditional high school guidance and counseling system be supplemented by the use of mentors—older people in the community whose experience and insights would be helpful to young people in dealing with a wide range of personal problems.

Youth Policy

Alfred Kahn stressed the need for reform of the present policy framework for dealing with youth, and for the formation of a new policy that would view young people as an asset to society rather than a problem. "We have to develop a new kind of agenda," he said. "How can we expect youth to be responsible unless we treat them as responsible individuals?"

Margaret Hastings, one of the auditors, pointed out that policy requires consensus and commitment. "Those are what we really need," she said. Sheila Kamerman agreed. "The problem is how to get political support for what we want to do," she said. "It's a problem of constituency building."

Willard Wirtz and others argued that youth policy should originate at the community level, because it is there that young people can participate most effectively in the formulation of policies dealing with their needs and concerns.

Robert Aldrich said the United States could learn a lot from the Canadian experience in developing a national youth policy. One of the secrets of the Canadian success, he said, was active participation by young people in the formulation of youth policy at all levels of government.

John Schweppe suggested that one element of a national youth policy should be the requirement that all young people, at the age of 21 or 22, be required to devote a year to some kind of national service, either military service or community service of the type performed by the Civilian Conservation Corps in the 1930's.

Willard Wirtz agreed that some sort of national service program might be a good idea but he warned that it would have to be voluntary and that it should have within it some mechanism for persuading young people that work is worthwhile. He also said that although such a program would have to be partly funded by the federal government, part of the cost should be defrayed by those who benefited from the work that was performed, and that the administration of the program should be vested in local communities rather than in the federal government. "People participation, the key to the success of such a program, is only possible at the local level," he remarked.

Federal Intervention

Several of the participants expressed apprehension about the growing intervention of the federal government in community affairs. The consensus seemed to be that youth programs originating at the community level should be administered locally, even if federal funding is necessary to get them off the ground.

Health

Donald King said there was a great need to educate young people about the dangers involved in the excessive use of prescription drugs and other agents that are used as "uppers and downers." He also said that much more needs to be known about substances that contaminate the environment, but he warned that there is a danger of overregulation with respect to some contaminants which have not been demonstrated to be dangerous. "We can't eliminate all of them," he said. "We're going to have to live with some of them."

Edmund Pellegrino agreed with Dr. King about educating the young to the dangers of certain prescription drugs. "We've got to do something about this incurable itch to take a pill," he remarked.

Robert Aldrich argued that there is a need for more and better courses in personal hygiene in the schools. June Christmas agreed, adding that she would go beyond that to include the promotion of healthful ways of living.

Leonard Duhl said he thought medicine ought to pay more attention to Oriental ideas about self-healing. "Doctors diagnose and prescribe," he said, "but they do not involve themselves in the healing process."

Sexuality

Sidney Werkman said he thought adolescents should have the opportunity to develop their own sexuality without outside interference. "We need to give adolescents a better chance to find themselves," he argued.

Philip Blumstein agreed. "Sexuality is a malleable thing," he said. "We have to be prepared to accept the fact that it unfolds during a lifetime."

With respect to sex roles, Dr. Blumstein said he thought men should be encouraged to develop expressiveness, even though this is generally regarded as a female quality. He also said society ought to come up with new ways of helping homosexuals deal with their situation and with public reactions toward it.

Cults

Harold Visotsky maintained that society's most effective response to cults is not to try to abolish them but rather to come up with competing groups that will fill the needs that cults are now serving, in a manner consistent with socially accepted values.

The Third World

Aklilu Habte found our educational system wanting with respect to teaching about other peoples and other cultures. He also recommended what he called "a people exchange," with the requirement that every visitor must live at least one year with a local family.

Paul Bohannan agreed, saying it was not enough for social scientists to study other cultures; they must also live in them in order to really understand them.

The Good Society

Early in the conference, Leonard Duhl and Alfred Kahn ventured the opinion that what was really needed in dealing with the problems and concerns of youth was a public debate on "the good society." Several other speakers, notably Dr. Pellegrino, agreed with this suggestion, and there was considerable discussion about it.

Robert Aldrich called for a national dialogue similar to that which laid the foundations for youth policy in Canada.

Willard Wirtz said he agreed with all of these comments. "I'm in favor of redesigning this society," he said. "I think it's our last best hope. What it will require is a new politics, a new economics, and a new set of values."

A. J. Kelso emphasized that the task of building the good society must begin with each individual person.

Selected Bibliography

Books:

Aries, Philippe. *Centuries of Childhood.* New York: Alfred A. Knopf, 1962.

Cox, Harvey. *Turning East: The Promise and Peril of the New Orientalism.* New York: Simon & Schuster, 1977.

Egan, Gerald. *The Skilled Helper: A Model for Systematic Helping and Interpersonal Relating.* Monterey, California: Brooks Cole Publishing Co., 1975.

Eurich, Alvin, C. (ed.) *Campus 1980.* New York: Delacorte Press, 1968.

Eurich, Alvin C. *Reforming American Education.* New York: Harper and Row, 1969.

Herr, Edwin L. *Guidance and Counseling in the Schools: Perspectives on the Past, Present and Future.* Falls Church, Virginia: American Personnel and Guidance Association, 1979.

Jencks, Christopher, et al. *Who Gets Ahead? The Determinants of Economic Success in America.* New York: Basic Books, 1979.

Kett, Joseph. *Rites of Passage.* New York: Basic Books, 1977.

Maccoby, E. and Jacklin, C. *The Psychology of Sex Differences.* Stanford, California: Stanford University Press, 1974.

Masters, William H. and Johnson, Virginia E. *Human Sexual Response.* Boston: Little, Brown, 1966.

Morris, Desmond. *The Naked Ape.* New York: McGraw Hill, 1967.

Parsons, Talcott and Platt, Gerald. *The American University.* Cambridge, Massachusetts: Harvard University Press, 1973.

Schmuck, R.A. and Schmuck, P.A. *Making the School Everybody's House.* Palo Alto, California: National Press Books, 1974.

Slater, P. *The Pursuit of Loneliness.* Boston: Beacon Press, 1970.

Tyler, Leona E. *The Work of the Counselor.* New York: Appleton-Century-Crofts, 1961.

Wittig, M.A. and Peterson, A.C. (eds.) *Sex-Related Differences in Cognitive Functioning.* New York: Academic Press, 1979.

Reports and Monographs:

Beck, Bertram M. *The Lower East Side Family Union: A Social Development*. New York: Foundation for Child Development, March 1979.

Fox, Robert S. *et al. School Climate Improvement: A Challenge to the School Administrator*. Bloomington, Indiana: Phi Delta Kappa, 1975.

Goodall, Jane van Lawick. *The Behavior of Free-Living Chimpanzees in the Gombe Stream Area*. Animal Behavior Monograph 1:161-311, 1968.

The Johnson Foundation. *Adolescent Sexuality and Health Care*. Report of a Wingspread Conference. Racine, Wisconsin, April 12-13, 1974.

Miller, J. Dean. *Developmental Theory and its Application in Guidance Programs: Systematic Efforts to Promote Personal Growth*. St. Paul, Minnesota: Minnesota State Department of Education, 1977.

Olsen, Laurie. *Lost in the Shuffle: A Report on the Guidance System in California Secondary Schools*. Open Road Issues Research, Citizens Policy Center, 1979.

U.S. Bureau of the Census. *Current Population Reports*. Series P-25, No. 800, "Estimates of the Population of the United States by Age, Sex and Race: 1976 to 1978. Washington, D.C.: U.S. Government Printing Office, April 1979.

U.S. Department of Health, Education and Welfare. *Facts of Life and Death*. DHEW Publications No. (PHS) 79-122. Washington, D.C.: U.S. Government Printing Office, 1978.

World Bank. *Education Sector Policy Paper*. Washington, D.C.: World Bank, 1980.

World Bank. *Health Sector Policy Paper*. Washington, D.C.: World Bank, 1975.

World Bank. *World Development Report*. Washington, D.C.: World Bank, 1979.

Articles:

Bem, S.L. "The Measurement of Psychological Androgyny," *Journal of Consulting and Clinical Psychology*, 42:155-162, 1974.

"Cults in America," a series of three articles published in *The New York Times*, January 21-23, 1979.

Demos, John and Virginia. "Adolescence in Historical Perspective," *Journal of Marriage and the Family*, Vol. 21, No. 4, November 1969.

Egerton, John. "The Issues Medical Schools Have Ignored," *The Chronicle of Higher Education*, October 16, 1978.

Fox, Robin. "Alliance and Constraint: Sexual Selection in the Evolution of Human Kinship Systems," in Bernard Campbell, *ed., Sexual Selection and the Descent of Man,* (Chicago: Aldine, 1972), pp. 282-331.

Lancaster, Jane B. and Lee, Richard B. "The Annual Reproductive Cycle in Monkeys and Apes," in Irven DeVore, ed., *Primate Behavior: Field Studies of Monkeys and Apes,* (New York: Holt, Rinehart and Winston, 1965).

McGuinness, D. "How Schools Discriminate Against Boys," *Human Nature,* February 1979, pp. 82-88.

Murphy, Robert F. "Man's Culture and Woman's Nature," *Annals of the New York Academy of Sciences,* 293:15-24, 1977.

"Residential Programs for Single Adolescent Mothers," The Johnson Foundation, Racine, Wisconsin, January 1979.

Udry, J.R. and Morris, N.M. "Distribution of Coitus in the Menstrual Cycle," *Nature,* 220:593-596, 1968.

"Where the Jobs Will Be for '79 Graduates," *U.S. News and World Report,* December 18, 1978.

Yarrow, Leah. "Everything You Want to Know About Teenagers (But Are Afraid to Ask)," *Parents* Magazine, February 1979.

Yinger, J. Milton. "Countercultures and Social Change," *American Sociological Review,* Vol. 42, No. 6, December 1977.

Part III
Young Adulthood

John Scanlon

29 Introduction to Part III

For five days in the summer of 1978, scholars and practitioners from nineteen different disciplines met in Aspen, Colorado, to discuss the multifaceted aspects of young adulthood — the stage of the human life cycle when the preparation for living has ended and the great adventure of living has begun.

These are the years, from 25 to 40, when most young adults establish families, raise children, develop lifestyles, embark on careers, and plan for the future. They are years of high hopes and lofty aspirations, but they are also years of disappointment and frustration. They are relatively healthy years, but accidents, cancer, and heart ailments take a heavy toll.

Much has been written about this critical period between adolescence and middle age, in novels as well as in scholarly treatises, but much remains to be learned. The charge laid upon the Aspen conferees was to examine young adulthood from a multidisciplinary point of view, to identify its special problems and concerns, to explore its relationships with earlier and later stages in the human life cycle, to shed more light on its particular stresses and strains, and to recommend programs of action that might improve the quality of life for people going through that stage of development as human beings and as members of society.

The conference, held at the Aspen Institute for Humanistic Studies, was the third in a series sponsored by the

Academy for Educational Development with grant support
from the Schweppe Research and Education Fund. The first
conference, held in the summer of 1976, examined the human life cycle as an entity and the critical transition periods
from one stage to another. Its findings and recommendations were published by the Academy in a report entitled
Major Transitions in the Human Life Cycle. The focus of
the second conference, which was held in the summer of
1977, was on the middle years in the life cycle—the period
from 40 to 65. The conference report, entitled *The Middle
Years: A Multidisciplinary View,* was published by the
Academy in the Spring of 1978.

The third conference, like its predecessors, was characterized by a multidisciplinary approach and a cross-
disciplinary dialogue. The participants included experts in
anthropology, biochemistry, demography, genetics, hema-
tology, history, housing, internal medicine, manpower policy, pathology, pediatrics, politics, preventive medicine,
psychiatry, psychology, religion, sociology, and urban studies. They are listed in Appendix A.

The major themes discussed at the conference were the
history of the human life cycle, the biological and demographic characteristics of the 25-to-40 age group, health
problems and biological changes that occur in the young
adult years, emerging trends in work force participation,
changing family structures and roles, housing and family
income, psychological needs and problems in the family, the
ethics of intervention, and priorities for the future. The conference agenda is reproduced in Appendix B.

At each session of the conference, one of the participants
presented a paper or a talk on the designated theme. Then
two, or sometimes three, panelists commented on the
speaker's remarks from the viewpoint of their own disciplines, and a general discussion followed.

The remarks made at each session were recorded and

transcribed. This made it possible for the conference reporter to let the participants speak for themselves, instead of trying to present the essence of what they said in a series of paraphrased essays. As a result, this report reflects much of the flavor of the discussions — the liveliness of the interdisciplinary byplay, the spontaneity of the remarks, the humor of some of the exchanges, and the informality of the spoken word.

Although the conference was not designed to produce formal recommendations — formal in the sense of being officially adopted by the conferees after discussion and debate — individual participants did make more than a score of suggestions and recommendations for dealing with the needs, problems, and concerns of people in the 25-to-40 age group. Most of these recommendations are initially set forth in the context within which they were made during the conference sessions, and all of them are summarized in the final section of this report.

30 Demographic Profile of the 25-to-40 Age Group

The marriage patterns and living arrangements of young adults have changed significantly in the past twenty-five years.—Arthur J. Norton

According to population estimates for 1977, published by the Bureau of the Census in April 1978, people in the 25-to-40 age group number about 47.7 million, or roughly 22 percent of the total population of the United States. Numerically, they comprise the second largest comparable segment of the total population, exceeded only by the generation that is coming along immediately behind them, the "baby boom" age cohort born in the late 1950s.

The chart on the following page, reproduced from "Estimates of the Population of the United States by Age, Sex, and Race: 1970 to 1977," published by the Bureau of the Census in April 1978, shows the age-group composition of the U.S. population and how it has changed since 1970. It looks much like a narrow-waisted Christmas tree, tapering upward from the broad base of children under ten years of age to the pointed tip of people over ninety. The narrow waist is made up of people between the ages of 30 and 40

311

Distribution of the Total Population
by Age and Sex: April 1, 1970 and July 1, 1977

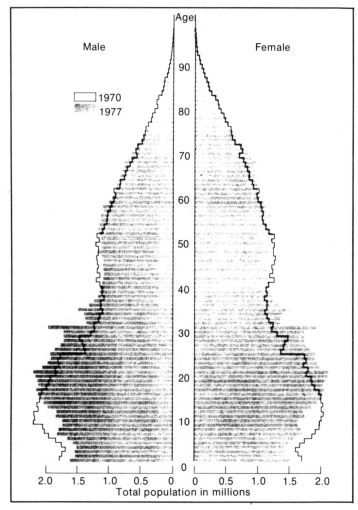

Source: "Distribution of the Total Population by Age, Sex, and Race: 1970 to 1977," *Current Population Reports*. Series P-25, No. 721, Bureau of the Census, Issued April 1978.

— the "slim" generation of the 1930s and 1940s — and it includes the young adults who are in the upper ranges of the 25-to-40 age group. The solid bars, representing the 1977 population profile, illustrate the fact that for the next ten years the number of people entering the 25-to-40 age group will be larger than the number of those leaving it at the upper end, which means that in the 1980s the 25-to-40 age group will constitute the largest segment of the U.S. population.

Sex and Race

Of those in the 25-to-40 age group, 23.6 million are men and 24.1 million are women, a reflection of the female's historic edge over males in the matter of survival. Classified according to race, 41.6 million are white and 6.1 million are black or members of other races. The numerical superiority of women over men in the black and other races is even greater than it is in the white race.

Marital Status

As the following table indicates, the overwhelming majority of the people in the 25-to-40 age group are married and living with their spouse.

| | Age Group | | | | | |
| | 25 to 29 | | 30 to 34 | | 35 to 39 | |
Percent, All Races	Male	Female	Male	Female	Male	Female
Single	26.1%	16.1	12.2%	7.0%	7.3%	5.4%
Married, Spouse Present	65.6	70.4	78.5	77.5	84.2	78.5
Married, Spouse Absent	3.0	5.4	3.5	5.3	2.7	5.0
(Separated)	(2.2)	(4.3)	(2.8)	(4.6)	(2.1)	(4.2)
(Other)	(0.7)	(1.1)	(0.7)	(0.7)	(0.5)	(0.8)
Widowed	0.1	0.4	0.1	1.0	0.3	1.7
Divorced	5.3	7.6	5.7	9.2	5.6	9.4

(*Note:* Because of rounding, not all columns add to 100%.)
Source: "Marital Status and Living Arrangements," Bureau of the Census, April 1978.

Divorce

As the following table indicates, more than 50 percent of all U.S. divorces occur among husbands and wives in the 25-to-40 age group. The divorce rate is highest among couples in the lower range of the age group — those between 25 and 29 — and then it begins a steady decline.

Distribution of Divorces
By Age at Time of Decree
(Based on 1974 data from 29 states)

Age at Time of Decree	Percent Distribution	
	Husbands	Wives
Under 20	1.2%	4.6%
20-24	15.7	23.0
25-29	24.2	24.3
30-34	18.2	16.1
35-39	12.4	10.7
40-44	9.6	7.9
45-49	7.5	5.9
50-54	5.2	3.6
55-59	2.8	2.0
60-64	1.7	1.0
65 years and over	1.6	0.8

Source: *Vital Statistics of the United States, 1974,* Vol. III, "Marriage and Divorce," National Center for Health Statistics, Government Printing Office, Washington, D.C., 1977.

The number of divorces in the United States has been increasing steadily for the past twenty years, rising from about 350,000 in 1958 to about a million in 1975. The number of children affected by divorce has increased at about the same rate, rising from about 400,000 in 1958 to well over a million in 1975. The trends are shown in the chart on the following page.

Divorce and Children

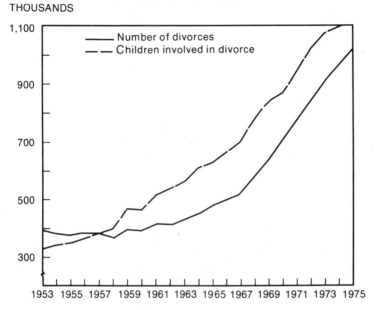

THOUSANDS

Reproduced from "Toward a National Policy for Children and Families," 1976, with the permission of the National Academy of Sciences, Washington, D.C., page 19.

Education

The intensive effort the United States has made in recent years to provide more of its children with more years of schooling shows up clearly in the educational attainment figures for people in the 25-to-40 age group. People in the upper range of the age group — those between 35 and 40 — have had more years of schooling than any generation that preceded them, but those in the lower range — 25 to 30 — have had even more. The trend is illustrated in the chart on the following page.

The chart illustrates not only that today's young adults are better educated than their forebears, but also that the

historic gap between male and female educational attainment — as measured by completion of one or more years of
college — has been closing in recent years. This trend is
illustrated by the close proximity of the solid and dotted
lines at the left side of the chart.

Educational Attainment in the United States:
March 1977

Percent of Persons 20 to 65 years old who were high school
graduates and percent who had completed 1 or more years of
college by sex: March 1977

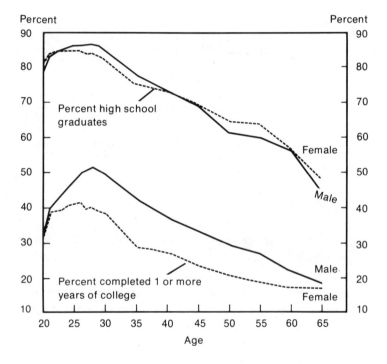

Source: "Educational Attainment in the United States: March 1977 and
        1976," *Current Population Reports,* Series P-20, No. 314, Bureau
        of the Census, Issued December 1977.

Income -

Although not all of the people in the 25-to-40 age group have reached the peak of their earning power, a great many have, and those who have not are approaching it. The following table, based on data from "Money Income in 1976 of Families and Persons in the United States," published by the Bureau of the Census in July 1978, shows the median income for individuals in various age groups:

Median Income By Age Group

	Age Group			
	25-29	**30-34**	**35-39**	**40-44**
Both sexes, all races	$ 8,258	$ 9,643	$ 9,953	$9,832
Whites	8,566	10,134	10,427	10,431
Blacks	6,397	7,256	7,267	6,853
Males, all races	10,534	13,062	14,224	14,438
White males	10,848	13,474	15,007	15,091
Black males	7,770	9,029	9,434	9,833
Females, all races	5,562	5,124	5,148	5,169
White females	5,660	4,942	4,996	5,237
Black females	5,173	5,685	5,876	4,943

Where They Live

Like most Americans, people in the 25-to-40 age group tend to live in urban areas. According to data from the 1970 Census, published by the Census Bureau in 1973, nearly 60 percent of the people between the ages of 25 and 39 (the age groups used in the Census data) lived in urban areas, and only about 30 percent in towns with less than a population of 2,500. Urbanization was more pronounced among people in the lower range of the age group (25 to 29) than in the higher range (35 to 39).

Life Experiences

The lives of people in every generation are affected to some extent by the historic events they live through, and the young adults who make up today's 25-to-40 age group have experienced a great many. Even the youngest of them — those born in 1953 — have lived through a period of history that has brought with it:

- The development of television as a worldwide medium of mass communication.
- The development of the hydrogen bomb and missiles capable of delivering it anywhere on earth.
- The McCarthy era.
- The development of space vehicles, the landing of the first man on the surface of the moon, and the exploration of Mars and other planets.
- The war in Viet Nam.
- The assassinations of John F. Kennedy, Robert Kennedy, and Martin Luther King, Jr.
- The campus turmoil of the late 1960s.
- The racial explosions in Los Angeles, Detroit, New York, and other major cities.
- The Civil Rights movement and the Women's Rights movement.
- The worst recession since the Great Depression of the 1930s, followed by double-digit inflation.
- The Watergate break-in and its aftermath, including the Senate hearings, the impeachment proceedings against President Nixon, and his subsequent resignation.
- Dramatic advances in medical science, symbolized by the successful use of open-heart surgery, organ transplants, and the birth of the first baby conceived outside the womb of its mother.
- Profound changes in religious life, symbolized by the Vatican Councils, the mass exodus of priests and nuns into secular life, the ordination of the first woman as a priest

in the Episcopal Church, and the election of the first non-Italian pope in 455 years.

The list is illustrative rather than exhaustive, but it does serve as a reminder of the momentous changes that have occurred in the lives of people in the 25-to-40 age group.

Arthur J. Norton, Chief of the Census Bureau's Marriage and Family Statistics Branch, described the years from 25 to 40 as "the age of application" in the human life cycle. "Formal education and training are usually over," he pointed out, "and young adulthood is the time for applying the knowledge and skills that have been learned. Occupational choice usually takes place at the very outset of this age span, and so does marriage and family formation. Psychologically, the years bring changes of perspective, from the initial optimism one has about goal attainment — the resolve to go out and make one's mark in life — to a gradual drift toward realism, and in some cases pessimism — an awareness of one's limitations and for most people an acceptance of one's lot in life."

He then went on to make several demographic comparisons between today's young adults and those of a generation ago, paying particular attention to the areas of his specialty — marriage and the family.

"Marriage," he said, "is and has been and probably will continue to be nearly universal in our society. Most first marriages occur before the age of 25. The most typical age for divorce is 29 for men and 27 for women. About 50 to 60 percent of the men who divorce ultimately remarry, usually by the age of 33. If they redivorce, they usually do so within six years afterward. About three-quarters of all women who divorce also remarry, usually by the age of 30. If they divorce again, they do so before the age of 40. In other words, most married people who divorce and remarry do so between the ages of 25 and 40, the age span we are discussing at this conference."

Dramatic changes in marital behavior are indicated by statistics spanning the past quarter century, Mr. Norton continued. In 1950, nearly 80 percent of the people in the 25-to-40 age group were married and living with their spouse. Only 2.5 percent were divorced. Today, only about 75 percent of the people in the 25-to-40 age group are married and living with their spouse, and for younger members of the age group the proportion is only about 70 percent. About 8 percent are divorced. "The American family," he observed, "has become increasingly physically fragmented and isolated over the past twenty-five years."

Accompanying the changes in marital status have been significant changes in the pattern of living arrangements. In 1950, nearly 95 percent of the men and women in the 25-to-40 age group who were maintaining households were doing so as married couples. By 1970, the proportion had dropped to 80 percent, and there was a corresponding increase in the proportion of young adults who were living alone or with roommates.

Mr. Norton mentioned two other significant changes in the family life of young adults that have occurred during the past twenty-five years: (1) the size of the average family has grown smaller, and (2) more wives are working outside the home.

In discussing the people in today's 25-to-40 age group, the experiences they have had and their prospects for the future, Mr. Norton made a distinction between those in the upper range of the age group (35 to 40) and those in the lower range (25 to 30). The former, he pointed out, were born before the post-World War II baby boom and, being members of a relatively small age cohort, have not had much competition for jobs and for positions of leadership in society. The latter, however, being members of the "baby-boom" generation, have had and will continue to have stiff competition from their peers, simply by virtue of their numbers. "The vanguard of the baby-boom generation have just about finished college and they're already finding it hard to get jobs suitable to their talents," he said. "This is creating all kinds of problems for them — psychological as well as economic. Many of them are not faring as well as their parents did at the same age. This is a blow to their pride. To them, it's a failure to maintain the American tradition of constant improvement from one generation to the next. It's no wonder that they are deviating from past marriage patterns and creating new variants of the traditional family structure."

According to Mr. Norton, life will be much less competitive for the next generation of young adults — the people born during the "baby bust" years since 1961. Because of their smaller numbers, he said, their employment prospects will be better and this, in turn, may lead to an increase in the marriage rate and a decline in the divorce rate.

Mr. Norton concluded by saying that the available socio-economic data about people in the 25-to-40 age group is not sufficient, and needs to be supplemented by more intensive investigation of the dynamic interaction between parents and children, wives and husbands, and members of peer groups. Especially needed, he said, is the type of information that will shed more light on the types of exchanges that provide security within the family. He recommended public support for federal surveys that would:

- Identify the kinds of kin-structure systems that transcend the physical boundaries of a housing unit, as a means of providing answers to questions relating to the impact of fragmentation and isolation on the social, economic, and psychological needs of family members.
- Explore in greater detail the dynamics and effects of marital dissolution.
- Appraise the overall economic well-being of families by studying the gross inflows and outflows of family income — from whom and to whom the exchanges occurred.

- Provide better data on health problems of people in the 25-to-40 age group and identify the most effective way of delivering health services.

In the discussion that followed, several of the conferees added information, insights, and speculations of their own.

Mr. O'Keefe pointed out that the younger members of the 25-to-40 age group came of age at a time when the greatest income shift in history began to take place—the worldwide transfer of some $60 billion a year to the oil-producing countries of the Third World. The United States has been the principal loser in this massive income shift, he said, with the result that we have lost our predominance in the world economy. Moreover, he continued, the OPEC countries have no effective consumer demand, because their populations are so small. They will purchase high-technology items and hold much of their new wealth in liquid investments— options that may not stimulate domestic job formation in the United States. "All of this is bound to have an adverse effect on the younger members of the 25-to-40 age group," he said, "because no matter what national economic policies we adopt young adults are going to experience trouble in the labor market. The income transfer and real resource constraints have put a heavy strain on our economy and have lessened our ability to stimulate growth internally."

Mr. Frieden, harking back to Mr. Norton's remarks about the decline in employment opportunities for the younger members of the 25-to-40 age group, said this might help to explain the rapid increase in their geographic mobility. "One of the ways in which people in the labor force adjust to the shortage of jobs is by moving to areas where the prospects of employment are better," he pointed out. "This has been true throughout the industrial age. We are now witnessing a modern version, in the form of migration out of slow-growth areas in the North and Northeast into the so-called Sun Belt states. This is quite unsettling for the people who move around a lot, and it's also unsettling from the standpoint of community stability, but in terms of our ability to compete with the rest of the world it's a good thing, because industry tends to move to localities where it can get quality production at the lowest cost."

Mr. Demos speculated that some of the demographic trends mentioned by Mr. Norton may well lead to an increase in the rate of childlessness among married couples. "The census figures already show that with women becoming more active in the labor force an increasing number of younger couples in the age group we're talking about are deciding not to have children at all, or to postpone having children until it's too late to have them. And ironically enough, people who don't have children may have more opportunities for advancement in their careers. The governor of our largest state, California, is a bachelor, and so is the mayor of our largest city, New York. This may be the sign of a trend, not only in government, but also in industry and the professions. We may well arrive at a situation where people without children or without families of the conventional kind are making important decisions that affect families with children. That kind of situation would constitute a real cultural shift, and its impact on family life would be enormous."

Mr. Norton said this was an interesting hypothesis, but he cautioned that decisions regarding child-bearing are often unpredictable. Even the demographers were surprised at the magnitude of the postwar baby boom, he remarked.

The discussion then turned to the cost of raising children as a determinant of family size. One of the conferees pointed out that when the United States was primarily an agricultural nation, children were an important economic asset because they relieved their parents of time-consuming chores on the farm, whereas today it costs a family about $50,000 to raise a child from birth to age 18.

Mr. O'Keefe said he doubted whether, in real terms, it costs more money to raise a child today than it did in earlier years. "The $50,000 that is spent on raising a child today becomes income to the people who are providing the necessary goods and services the child needs," he pointed out. "In fact," he continued, "with our present social support system and the income redistribution that occurs because of improved medical and educational services, the cost of child-rearing may actually be less in real terms. A substantial portion of the $50,000 is an investment that eventually enhances the child's ability as a wage-earner."

One of the other conferees said that although he did not quarrel with Mr. O'Keefe's economic analysis, it did not cover the entire situation. "What about the psychological attitudes of the parent and the child?" he asked. "The farm boy felt important, and was regarded as important, because he was helping his father produce income for the family. Not many boys play that role today."

Robert Aldrich called attention to a recent United Nations report on the economic value of children in Third World countries. "The data show that they are an economic necessity," he said, "even at an early age. They fetch water, care for the chickens and ducks, tend the goats and cattle, and help with the harvesting of rice."

Mr. O'Keefe pointed out that the Third World countries are highly labor-intensive, a stage in economic development that the United States has moved away from, and that by using children as income producers they are failing to develop the full economic potential of the children, including the ability to use machinery that can make farming more productive. "You have to remember," he said, "that when we keep our children in school we are making an investment in human capital. We are enabling them to develop the skills that they're going to need to generate income for all of us."

31

History of the Human Life Cycle

There is a certain sense in which our own time has become far more self-conscious about the life cycle than was ever true in the past.—John Demos

According to John Demos, Professor of History at Brandeis University, the historical study of the human life cycle is a relatively new, and in many ways incomplete and imperfect enterprise.

"Twenty years ago," he said, "and possibly even more recently than that, you would have found nothing that historians of any sort could contribute to a conference of this kind. I don't think you could have found an historian to invite. Today, there are quite a number of historians who would have something to contribute, but those of us who devote a major portion of our time to the study of the human life cycle are still a tiny minority within our professional brotherhood."

Mr. Demos attributed the recent wave of historical interest in the human life cycle to a landmark book by the French cultural historian Philippe Aries, written in 1960 and published in English in 1962 under the title *Centuries of Childhood*.

325

"In that book," he said, "Aries threw open a whole new territory of historical materials and questions — all of them pertaining more or less to childhood throughout the ages. But in addition to that, I think he opened up to many historians the notion that there might be something useful in the study of the life cycle as a whole. Since that time, the historical study of childhood has become a fairly substantial and flourishing enterprise. From childhood it was only a small and logical step to adolescence, and within the past few years there has been a mini-wave of studies of old age. So we have a fairly rounded set of information about the early and late parts of the life cycle as they have been experienced in the historical past, but very little on the middle parts of life."

Mr. Demos developed an interest in midlife as a stage in the human life cycle as a result of research he initially undertook on old age in early America.

"I got into various materials on old age rather deeply," he said, "and I soon began to realize that there was something missing in Colonial perceptions of the life cycle. The people I was studying, mainly the people of early New England, had fairly clear and sharply articulated views of old age, and they also had fairly clear views of childhood and what they called youth, but they had scarcely a term, let alone a full-blown concept, of middle age. There seemed to be a kind of gap there. The whole life period from age 25 or 30 to age 60 was thought of not as a stage or a phase in an ongoing sequence of events, but simply as the full realization of personhood."

"This was the time when a man was fully a man, or a woman was fully a woman. Everything that came before that middle period of life was essentially preparatory and by definition incomplete. Everything that came after that period was in some sense an aspect of decline, a falling away from the peak of personhood which had already been realized."

Mr. Demos hastened to add that there were exceptions to his generalization about the segmented view of the life cycle that prevailed in early times. "Playwrights like Shakespeare talked about the life cycle in a way that seems quite modern and sophisticated by our lights," he said, "but as a social historian I must point out that in those days there was a vast gulf between what you might call the high and the popular points of view. The high-culture exemplars like Shakespeare thought about life in a way which cannot be regarded as representative of the people of their time."

Possibly one of the largest differences in our experience in growing up and growing older from that which obtained in pre-modern times, Mr. Demos said, has to do with the whole range of choices that confront us in many sectors of our lives — choice of careers, choice of marriage partners, choice of friends, choice of values, and choice of religious systems.

"The average young person growing up in Colonial New England," he continued, "did not experience anything like the same range of choices. That difference has important implications for several parts of the life cycle — most conspicuously for adolescence, which is in a certain sense the choice-making time today, but also perhaps for midlife, when many people review earlier choices in an attempt to live more comfortably with them or else to change them."

Pointing out that cohort experience plays an important part in shaping the lives of people in any age group, Mr. Demos then turned to an examination of the historical context in which today's young adults have grown up and matured.

"As a shorthand way of doing that," he said, "I suggest taking 32 or 33 as the average age of people in the 25-to-40 age group. That would mean birth just after the Second World War, when the postwar baby boom was just beginning, childhood in the late Forties and early Fifties, adolescence in the late Fifties and early Sixties, college — for that part of the age group that went to college — during the latter part of the Sixties, and career choices in the early years of the Seventies. It goes without saying that in some very formative sectors of that hypothetical life course, powerful cohort experiences impinged. The beginning of college for these people would have just about coincided with President Kennedy's assassination — the first of the several major assassinations that took place in the Sixties. It also was a period of rapid increase in college enrollments, the war in Viet Nam, and the campus protest movement. That particular set of historical circumstances has shaped and indeed scarred people in the 25-to-40 age group in a way that may well set them aside as somewhat special."

Turning back to his studies of life in early New England, Mr. Demos said he was struck by the demographic characteristics of the principal figures involved in the witchcraft hysteria. Most of the alleged practitioners were women in the 40-to-60 age group, and many of the professed victims were teenaged girls, but by far the most numerically important victims were young men between the ages of 25 and 40. Mr. Demos confessed that he was not positive about the reasons for this phenomenon, but he speculated that they may have grown out of the particular circumstances of early New England life. At that time, he pointed out, men between the ages of 25 and 40 were expected to have achieved their economic independence, so that they could move out of their families of origin, marry, and establish households of their own. However, the prerequisites of economic independence — and particularly land — did not come easily to all young men of that era. Some had to count on inheriting land from their fathers, and not all fathers were inclined to be generous. They gave rather meagerly, or on terms that fell short of outright ownership.

"All of this suggests to me that there might have been good reasons why young men in early New England had their own particular anxieties," Mr. Demos continued, "and there may well be certain parallels between their plight and the plight of young men in our own times. I'm not sure."

In the discussion that followed, Alex J. Kelso, Professor of Anthropology at the University of Colorado, was the leadoff speaker. From the viewpoint of a cultural anthropologist, he underscored what Mr. Demos said about the ever-widening range of choices that confront human beings in various stages of their lives and differentiate members of one age group from another. It is this phenomenon, produced by economic, technological, and cultural change, that leads to the so-called generation gap, he said. By way of illustration, he pointed to the multiplicity of choices that confront today's purchasers of breakfast cereals and automobiles as compared to the rather modest range of alternatives available a generation ago. "Today," he continued, "the biological differentiations between members of the same age group in various cultures throughout the world are far less than the value differentiations between members of various age groups in an industrialized society like ours." He went on to point out that the gradual shift to the metric system in the United States is another example of cultural change that differentiates members of one age group from another. "Our children are already learning the metric system in school," he said, "and it won't be long before that becomes one more area where parents and children have nothing in common."

Arthur Robinson, a medical geneticist, contributed several observations to the discussion. He said that in discussing people in the 25-to-40 age group the conference participants should guard against sweeping generalizations. "Personally," he continued, "I am much more impressed by the genetic uniqueness of each individual than by the genetic homogeneity of an age group." He also made the point that no examination of the human life cycle can be complete without some consideration of the periods of conception and gestation. "There is growing evidence that the behavior of adults — including those in the 25-to-40 age group we are discussing here — is profoundly affected by genetic errors occurring at the very beginning of the life cycle."

Genetic compatibility is already becoming a consideration in the choice of marriage partners, he continued, and the awareness of genetic hazards is beginning to affect child-bearing decisions. "Women over 35 are becoming increasingly aware of the fact that they run more risks of producing chromosomally defective children than younger women," he pointed out.

Harold Visotsky said he agreed with John Demos' statement that people today are much more age-conscious than their forebears, and that age-grouping is much more common than it used to be. "There are a whole series of rules, regulations, laws, and customs which establish age thresholds in modern society," he pointed out. "We are required to state our age when we start school, apply for a driver's license, drink at a bar, enter college, look for a job, get married, take out life insurance, pay our income tax, and collect our Social Security benefits." He went on to add that demographers are not the only ones who divide people into age groups. "Advertisers do it all the time," he said. "In fact, the whole concept of marketing is age-specific as well as income-specific. The advertisers put us into cereal groups, soft-drink groups, clothing groups, and even automobile groups. There's a tremendous amount of age-grouping going on in modern society, and it's increasing all the time. There's also a significant amount of inner conflict as we move from one age threshold to another. Each move confronts us with a new array of choices and decisions that have to be made. When we reach 30, 40, 50, and 60 we suffer from birthday depressions."

Arthur Robinson pointed out that because of genetic differences people vary in their ability to withstand the stresses of life. There are genetic subgroups that are more susceptible to various kinds of social and environmental stresses than others. "We need to know more about this whole area of genetic susceptibility," he said.

John Schweppe agreed. "I think we need to know a lot more about how genetic characteristics affect human behavior," he said. "I also think we need to understand the relationships between the biological determinants of human behavior at various stages of the life cycle — the age-related determinants — and those determinants attributable in large part to the social and cultural milieu."

32

Biological Characteristics of Young Adulthood

It's unfortunate that not more than ten percent of those who would benefit from genetic counseling are currently receiving it.—Arthur Robinson

Dr. Arthur Robinson, Professor of Genetics and Pediatrics at the University of Colorado Medical Center and Director of Professional Services at the National Jewish Hospital and Research Center in Denver, described the significance of chromosomal differences between males and females, discussed the impact of genetic diseases on people in the 25-to-40 age group, and emphasized the importance of genetic counseling for young adults who are planning to have children.

In human biology, he pointed out, the sexes are usually distinguished by observation of morphological and cultural attributes — the anatomic and psychosocial aspects of sex. "However," he continued, "the ultimate difference between the sexes in mammals is controlled by a sex chromosomal difference which starts a train of events that leads to the gonadal, somatic, hormonal, social, and reproductive aspects of sex.

333

"The female has two X chromosomes, and the male—the heterogametic sex — has one X and a Y chromosome. Of significance is the fact that although the X chromosome has at least a hundred known genes, the Y chromosome has only one known gene—the gene for the H-Y antigen that is presumably responsible for the development of the male gonad. In addition, the female, with approximately four percent more biosynthetic material than the male, is a mosaic of X chromosomal activity. In each cell of the early female embryo, a random decision is made as to which X chromosome—the paternal or maternal—will be genetically active and which inactive, and all descendants of these cells will maintain this distinction, a form of fixed differentiation.

"Presumably, the four percent of additional biosynthetic potential, plus the increased flexibility inherent in the mosaicism of X chromosomal activity I referred to earlier, increase opportunities for gene expression denied to males and are involved in the lower susceptibility to disease and the increased longevity of the human female. The male has only one X chromosome and hence has no homologue for a significant number of genes on the X chromosome. As a result, he will express harmful, recessively-inherited traits such as hemophilia and muscular dystrophy, whereas the heterozygous female — with one mutant and one normal homologue—with rare exceptions, will not. This biological advantage is partly responsible for the disparity in the male-female survival rate. At birth, the male-female sex ratio is 1.06 to 1.0, but a surplus of females is evident by the fourth decade of life, and it increases thereafter."

Young adult males are well aware of the female biological advantage, Dr. Robinson continued, and many of them try to compensate for it through exercise, careful attention to diet, and other practices that offer hope of increasing their longevity.

Of increasing importance to people in the 25-to-40 age

group who are planning on raising families, Dr. Robinson said, has been the relatively recent understanding of the impact of genetic diseases upon society. He cited the following statistics:

- 50 percent of spontaneous abortions involve chromosomal abnormalities.
- 10 percent of stillbirths and neonatal deaths involve chromosomal abnormalities.
- 10 percent of all newborn infants will have a genetically determined disease.
- 30 percent of the patients in children's hospitals have a genetically determined disease.
- 12 percent of the adult patients in hospitals have a genetically determined disease.
- 15 percent of all Americans are victims of birth defects.

"This new information," he continued, "comes at a time when a major change is developing in the orientation of the primary-care physician, who aggressively seeks out those with genetically determined diseases. Care is continuous and not episodic, and preventive medicine has become an important part of family medicine. The early identification of patients with familial maladies is of particular importance to people in the age group we are discussing. In addition to being concerned about their own personal health, couples are also concerned about the health of their present or future children. They are increasingly requesting premarital counseling for Rh status, rubella immunity, and carrier state for a variety of diseases such as Tay-Sachs disease and sickle-cell anemia, and they are increasingly turning to the genetic counselor for information about the likelihood of prospective children being affected by some condition frequent in their family or ethnic group. They want to know the odds against having a defective child. They want to decide for themselves whether to take a chance. They also want to know the alternatives available to them. In short, both prenatal and postnatal genetic counseling have become very important for couples in the 25-to-40 age group. And it's unfortunate that not more than ten percent of those who would benefit by genetic counseling are currently receiving it."

Pointing out that the moral and ethical implications of genetic counseling and the steps that couples take to avoid producing genetically deficient offspring are being debated with growing intensity, Dr. Robinson went on to say: "I would like to turn around the question I often hear, and ask in return, is it moral to permit the preventable birth of a severely defective child?"

But couples who are at a significant risk need no longer refrain from conceiving, he continued. Artificial insemination is now fairly common, and females who are sterile need not necessarily remain so. By way of example, he cited the recent case of the woman in England whose child was conceived *in vitro*. "This," he said, "was a uniquely human endeavor and not, as some people have claimed, an inhuman experiment." He went on to point out that the technique used by the British doctors also has several other potential applications that might be beneficial to women with particular problems. "For example," he said, "a woman who cannot produce ova of her own could borrow an ovum from another female, have it fertilized *in vitro* by her husband's sperm, and implanted in her own womb. Then she would have the wonderful experience of gestating a child and bearing it. The same technique could be used in cases where the wife is the carrier of a severe genetic defect and her husband is not. And finally, of course, there is another application. It may seem rather far out at the moment, but it's entirely within the realm of possibility. I refer to the case of the professional woman who does not want to have her career interrupted by nine months of gestation. She could hire a surrogate mother to gestate her fertilized egg.

There's no reason why the female should be discriminated against in artificial insemination."

Recent advances in the field of medical genetics also make it possible for couples to avoid bearing genetically defective children, Dr. Robinson continued. A new technique now being used in Britain can identify women who are at a genetic risk of having children with neural tube defects, and it is now possible to screen both prospective parents for a genetic defect that predisposes males and females to the early onset of emphysema, particularly if they smoke or work in a polluted environment.

Dr. Robinson went on to observe that young adults in the 25-to-40 age group are becoming more and more knowledgeable about diseases that can be passed on to their children, and are taking steps to avoid such eventualities whenever possible. As examples, he mentioned the caution exercised by people who have a family history of lung cancer or breast cancer, and the growing awareness of prospective mothers about the dangers of smoking and drinking during their pregnancy.

"This leads me to believe that more effective genetic education is one important way of helping people in the 25-to-40 age group," he said. "Much is being done now to bring young adults up to date on recent advances in genetics, but much more remains to be done. Also, we in the medical profession need to learn a lot more about the biological processes underlying the normal development of people of both sexes, as well as the interrelationships of these processes with environmental influences."

Returning to the subject of genetic counseling, Dr. Robinson said it would not be prudent to expect maximum effectiveness. "Let's not forget the matter of patient compliance," he said. "Many patients refuse to act logically on the information they receive. As a result, genetic counselors don't expect 100 percent compliance. Their job is to provide the information the patient needs in order to make an intelligent decision. But the decision patients sometimes make is very different from the one that might be expected."

He ended his remarks with the reminder that education and research are slow processes, and cannot produce instant results. This led him to the observation that the vagaries of federal funding are a constant source of worry to medical researchers, and that interruptions in the flow of funds can be costly. He recommended that Congress and the National Institutes of Health work out a more rational method of allocating funds for long-term medical research and service.

In the discussion that followed, several of the conferees supplemented Dr. Robinson's remarks with additional observations, insights, and ideas.

John Schweppe described the intricate process by which cells receive and pass on genetic instructions from one generation to the next, and ventured the observation that in one sense the human being can be regarded as "a gene machine — a carrier of life."

"This is a very mechanistic point of view," he said, "but it does illustrate the importance of genetic characteristics in determining human development and human behavior. The environment also plays an important role, of course, but the genetic uniqueness of each individual is the basic determinant."

One of the conferees asked if genes are affected by the environment and, if so, to what extent. Donald King said that during a person's lifetime there are a whole host of environmental agents that produce alterations in the gene structure, including chemicals, drugs, viruses, and nutritional deficiencies. Arthur Robinson agreed. "The notion of programming in the genetic process leads to an assumption of immutability," he said, "and that's not true. We may not be able to change the genes, but we can change the environmental factors that affect the genes. That's what preventive medicine is all about. There's a perfect example of this in the treatment of the genetic disease called galactosemia — an enzyme defect which doesn't permit babies to metabolize the sugar galactose which is present in human milk. The treatment for this terrible disease, which causes mental retardation and death, is to diagnose it early and take human milk out of the baby's diet."

Robert Aldrich, Professor of Preventive Medicine and Pediatrics at the University of Colorado Medical Center, said he had been interested for a long time in the biological changes that make a difference at different stages of life and had been struck by the fact that there has been no systematic follow-up of these changes in cohorts of people as they advance from childhood through adolescence to young adulthood. "When a branch of medicine called pediatrics began to be concerned with the special health problems of children," he continued, "it began to develop a considerable body of scientific knowledge about how to keep children healthy. But very little thought was given to the period that came after childhood, which we call adolescence, or the period we are discussing at this conference, young adulthood. In other words, we know much more now than we used to know about how to raise healthy children, but we also know that healthy children do not always become healthy adults. I don't find in the public health statistics any evidence of a dramatic improvement in the health of people in the 25-to-40 age group in recent years. What I'm suggesting is that maybe we need a new branch of medicine that might be called 'mediatrics,' which would be concerned with making follow-up studies of age-group cohorts as they move from childhood through adolescence to young adulthood. If we had people in medicine who began to specialize in these follow-up studies, we would gradually build up a new body of knowledge comparable to what we now have in pediatrics."

Dr. Aldrich also expressed strong support for Dr. Robinson's recommendation for increasing the amount and regularizing the flow of federal funds devoted to the study of the embryo and the fetus. "I've been watching the federal budget for NASA grow to astronomical proportions," he said, "and I have wondered whether those funds — in terms of their impact on human affairs — are as important as funds that could be spent on what I would call a 'matronaut' program — an intensive study of the prenatal period of human life. As Arthur Robinson has pointed out, this is a critical period in human development, and what happens during that period helps to influence what happens during the subsequent stages of human life. We need to know much more about the prenatal period."

Dr. Aldrich concluded his remarks by raising a question about the ethical standards followed by doctors throughout the country in cases involving artificial insemination. He cited the results of a study one of his graduate students had made of the procedure used by physicians in Colorado in screening male donors. "After reading her findings," he continued, "I found myself wondering what kinds of standards we have set up, if any, and whether or not they are adequate if artificial insemination and laboratory fertilization get going on a large scale."

Donald King asked Dr. Robinson whether there was any evidence of genetic susceptibility to certain diseases. Dr. Robinson said there was, and by way of example he cited cleft lip, cleft palate, congenitally dislocated hips, neural tube defects, and arteriosclerosis. Someone objected that arteriosclerosis was not a good example because it is so common, but Dr. Robinson said he did not agree. "I think it's a good example," he said. "If you look at the age of death of twins suffering from arteriosclerosis, the concordance is almost 100 percent."

33 Biological Changes

The gradual loss of T-cells may well be at the heart of the aging process.—Harry Ward

Every phase of the human life cycle is marked by profound biological changes that stem from prior events and lead to subsequent events. Dr. Harry Ward, Dean of the University of Colorado's School of Medicine, described some of the biological changes that take place during young adulthood.

He began with the hematopietic system — bone marrow. At birth, he said, the myeloid cells that make up bone marrow — red cells, white cells, and platelets — are present in all of the bones throughout the body. Then, as a person grows older, bone marrow begins to disappear in the extremities and centralize in the ribs, sternum, spine, and pelvis. By age 60, there is often an absence of myeloid cells in the extremities, but a lot in the central skeleton.

At the same time the bone marrow is moving centrally, he continued, the amount of cellularity of the bone marrow is also changing. At birth, 80 percent of the bone marrow is made up of cells, and 20 percent is fat. By approximately age 20 the proportions have changed to about 60 percent cells and 40 percent fat, and by age 50 to 60 they have shifted again to about 30 percent cells and 70 percent fat.

341

"Despite these major changes in the bone marrow — the factory — we really see no changes in the circulating cells of the blood during the age span from 25 to 40," he said. "This is another illustration of what has been pointed out earlier — that the period from 25 to 40 marks the beginning of various disease processes that will eventually lead to our destruction."

Dr. Ward then went on to discuss unique diseases of the hematopietic system that afflict people in the 25-to-40 age group. For the red cells, the most common is iron-deficiency anemia. In males, it is usually caused by ulcers. In females, it usually results from a significant loss of blood during the menses. Then there are the anemias associated with autoimmune diseases, like rheumatoid arthritis and systemic lupus erythematosus. These affect women in particular. Finally, there are abnormalities of the red blood cells caused by the use of drugs. "Virtually every drug you can think of produces toxic reactions," he said, "and these reactions are fairly common among people in the 25-to-40 age group."

There are no noteworthy diseases that affect the white cells during the age span from 25 to 40, but there are several that affect the platelets — the cells that help the blood to clot. Chief among these is a deficiency in the number of platelets, which is a significant problem among women. This deficiency is often caused by drugs, but it may be associated with autoimmune diseases.

Dr. Ward then turned to the lymphatic system, the system that serves as our surveillance mechanism against invaders. It forms antibodies that protect the body from outside invaders, such as bacteria, and from internal enemies such as cancer cells. In order to simplify his explanation, Dr. Ward showed a slide, divided into two parts, illustrating how the lymphatic system functions. The upper part of the slide showed the thymus, an organ located in the chest. The lymphocytes that make up one part of the lymphatic system pass through the thymus, and there they are converted into T-cells, which are programmed to identify and reject harmful cells, such as cancer cells or transplanted cells that are incompatible. The lower part of the slide illustrated the formation and circulation of the so-called B-cells, which synthesize the antibodies that provide protection against infectious agents.

The two parts of the lymphatic system should not be looked at as totally separate, Dr. Ward explained, because they work together as mutually-supporting parts of the overall lymphatic system.

Dr. Ward then turned to a discussion of the diseases of the lymphatic system that affect people in the 25-to-40 age group.

With age, he explained, the lymphatic system begins to degenerate. As the immune function begins to break down, there is an increasing incidence of various immunological diseases, and particularly the autoimmune diseases, so-called because the body begins to react against itself. These diseases, he said, are reflections of the process of self-destruction which is going on within the human body. The lymphatic system begins to produce various healing substances that attack internal abnormalities, and the healing substances themselves cause a systemic disease.

The presence of cancer cells in young adults may well be an indication that their T-cells are not identifying the cancer cells and rejecting them, Dr. Ward continued. "In fact," he said, "the gradual loss of T-cells may well be at the heart of the aging process."

Another disease of the lymphatic system is infectious mononucleosis. Dr. Ward said it is caused by a virus, and is common among people in the 25-to-40 age group. "It looks like a malignancy, but it's self-limited," he reported. "The patient usually survives, if the physician is smart enough

not to rupture the patient's spleen." He also called attention
to an increase in the incidence of lymphoma, or Hodgkin's
disease, among young adults. "This is really a manifestation
of the breakdown of the lymphatic system, which begins at
about age 20. The major incidence of Hodgkin's disease
occurs at about age 30 to 35."

During the discussion that followed, Dr. Pellegrino emphasized the importance of early diagnosis in dealing with
the diseases Dr. Ward had mentioned. "There are obviously
some things we can't do anything about," he said, "but
there are a lot of others where we can be helpful. With
respect to the coronary diseases, for example, we know that
there are seven or eight risk factors to look out for, and
from them we can construct a probability index that can be
very useful."

He went on to say that one of the major problems in
medicine was how to make prudent decisions in the face of
uncertainty. "That's the situation we face most of the time,"
he continued. "We deal with probabilities rather than certitudes, and in some cases if all we can do is reduce the
probabilities of a particular disease, we ought to be content
with that. I think we've got to be careful that preventive
medicine doesn't become a religion, a panacea. A sensible,
prudent kind of preventive medicine is the best we can hope
for."

Someone else stressed the importance of helping to
strengthen the patient's ability to cope with disease. "We
may not be able to predict in advance whether a particular
person will have a given disease," he said, "but if we detect
a susceptibility we can strengthen the person's coping capacity and thereby reduce the amount of disability. We
know, for example, that angry people are highly susceptible
to rheumatoid arthritis. If we can get them to understand
that, and do something about it, we can get a fairly significant reduction of not only the symptoms but even of the
disease level."

Dr. Schweppe asked if there was anything that could be done to make the well person in the 25-to-40 age group "weller," so that the mortality curve could be pushed further to the right.

"I wish I knew," Dr. Cahill replied. "In biology we've just about peaked out in terms of what we can do for the physical well being of the individual, and I honestly think that from here on in the major emphasis has got to be on helping people to cope with disease."

Thomas Edmunds suggested that the medical profession develop special courses in biology that people in the 25-to-40 age group could take in order to get a better understanding of the body systems, how they function, and what can be done to keep them functioning effectively. "I think there's an eagerness to know more about the body," he said. "Women have shown a particular interest in their bodies in recent years, and I think courses in biology set up within a community structure would be very helpful."

Susan Klein said she thought that health education should begin much earlier than among people in the 25-to-40 age group, and that it should focus on developing a sense of responsibility about one's own health. "That's where the action's at," she remarked.

Several of the conferees provided fascinating examples of the gradual but inexorable process by which human beings begin to die from the day they are born.

As Dr. King put it, "We have within us the seeds of our own destruction, and these seeds mature throughout our lifetime. The rate of deterioration is not the same for all individuals, or for all the biological systems within us, but it continues like the ticking of a clock."

"Every human being is born with a fantastic reserve of tissues and cells, and this reserve is gradually depleted with age. Some cells, notably those that make up muscles and the central nervous system, have a limited replicative potential. Others, like the liver and kidney cells, are constantly replenishing themselves. But all of them have limited lives, and the older a person gets the weaker the regenerative process becomes. For example, if you take a random cell — and one of the easiest random cells is one of the connective tissue cells from a person's skin — and put it into a tissue culture, it will do very well over a period of time. But what's fascinating about this process is that the tissue cells of a 20-year-old will do much better than those of a 70-year-old. There seems to be a built-in time clock that governs the rate of self-destruction."

Dr. Cahill pointed out that there is a growing body of evidence suggesting that a limited reproductive capacity can be inherited. "So," he continued, "it looks as though we are working against a kind of genetic destiny, and the question is how can we modify the environment in ways that will change the time clock? It's going to take us a long time to change the sperm and the egg that carry the genetic instructions. I honestly think that none of us here, nor our children, will live beyond the ten decades that currently seem to be the upper limit of the lifespan."

Dr. Pellegrino pointed out that the same loss of cellular reserve that Dr. King mentioned also takes place at larger levels of physiological functioning, such as the kidneys, the lungs, the heart, and the brain, and that the diminution process in these areas sometimes begins before age 40.

Dr. Cahill said that four or five of our systems are falling apart simultaneously, and if a person is lucky enough to be on the normal distribution curve for the deterioration of all the systems he or she is likely to live to be a hundred. The systems he mentioned were the brain, circulatory system, skeletal system, and the kidney system. "If they are genetically programmed to last 100 years, you'll live that long," he said. "I think that's the explanation for why some people live to be a hundred and others don't."

Someone asked how many people were that lucky, and

Merrell Clark said there are about 15,000 people in the country today who are over a hundred.

Someone else asked if there was anything young adults in the 25-to-40 age group could do to prolong their lives. Dr. King replied that sustained, moderate exercise can be very helpful, so long as it's not too strenuous.

Dr. Cahill remarked that moderate exercise is especially important for women in the 25-to-40 age group. "Caucasian females seem to be particularly susceptible to osteoporosis," he said. "Their bones fizzle out. We know that physical exercise can help with that, but too much exercise is bad for the joints, so we have to urge moderation."

Dr. Pellegrino agreed. "For the post-menopausal woman," he said, "regular, gentle exercise is an excellent safeguard against osteoporosis. It's more important than the estrogens or vitamins, because it strengthens the muscles that cover the bone structure. But we do have to recommend caution. Too many Americans go off the deep end when it comes to exercise. They put too much stress and strain on an aging set of joints."

Dr. Robinson agreed. "There are a lot of middle-aged-women — and men — who are jogging their joints off," he said.

34 Health Problems

By the time we reach young adulthood, the lesions of the diseases that will prove catastrophic later on are already present.—Donald W. King

Donald W. King, Delafield Professor of Pathology and Chairman of the Department of Pathology, College of Physicians and Surgeons at Columbia University, and George F. Cahill, Jr., Professor of Medicine at the Harvard University Medical School, led the discussion of health problems common among people in the 25-to-40 age group.

Dr. King began by saying that although young adults are relatively healthy, many of them are beginning to develop symptoms of certain diseases that will affect them with full force at later stages of their lives. He called these "the silent diseases," and among them he cited the following:

- Rheumatic fever
- Atherosclerosis
- Emphysema
- Lung cancer
- Cirrhosis of the liver
- Kidney diseases
- Arthritis

"These are called silent diseases," he explained, "because, although they have not begun to have their full effect, all the early morphologic lesions are there."

By way of illustration, he showed a series of slides depicting the early stages of rheumatic fever, atherosclerosis, emphysema, cirrhosis of the liver, polycystic kidney, and osteoarthritis in young adults.

In addition to the "silent" diseases, Dr. King continued, young adults in the 25-to-40 age group also suffer from a number of other diseases, some of which are relatively rare and others relatively common. Some of the rare diseases are of genetic origin, including cystic fibrosis, sickle cell anemia, diabetes, and Huntington's chorea; in others, including multiple sclerosis and rheumatoid arthritis, the etiology is completely unknown. The relatively common diseases are often associated with stress. This is especially true of hypertension, the abuse of drugs and alcohol, mental depression, ulcers, and obesity. Injuries caused by home and automobile accidents are also fairly frequent among young adults, and some women in the 25-to-40 age group suffer from pregnancy complications. Finally, of course, large numbers of young adults suffer from the common cold and other infections, including venereal disease.

Dr. King said that the indiscriminate use of drugs is a serious health hazard to people in the 25-to-40 age group. He estimated that there are anywhere between 15,000 and 30,000 different kinds of drugs on the market, and went on to say that many people are unaware of their effects on the human system.

What can be done about the diseases that afflict young adults?

Early diagnosis and improved methods of treatment can be helpful in many cases, Dr. King said, but an intensive education program designed to acquaint the public with early symptoms and late consequences, as well as preventive measures, is probably the best long-run answer. "However," he continued, "education is not a panacea. The problem of compliance must be taken into account. Public knowledge about the dangers of smoking is now fairly wide-spread, thanks to the intensive education campaign that has been carried on in recent years, but there are still millions of people who haven't stopped smoking." He also cited an early campaign to wipe out carcinoma of the cervix in a large Southern city. Although the women of that city were assured that this form of cancer could be completely cured, and that all they had to do was to show up for pap tests every six months or a year, the turnout for the tests dropped steadily from a high of 80 percent of the white women and 60 percent of the black women during the first year to less than 60 percent of the white women and 40 percent of the black women during the third year.

"We have a well-educated, highly sophisticated popula-tion," he remarked, "but education hasn't worked as well as we hoped. Maybe we need a new type of education."

Dr. Cahill said that although disease among young adults is minimal, the 25-to-40 age span is the time when the pat-terning is set for the degenerative diseases that become cat-astrophic later on.

Despite the dramatic advances that have been made in the field of medical science in recent years, he said, not much has been done to improve the well-being of the people in the 25-to-40 age group. "All we are doing," he said, "is permitting a larger proportion of the population to achieve longevity."

By way of illustration, Dr. Cahill drew on the blackboard a chart comparing mortality rates of a few centuries ago to those of today. He then showed a series of slides illustrating mortality rates among different age groups.

Dr. Cahill went on to say that all the longevity and mor-tality data indicate that human beings are "programmed" to live not much more than 100 years. Recent articles in pop-ular magazines about people in some parts of the world who claim to be well over a hundred must be taken with a grain of salt, he continued, because the birth records in those countries are not very accurate. He conceded, however, that stories concerning former slaves in Southern communities living to be 110 or 115 years of age are probably true. "Black people, once they get beyond a certain critical point of life do live longer than most white people," he said. "They do tend to die younger, but those who make it to old age tend to live longer. It's a very interesting phenomenon."

Turning specifically to the diseases most common among people in the 25-to-40 age group, Dr. Cahill mentioned the following:

- Respiratory
- Accidents and violence
- Joints and muscles
- Digestive
- Genitourinary
- Circulatory system
- Infection
- Nervous system
- Mental
- Skin
- Neoplasm
- Endo-Metabolism
- Blood
- Childbirth complications

Dr. Cahill said that respiratory infections, including the common cold, were probably the most prevalent. "They cause more personal discomfort, and result in more time lost from work, than any of the other diseases that affect the people we are discussing," he said. "And the amazing thing is that in the last hundred years we haven't advanced one iota in conquering the common cold. It's the same problem it was a hundred years ago, and probably ten thousand years ago. Every time we develop a new vaccine, we're vaccinating people against last year's virus. We just don't know which type of virus is going to strike next."

Dr. Cahill recommended two simple preventive steps for people in the 25-to-40 age group — pap smear tests for females, and blood-pressure checkups for males and females. "With respect to obesity," he said, "education becomes extremely important. It is a disease of the over-nourished. So is maturity-onset diabetes. I think 80 to 90 percent of the people who are diagnosed as diabetic would return to normal sugar metabolism if they lost twenty or thirty pounds. So education is important here, too. But we must be careful not to overshoot the mark, or we might turn the American people into a bunch of hysterical hypochondriacs."

Dr. Cahill concluded his remarks with the warning that medical science may be entering a phase where the technology is outracing practicality. "The technology is becoming extremely expensive," he said, "and it may not always produce benefits that justify the cost." As an example, he cited expensive new methods of identifying genetic "blueprints" that predispose people to certain diseases like rheumatoid arthritis and multiple sclerosis, about which physicians can do very little at the present time.

William M. Marine, Chairman of the Department of Preventive Medicine at the University of Colorado Medical Center, focused on what he called "strategies of intervention" for improving the health of people in the 25-to-40 age group.

He reported that Marc Lalonde had developed a promising new statistical technique for determining which strategies of intervention might be most effective. In a recent health study of Canadians, Lalonde decided that instead of looking at mortality rates by disease categories (see accompanying table) he would classify them in terms of the number of years of life lost from particular diseases.

Ten Leading Causes of Death
Among People 25 to 44
(1976)

	Rate per 100,000 People
Accidents	40.8
Malignant neoplasms (cancer)	30.0
Heart Diseases	26.2
Suicide	16.1
Homicide	15.6
Cirrhosis of the Liver	9.2
Cerebro-vascular diseases (stroke)	6.8
Influenza and pneumonia	3.7
Diabetes mellitus	2.7
Congenital anomalies (birth defects)	1.3

Source: Mortality Statistics Branch, Division of Vital Statistics, National Center for Health Statistics, DHEW, Hyattsville, Md.

"In other words," Dr. Marine explained, if a person dies of cancer at the age of 65 and the normal life expectancy of that person was 72, then seven years of life were lost because of that disease. By looking at the mortality statistics in those terms, Lalonde found that the four leading causes of death in Canada — and in the United States — were, in the order of importance, auto accidents, all other accidents, heart disease, and suicide."

Dr. Marine went on to say that this technique might be a helpful way of looking at the principal causes of death among people in the 25-to-40 age group because the toll they take in terms of the number of years of life lost is very high.

"Therefore," he continued, "I agree with George Cahill that education is the most promising strategy of intervention — even in the case of heart disease. We can't use the fact that education doesn't always work as a cop-out. I think a lot of effort has got to be put into finding out what it does take to change people's minds and attitudes about their health and about the steps they can take to improve it. But I also think we have to respect the rights of people who choose not to take our advice as physicians."

Dr. Marine went on to suggest that people other than physicians, and especially people with highly developed communication skills, ought to undertake the educational program. "In fact," he said, "we really need to have two kinds of health care systems in this country — one to take care of sickness and the other to educate the public." Harking back to what had been said earlier about the importance of annual physical examinations, he also recommended that serious thought be given to the idea of a lifetime health-monitoring program, advanced a couple of years ago by Lester Breslow and Anne Somers in the *New England Journal of Medicine*.

Dr. Marine concluded his remarks by saying that although he agreed with Dr. Cahill's concern that medical technology may be outrunning practicality, he nevertheless felt that the intervention approach should be pursued. "I don't think preventive medicine should be sold as a cost-effective measure," he said, "but rather on the basis that it can contribute to improving the quality of life."

Dr. Aldrich remarked that in considering the diseases that affect people in the 25-to-40 age group it was important not to lose sight of the fact that parents often contract infectious diseases from their children. "I've seen my share of mumps in mothers and fathers," he said, "and I've also seen several cases of measles and whooping cough." Even pets can be a source of infectious disease, he pointed out. "I have seen

several cases of psittacosis, a viral disease that affects parakeets and other members of the parrot family. It's highly
contagious, and it's pretty serious when humans get it."

Dr. Aldrich went on to say that parents in the 25-to-40
age group also suffer from severe emotional stress when
they have mentally retarded or physically handicapped children, when a teen-age daughter becomes pregnant, when a
teen-age son gets a girl pregnant, or when one of their children indulges in delinquent behavior — particularly when
the behavior is associated with the heavy use of drugs or
alcohol. "The degree of disruption and stress that these
things create within a family is usually quite severe," he
said, "and a doctor might find himself spending a lot of time
with the parents, trying to help them through their ordeal."

Dr. Pellegrino focused on the socio-cultural factors that
contribute to disease. Pointing out that coronary disease is
on the rise among people in the 25-to-40 age group, he said
that smoking and poor diet might well be associated with
the increase. He also called attention to the fact that the
early symptoms of coronary disease are beginning to be
found with greater frequency among women in that age
group, who were once thought to be protected in some way
by their sex, and he speculated that this phenomenon might
well be related to the wider use of contraceptive pills.

Longitudinal studies of ethnic groups in the United States
strongly suggest that socio-cultural factors play an important role in the incidence of coronary disease, Dr. Pellegrino
said. One of the studies he cited dealt with two different
groups of immigrants who settled in a small town in Pennsylvania. One group, which maintained the customs and
lifestyle it had brought with it, continued to show a relatively low incidence of coronary disease while the other,
which had quickly become Americanized, experienced a
significant rise in coronaries.

Dr. Pellegrino ended his remarks by recommending that

doctors periodically engage in what he called a "lifestyle review" with their patients, on a one-to-one basis. "Somewhere between 25 and 40," he said, "most people become acutely aware of their own mortality, and this realization, painful though it may be, does prompt some people to start taking better care of themselves. They start getting annual physical examinations, and that's a good thing. The examinations permit early diagnosis of what Donald King called the silent diseases, and the results of the examinations are often important in the kind of preventive education we've been talking about. But even more important than the physical exam is the opportunity to evaluate the patient's lifestyle and then recommend preventive measures designed for the individual."

Dr. Pellegrino said his own practice has been to use the periodic examination to review the patient's habits with respect to such things as smoking, diet, exercise, use of alcohol and medications as coping mechanisms, family problems, job satisfaction, marital situation—all of the factors which can affect health.

"The aim," he continued, "is to detect potentially damaging personal habits and attitudes, and to write an individualized 'prescription' for prevention for *this* patient. This prescription takes cognizance of lifestyle aberrations, as well as physical risk assessment in *this* person—those factors we know affect the probability of illness—blood pressure, blood lipids, family history, body type, type personality and the like. The resulting prescription could be designed to lessen the risk factors, and emphasize prevention by appropriate changes in lifestyle. The whole thrust is in the direction of hygienic living in the original sense of that word — as used by the Greeks to mean a health-promoting way of life with due attention to body, mind and spirit."

Dr. Pellegrino emphasized that this individualization of

the preventive prescription was far more effective and meaningful than educational programs in schools or in the media aimed at the general rules of good health. "The prescription makes the recommendations concrete, special, and personal," he said. "It is *your* health and *your* prescription. Disciplining one's life because of some general probability of illness in the entire population is far less effective than doing so because you know the probability of risk for you. Periodic examinations serve to update, reenforce, and clarify the prescription and assist the patient's motivation —or at least compel him or her to make conscious choices."

Calling attention to the rapid spread of self-help and self-improvement groups throughout the country, Dr. Visotsky said these might well represent a new trend in preventive medicine. "People seem to be more and more interested in improving themselves as human beings, and we ought to take advantage of that. This is a good time for us to dip our oar in and give them some direction. If they want it they'll take it, and as long as we don't insist that they must take it we may be able to make an impact."

Arthur Robinson said physicians need to respect the individual's ability to cope with medical problems by giving patients the facts and then discussing what they can do to help themselves. "I agree strongly about the importance of the annual physical," he continued. "It's a perfect time for the internist to be involved in family medicine and genetics. Many internists never take the time to sit down with their patients and talk about the things Dr. Pellegrino was talking about."

Dr. Robinson also said that Dr. Pellegrino's remarks underscored the importance of recognizing the genetic uniqueness of the individual. "When we sit down with a patient we shouldn't talk about the diseases that afflict young adults as members of an age group. We should talk about the patient's own genetic makeup and what his problems are likely to be in the future."

Dr. Robinson went on to mention the growing incidence of alcoholism among women as a problem physicians should be concerned about. "If the alcoholic woman happens to be pregnant, the risk to the unborn child is very high. The figures are not precise, but they show that in somewhere between 10 and 30 percent of such cases, the result is some form of intellectual disability. That's also a matter that can be talked about at the annual physical."

Bernard Frieden said it was very important to try to distinguish between those educational approaches to preventive medicine that pay off and those that don't. He went on to point out that policy analysts in areas other than medicine have made some interesting discoveries about educational approaches. In the field of highway safety, he said, studies that compared the results of driver education programs with the results of changes in automobile design concluded that the payoff in terms of injuries avoided or lives saved was many times greater from the approach of changing the automobile than from the approach of trying to educate the driver.

Mr. Frieden then cited an example of an education program that backfired. Several years ago, he said, New York City, in an effort to reduce the incidence of drug abuse among high school students, hired some ex-addicts to visit the high schools and discuss their own experience with drugs as an example of the consequences of addiction. "The results," Mr. Frieden said, "were directly opposite to what had been expected. The kids in the ghetto schools were thrilled by the stories of life in the streets, and later on, when asked in an evaluation questionnaire what they would like to be when they grew up, a significant number said 'ex-addicts.' They were under the impression that this was a new kind of career position in the city government."

Mr. Frieden hastened to add that examples of the kind he had cited should not be taken as evidence that education would not be effective in the field of preventive medicine, but rather as illustrations of the importance of determining the most effective kind of educational approach.

In a jocular vein, he went on to say that the campaign to educate the public about the possible dangers of certain food additives has had unexpected results. "Who would have guessed ten years ago," he asked with a smile, "that one of the hottest middle-class political issues today would turn out to be 'What's in our hot dogs?' "

John Demos said he was not pessimistic about the potential of education in the field of preventive medicine. By way of example, he said that although the figures on the number of young people starting to smoke were alarming, there were also a lot of people who would have become smokers if it had not been for the education program conducted by the Surgeon General, the American Cancer Society, and other groups. "I'd be one myself," he confessed. He went on to express the hope that in many cases teenage smoking may be merely a transitory manifestation of adolescence that will be outgrown when the young smokers mature.

He also said he liked Dr. Pellegrino's suggestion of a one-to-one lifestyle review. "I think it's enormously constructive," he said. "Speaking from my own experience as a patient, I think it's rather disappointing when the doctor is at a loss as to what to talk about in the annual physical examination. Maybe the doctor is simply uncomfortable about that kind of discussion because he hasn't been prepared in how to go about it. Maybe the medical schools need to pay more attention to this aspect of preparation."

Dr. Schweppe said he would like to add back problems to the list of diseases common among people in the 25-to-40 age group. "They seem to be on the rise, at least among the people I see," he said.

He went on to suggest that the early identification and treatment of young adults who have a family history of heart disease, diabetes, arthritis, emphysema and other diseases of genetic origin may well be a way of lowering the cost of their medical care in later years, when these diseases have their greatest impact.

Mr. Eurich said he wanted to underscore the importance of what had been said about treating young adults as genetically unique individuals rather than as members of a specific age group. "Let's not forget," he said, "that the age group we are discussing is made up of individuals. I have frequently thought that one of the major weaknesses in our educational system is that we classify students in terms of grade levels, rather than treating them as individuals with varying interests, capabilities, and rates of development."

35 Emerging Trends in Work Force Participation

It is during the years 25 to 40 that men and women place themselves on a trajectory that will determine their career paths, but we don't know what influences their choices. What we do know is that there is considerable labor market mobility among this cohort.—Patrick J. O'Keefe

Men and women in the 25-to-40 age group form the backbone of the U.S. labor force. Nearly 77 percent of them are either employed or actively seeking employment, and taken together they represent nearly 40 percent of the total labor force, a larger segment than that for any other comparable age group.

According to Patrick J. O'Keefe, Deputy Director of the National Commission for Manpower Policy, nearly 98 percent of the men and 60 percent of the women in the age group participate in the labor force, and many of them hold down two jobs. The figure for women is particularly striking, he said, because it reflects the fact that the rate of participation in the labor force by adult women has just

363

about doubled since 1950. He also noted that the rate of participation by black and other minority women is higher than that for white women — 63.7 percent as against 58.9 percent—and that minority males are withdrawing from the labor force faster than white males. All of these trends, he said, have important implications for family life in the United States.

Marriage has a significant influence on the rate of labor force participation. It tends to depress the rate for females, and raise the rate for males. Marriage also tends to lower the unemployment rates for both sexes. "Maybe one of the solutions to the high rates of unemployment among young people would be to require everyone to marry at the age of 16," he observed with a smile.

Mr. O'Keefe also pointed out that people in the 25-to-40 age group have a relatively high rate of educational attainment — an average of 13 years of schooling — which indicates that they are either holding or seeking jobs that pay fairly well. "These people want to be involved in the economic mainstream," he remarked. "The question is whether we can continue to generate the kinds of jobs that will fully utilize their skills and competencies. Society has made a substantial investment in their education, and they have, too."

To illustrate the importance of educational attainment to people in the job market, Mr. O'Keefe called attention to a chart showing 1976 unemployment rates for persons 25 years and over by the number of years of schooling completed. The chart showed that for people with 11 years of schooling or less, the unemployment rate was about 8 percent, but for high school graduates it fell to 6 percent and declined further for people who had completed one or more years of college.

From a labor economist's point of view, Mr. O'Keefe said, the people in the 25-to-40 age group are quite successful, because they rank very high in the kinds of jobs they perform and in the income they earn. They are also industrious. About 6.2 percent of them hold more than one job, as against 5.3 percent for the labor force as a whole. In most cases, the extra income they earn is spent on housing and consumer goods. They frequently get into debt, but they pay it off promptly. The males have a higher dual-job rate than the females, but the females are catching up rapidly. One reason is that there seem to be more part-time jobs available to women than to men, and many women take two part-time jobs in order to supplement the family's income.

Unemployment among people in the 25-to-40 age group is relatively low, Mr. O'Keefe reported. It is considerably lower than for younger members of the labor force, but somewhat higher than for older members. The principal reason for the latter phenomenon, he explained, is that after age 45 there is a relatively high rate of withdrawal from the labor force in response to job loss; hence fewer older people are counted among the unemployed.

As might be expected, the unemployment rate for women in the 25-to-40 age group is higher than that for men, because a high proportion of the women are re-entrants to the labor market after a period of raising children, and tend to experience an initial period of unemployment before they find a job that suits them. Nevertheless, many people in this age group who experience unemployment do so for a very short period of time — a week or two. "Much of the job-changing that goes on in this cohort happens without an intervening period of unemployment," Mr. O'Keefe said. "These people are occupationally mobile, and they move from one job to another without ever having to go out and test the job market by being totally unemployed."

Much of the unemployment in the 25-to-40 age group occurs among minority people, Mr. O'Keefe reported. "Very often they are located in poor labor markets, such as rural areas or the inner core of large cities. Also, they tend to be less well-educated than their white counterparts. This is a reflection of years of discrimination. The tragic part of it is that they are being scarred just at the time when they ought to be making the most headway. And once we have written them off, in terms of employment, we've written them off for however long they're going to live."

Turning to the rapid increase in female participation in the labor force since 1950 (see chart on Page 367), Mr. O'Keefe said it was not an aberration. "As a matter of fact, it has been much slower than might have been expected. Some of the Scandinavian countries, and particularly Sweden, have found that about 60 to 65 percent of the entire female population participates in the labor force, and for the age cohort 20 to 40 it is approaching 75 percent."

There is no reason to believe that female participation in the labor force will ever get as high as male participation, Mr. O'Keefe said, but there is some reason to believe that the male participation will continue to decline, and if there are institutional shifts to accommodate the entry of more females into the labor force, the gap in the current male-female ratios could narrow significantly.

Ironically enough, the members of today's 25-to-40 age group are unwittingly stoking the fires of inflation. "They are the most desirable group of workers," Mr. O'Keefe said. "Their labor-force participation is high, and so is their utilization rate. If employers continue to prefer them to younger and older members of the work force they will bid up their wage rates to the point where we will have labor-market-induced inflation."

Looking ahead fifteen years, Mr. O'Keefe said the total labor force will begin to decline as a result of the declining birth rate, that it will contain a higher ratio of minority members to reflect changes in the racial composition of the "baby bust" generation, and that the educational attainment level of its members will increase by nearly a year.

Percent of
population in
labor force

**Civilian Labor Force
Participation Rates, 1957 to 1977**

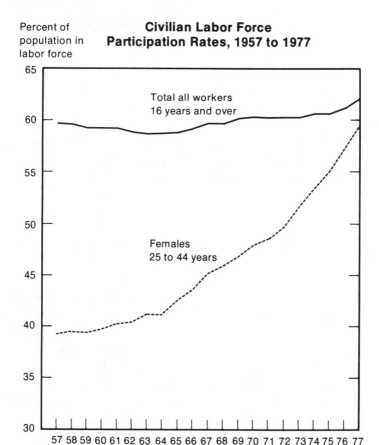

Source: *Employment and Training Reports of the President, 1978,* U.S.
Department of Labor, Washington, D.C.

"What we may find," he said, "is highly-skilled and well-educated labor pushing down into lower-level jobs. If that happens, we will be wasting the very large investment we are now making in their education, and their output will fall short of its potential."

Conceding that it is impossible to accurately forecast future trends in the U.S. labor force, Mr. O'Keefe went on to say that the following additional developments seem highly probable:

- Early in the 1980s, the cohort between 25 and 40 will become the largest single group in the labor force, accounting for more than 45 percent of the total.
- Women will continue to participate in the labor force, at an accelerated rate.
- The 25-to-40 age cohort will be increasingly frustrated by the growing gap between their expectations and the actual availability of highly skilled jobs.

"All of these developments will have significant social and economic consequences," he said. "The potential for job frustration and dissatisfaction will be enormous."

In the discussion that followed, Mr. O'Keefe was asked a number of questions about the economic attitudes and prospects of people in the 25-to-40 age group.

One of the conferees asked about their attitude toward early retirement. "I think they have pretty well set their sights on getting out early," Mr. O'Keefe replied. "Unlike older members of the labor force who were hard hit by the Depression, they've never known what it was like to be chronically out of work, and most of them are covered by pension plans as well as by Social Security, so they're not inclined to worry about the future."

Someone else asked if anything could be done to alleviate the job frustration and dissatisfaction that stems from the gap between workers' expectations and the limited number of highly skilled jobs currently available to them. Mr. O'Keefe mentioned work sharing, sabbatical leaves, a further shortening of the work week, and raising the rates for overtime work as possible remedies, but he warned that all of these alternatives are economically expensive. "Our best bet," he said, "is to keep on creating new kinds of jobs that require higher and higher levels of skill." He pointed out that many of the jobs that are being performed today did not exist ten years ago, and he cited forecasts by the U.S. Bureau of Labor Statistics that between 20 and 30 percent of today's jobs will disappear in the next ten years because of technological advances, capital shifts, and other economic developments. "This points to the need for extensive retraining programs that will help workers to upgrade their skills," he continued.

Susan Klein asked what might be done to make it easier for women to participate in the labor force. Mr. O'Keefe prefaced his answer by pointing out that of the 30 million new jobs created by the economy since 1930, a high proportion have been filled by women. "That has been a very fortunate development," he said, "because there just weren't enough men around with the level of education necessary to fill the kinds of jobs that were being created. Therefore, I think it's important to facilitate the entry of women into the labor force. Better job-training would be one way to go about it, and a better institutional response to the special needs of working women would be another." He then went on to say that if it were possible to find acceptable ways of reducing the number of men holding two or more jobs that, too, would open up more employment opportunities for women. He cautioned, however, that the average income of working wives does not add significantly to the family's total income. "Most working wives have been forced to accept a secondary income status," he said. "They're viewed as being in the labor force on a part-time basis, and they generally tend to have part-time jobs. As a result, a working wife adds only about $2,000 a year to the income of the family. The median income of families where only the husband works is about $14,000. When the wife also works it's only about $16,000."

These figures drew a whistle of surprise from several of the conferees. Mr. O'Keefe went on to say that the wage differential between men and women may be getting worse, because of the fact that women are tending to take the more marginal jobs available in the labor market, and that these jobs are generally the least remunerative.

The discussion then turned to the affirmative action laws that require equal pay for equal work. In an expanding economy, Mr. O'Keefe said, the effect of these laws on employment and income distribution is not noticed because the pie is getting larger, but if economic growth begins to fall off the redistribution of employment and income will be largely at the expense of white males. "That kind of situation could create political dynamite," he remarked.

When asked about the future employment outlook for people in the 25-to-40 age group, Mr. O'Keefe said he was optimistic. "Our economy is continuing to generate new jobs," he said, "and although productivity has declined in recent years because of the diversion of capital into pollution controls and the outflow of dollars to other countries, our labor costs in the international arena are lower than those of many other Western nations."

John Demos, citing the dehumanizing effect of monotonous, repetitive, assembly-line work, asked Mr. O'Keefe if manufacturing corporations were taking any steps to improve the quality of life among their employees. Mr. O'Keefe replied that many corporations have undertaken experiments aimed at improving job satisfaction, but not all of these programs have had the desired results. He also said that organized labor tends to be suspicious of such experiments, remembering the days when efficiency experts were hired to speed up production.

Alvin Eurich asked Mr. O'Keefe if he could identify some of the most critical manpower problems that lie ahead, and suggest ways of dealing with them.

"One of the most scary ones," Mr. O'Keefe replied, "has to do with the underutilization of skills. I've seen forecasts which indicate that 25 percent of our college graduates are

not going into white-collar jobs that demand a college education. Many of them will wind up in blue-collar jobs, or as
clerks in liquor stores. That would be unfortunate on two
counts. First, they themselves would be underutilized in
terms of their educational preparation, and second, by taking lower-skilled jobs they would be bumping people who
would be optimally utilized in those occupations. France
had that experience in the late 1960s, and has taken steps to
prevent it from happening again, through subsidies of various kinds and changes in the tax laws. I think continued
economic growth is the best answer."

John Schweppe asked whether better job counseling
might help. Mr. O'Keefe said it undoubtedly would, but
there were two difficulties with this approach. "Our job-
aptitude tests are not very good predictors of actual performance," he said, "and too many young people think in
terms of job titles without knowing anything about what
kind of work the job involves."

Harking back to the subject of job satisfaction, Harold
Visotsky said he was under the impression that surveys
showed a relatively high degree of job satisfaction among
American workers. Mr. O'Keefe replied that this was true,
but that a lot of the surveys were scientifically soft. "Some
ill-informed social scientist goes into a factory and asks the
workers if they like their jobs, without ever mentioning alternative occupations, and the response is usually yes.
That's what gets reported."

Bernard Frieden remarked that the high rates of labor-
force participation among men and women in the 25-to-40
age group seemed to indicate that job satisfaction was not a
very great problem. "In fact," he said, "when we look back
at some of the dire predictions that were being made ten
years ago about the boys and girls who now make up the
younger cohort of the 25-to-40 age group, we ought to be
grateful that those predictions didn't come true. If a confer-

ence like this had been held in 1968, it would have concerned itself with the disappearance of the work ethic, withdrawal of youth from American society, rejection of the notion of success, and the trend toward communal living and menial work. But a funny thing happened to these young people in the last ten years. They are now participating in the labor force in record numbers, and many of them are working at two jobs. That would suggest that whatever job dissatisfaction there may be among these young people, they have found ways of making compromises."

Citing the results of a recent study by Richard Coleman, based on a national sample of families who were interviewed from 1968 to the present, Mr. Frieden provided some interesting insights into why women enter the labor force, what happens to the money they earn, and how the extra money affects the family's lifestyle.

"The most striking finding," he said, "is that if you take two families of equal income — for example, $15,000, which is the median for the United States — and one family derives that income from the husband's earnings while the other derives it from a combination of the husband's and wife's earnings, those two families are likely to live very differently.

"The family that derives all of its income from the husband's earnings is likely to have a higher standard of living than the family whose income is a mix of the husband's and wife's earnings.

"In other words, the husband's income is the most stable part of the household income, and it tends to be the baseline that establishes the family's general standard of living. This is true of many different kinds of expenditures, and it's true even when the wife contributes a significant amount to what the family earns.

"One reason why this seems to be so is that families make a lot of their consumption decisions on the basis of what they regard as permanent income — not necessarily what they are earning today but what they expect to be earning over a lifetime. And even across a broad spectrum of families — not merely the well educated — permanent income means the husband's earnings. That's the base.

"With respect to what the wife's income does for a family, the study showed that in general the family does not move to a higher standard of living on the basis of the wife's income, but it does tend to live better within its own definition of family lifestyle. For example, if both husband and wife work at lower-level white-collar clerical jobs, but their combined income puts them into a higher income bracket, they still live like other people with clerical jobs. They may have the income of an upper middle-income family, but they don't live that way.

"Likewise, there are cases where husbands and wives both work at blue-collar jobs, but their combined income puts them into the top fourth of the country in terms of total earnings, yet what they tend to do is live better in the blue-collar style. They buy season football tickets, or they fix up a game-room in the basement rather than moving to a house in an upper middle-income neighborhood.

"As to what seems to bring women into the labor force, the motives identified in the study were very diverse and partly based on class. Upper middle-income women, highly educated and career oriented, tend to work full time. For them, career considerations are dominant. They enjoy what they do, they get a sense of self-satisfaction and self-realization from their work. The money is also important, because it enhances their feeling of self-worth. Among middle-class women, the motives seem to be different. Some are working mainly for the extra income, others because they feel it is a sign of higher status to be working rather than being a housewife. A few of the women in the study wanted to get out of the house. They didn't like housework, and thought working outside was more interesting than doing housework. But for many of the women, part of the motivation seems to be a desire to have a greater say in how the household spends its money.

"The study also found that the kind of work women chose said something about their special motivation for getting into the labor force. Some worked in school cafeterias or drove school buses because they wanted to be available when their children got out of school. Others, exhibiting a strong interest in improving their personal appearance, chose jobs in dress shops and jewelry stores.

"What the family wanted to do with the extra money also was class-related. For upper-middle-income families, the extra income often went for things like foreign travel, collecting art or books, sending their children to private school, or helping to pay for their children's college education. For lower-income families, the money was likely to go for home improvements, a place in the country, or weekend sports."

William Oriol said that one ironic aspect of the increasing participation of women in the labor force is that it is happening at a time when the need for non-institutional care of elderly people — the kind of informal family care usually provided by women — is on the increase. "We are just beginning to realize how important that kind of care has been, and how much it will cost to replace it," he said.

36 Housing Problems

It is no exaggeration to say that the women's liberation movement has been the salvation of the American housing industry. It now takes two incomes instead of one to finance home ownership. But if the cost keeps rising, young people may turn to polygamy!—Bernard J. Frieden

Owning a home has always been part of the American dream, and according to Bernard J. Frieden, Professor of Urban Studies and Planning at the Massachusetts Institute of Technology, people in the 25-to-40 age group have been achieving that goal in record numbers since the end of World War II.

Public opinion polls show consistently that the great majority of people want to own their own homes, and that of those now renting, a substantial proportion want to move in order to become homeowners.

Home ownership, according to Mr. Frieden, is even more popular during the child-raising period. Among husband-wife households with husbands aged 30 or over, 82 percent were homeowners by the mid-1970s.

Among people in the 25-to-40 age group, a suburban single-family home is the preferred setting for raising children. Life in suburbia or in small towns is vastly preferred to life in large cities. Since the 1950s, most families have had reasonable access to home ownership in the suburbs. Government policy and an expanding economy, together with the postwar baby boom, have had much to do with the population shift to the suburbs. By the mid 1970s, two thirds of the American people were living either in suburbs or in small towns, and about two-thirds owned their own homes.

Home ownership is the middle-income norm. It is widely regarded as the major symbol of middle-class success and status. Blue-collar workers are highly motivated to own their own homes. According to Mr. Frieden, 77 percent of AFL-CIO members (with median incomes slightly higher than the national average) own their own homes, and among those who are renters a majority want to achieve home ownership.

The poor and members of minority groups are much less able to become homeowners or to live in the suburbs. Among families with incomes below $7000 in 1975, Mr. Frieden reported, only 49 percent were homeowners, and this includes many retired people who had paid for their homes during their working years. Home-ownership rates among low-income people in the 25-to-40 age group are much lower. Only 44 percent of black families were homeowners in 1975. Nevertheless, Mr. Frieden said, there is a tremendous latent demand for home ownership among low-income families. This became clear during the short-lived federal program to encourage home ownership among the poor.

If the age span from 25 to 40 is the prime time for buying homes, it is also the time for moving from one part of the country to another. Almost all people aged 25 to 40 will move at least once during these years, Mr. Frieden said, and many will move more than once. Of the households headed by people from 25 to 34, only 18 percent will stay in the same house for as long as five years. But after this period of moving, most households settle down for a long time. Sixty-eight percent of the households headed by people in the 45-to-64 age group will stay in the same house for five years or longer.

Pointing out that members of the postwar "baby boom" generation have been entering the 25-to-40 age group in ever-increasing numbers since the late 1960s, Mr. Frieden said that the record number of new families being formed by these young adults is putting enormous pressure on the nation's housing markets — just as the same people overwhelmed college and university campuses during their earlier years. From 1950 through 1970, the number of households in the United States increased by an average of about a million a year. From 1970 to 1975 the increase averaged 1.5 million a year, and from now until 1985 it will average about 1.2 million a year.

The demand for new housing is intensified by the tendency of young families to migrate out of slow-growth, housing-surplus areas and into expanding areas with a very limited stock of older housing. Nationwide, the move has been from the Northeastern and North Central states to the Sun Belt states of the South and West. The South gained 5.1 million people from 1970 to 1975, and the West gained 2.9 million. Ten of the largest metropolitan areas in the Northeast suffered population losses between 1970 and 1974.

"In others words," Mr. Frieden continued, "the people in the 25 to 40 age group are a large and footloose population, and they need lots of new housing."

The major housing problem among people in the 25-to-40 age group is the rapid rise in costs. For houses of constant quality, prices increased 63 percent between 1970 and 1976. For all houses sold, the increase was 89 percent. The impact of this escalation has been greatest among low-income families, but it is now beginning to spread upward to middle-income families. "We used to think of housing problems as slum conditions, and as only affecting the poor, but neither is true any more," Mr. Frieden said.

The chart and the cartoon on page 379 illustrate the plight of prospective home owners.

There are several kinds of housing deprivation. The three principal ones, according to Mr. Frieden, are deterioration, overcrowding, and excessively high rental costs. "For the poor," he said, "the rent squeeze is rapidly replacing the dilapidated slum tenement as the number one housing problem. Physically inadequate housing is now a problem that occurs mainly outside of the metropolitan areas, but in the case of half of the renting households with 1973 incomes of less than $5700, their rent burden was excessive."

Minorities have been hardest hit by housing deprivation. Despite a decade of civil rights activity, the proportion of black families with some form of housing deprivation has held at twice the proportion for all families. There has been some progress, but the gap remains. In 1973, 37 percent of black families suffered from one or more of the three forms of housing deprivation, as against 18 percent for all families. Nearly two out of five black families are still housed inadequately.

Turning next to the question of what has been happening to the average family's ability to buy a home, Mr. Frieden said that despite a federal commitment that goes back to New Deal days, and the massive federal help stemming from that commitment, the average American family is losing purchasing power when it comes to home buying. The effects of rising costs are quite complicated, he said, because both consumers and suppliers can adjust to them in various ways. As a result, the present situation seems contradictory. Despite the continuing rise in costs, 1976 and 1977 were banner years for new single-family housing starts.

Who are the hardest hit by the cost increases? Families

Narrowing of the Market for New Houses, 1970-1976

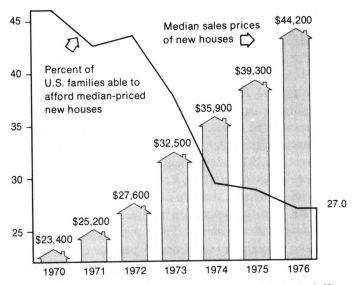

Source: Bernard J. Frieden, Arthur P. Solomon, *et al., The Nation's Housing: 1975 to 1985,* Cambridge, MIT-Harvard Joint Center for Urban Studies, 1977. Reprinted with permission.

"THIS WE CAN AFFORD — IT'S THE EXTRAS, LIKE THE ROOFING, SIDING, FLOORING, FIREPLACE, CHIMNEY, WINDOWS AND DOORS THAT PUTS IT OUT OF OUR PRICE RANGE"

By John Fischetti. © 1978 Field Enterprises, Inc. Courtesy of Field Newspaper Syndicate.

with an income of less than $15,000, first-time buyers, and those seeking to buy new homes.

How much ground has the average family lost? Since 1970, according to Mr. Frieden, increases in median sales prices for both new and existing homes have been much greater than increases in consumer income or in the cost of living in general. Full ownership costs—including fuel, utilities, mortgage interest, maintenance and repair, taxes and insurance—have more than doubled for the median-priced new home, while median family income has increased only 47 percent. In 1970, almost half of all families could keep the monthly costs of owning a median-priced home within 25 percent of income, the rule-of-thumb measure of affordability. By 1976, the proportion had dropped to a little more than a fourth of all families. For a median-priced existing home, the proportions fell from 45 percent in 1970 to 36 percent in 1976.

How are families coping with the cost increases? Mr. Frieden mentioned the following strategies:

Trading up

This means starting out with a cheaper house, probably an older one in a city neighborhood, and, through a series of sales and new purchases, winding up with a more expensive house in the suburbs. Each sale in a rising market generates cash for the down payment on the next house. This is the most common strategy used by young families, Mr. Frieden said. Data for 1976 and 1977 show that only 35 percent of the buyers of new homes were first-time buyers, and they tended to concentrate on lower-priced houses.

Pooling incomes

Of families that bought new homes in 1975 and 1976, about 43 percent had both husband and wife working. This was close to the national labor-force participation rate for married women. Among first-time home-buyers, the figure was 59 percent.

Stretching resources

In 1975-76, the homes bought by families of near-average income ($10,000 to $15,000) tended to be in the $30,000 to $40,000 price range. In view of the average operating costs for houses in that price range, those families took on a very heavy financial burden — nearly a third of their income for housing expenses. But to them it made sense, because they were thinking of long-term income and the likelihood of continuing price rises in the future. They regarded housing as an investment as well as a consumption item.

Buying older houses

Families of average income are withdrawing from the market for new homes. That market, once dominated by middle-income families, is now dominated by families in the top-quarter income bracket — $20,000 and higher. They bought 58 percent of all new homes sold in 1975-76. Many of them got into the top-quarter income bracket by pooling two incomes, another consequence of the increasing partic-ipation of women in the labor force. Much of what has been interpreted as a "back-to-the-city" movement is really a reflection of the fact that young renter families are buying their first homes in city neighborhoods.

Other possible strategies, such as renting longer and buy-ing smaller homes, are not being widely used. Mr. Frieden reported that the proportion of home-buyers to renters is climbing steadily year by year, and that the trend in new homes is toward more living space rather than less. "There has been no return to basics in housing," he said. "New homes are getting bigger and better equipped."

The rising cost of housing has profound implications for

people in the 25-to-40 age group, Mr. Frieden said. He cited the following as examples:

- Many young families are vulnerable homeowners. If the wives lose their jobs, or if family incomes fail to grow, they will find themselves seriously overcommitted.
- If the price increases continue, more moves will be necessary to achieve the suburban goal.
- Central-city housing will become more attractive, especially for couples whose children have not reached school age.
- As the costs of home ownership rise, other parts of the family budget will have to be cut to cover the increase.
- Home purchase will become an increasingly important part of family financial strategy and investment-planning.
- The cost of "household splits" is rising. From 1960 to 1975, the difference between the cost of maintaining separate living accommodations and the cost of one larger home with equal space fell substantially. It was cheaper for parents to separate, or grown children to leave home. Since 1975, however, the cost of this kind of "splitting" has been rising. Continued home ownership may become less possible after divorce, because of the big income gap between husband-wife families and female-headed families. Home ownership is becoming increasingly difficult for one-parent families.
- The increase in the size of the average home built during the past twenty years, coupled with the dramatic decline in the size of the average family, will create surplus space in single-family homes. One way of coping with increased ownership costs may be to take in relatives or roomers, or to create apartments out of the extra space.
- Since people who already own their own homes are the principal beneficiaries of the recent rise in housing prices, inter-generational conflict may develop. This is particularly true in the suburbs, which are already adopting

growth restrictions. People living in those communities have learned how to exclude families of average or lower income, or how to make them pay heavily to enter. Today's suburbanites got there when the entry fee was much cheaper, and when new families were welcomed. Now they are busy keeping newcomers out in order to protect the social and financial advantage they have achieved.

Mr. Frieden concluded his remarks by discussing some of the things that might be done to ease the housing problems of people in the 25-to-40 age group. Specifically, he recommended the following:

- A revival of the federal program for encouraging home ownership among the working poor. "They are good risks," he said, "and they deserve to be helped."
- Modification and simplification of the maze of municipal building codes and zoning ordinances which, he said, impose a "staggering" burden on the housing industry.
- A restructuring of the present system of financing home ownership.
- Special help for families that want to buy older homes in the central cities.
- Stabilizing the housing industry so as to eliminate the peaks and valleys in new housing starts.
- Eliminating some of the barriers to new home ownership in suburban areas, and particularly the zoning restrictions that keep moderate-income families out.

37

Changes in Family Structure and Roles

People in the 25-to-40 age group are continually moving in and out of various family forms, and we ought to help them cope with these transitions. —Susan Klein

One of the basic problems confronting married people in the 25-to-40 age group, according to Susan Klein, Assistant Professor of Preventive Medicine and Sociology at the University of Colorado Medical Center, is how to balance the needs of their families against their own indivdual needs for growth, development, and self-realization.

This is a problem for all married people, she said, but it is particularly acute for those in the 25-to-40 age group because this is the period in the human life cycle marked by maximum opportunities and highest expectations for achievement. Studies of changing goals in adulthood show that for married and single women, as well as for married men, there is a peak in external achievement needs in the thirties. Women of this age show a decreasing need to get married or to have a home.

The responsibilities of work and marriage may often be in conflict during young adulthood, she continued. The probability of divorce is highest during the first nine years of marriage, particularly for those who are under the age of 30 when they marry. The actual frequency of divorce is greatest among people between 20 and 40. This is the period when the needs of the family and the developmental needs of the individual adult are both at a very high level.

The greatest conflict in balancing individual needs against the demands of family life, according to Ms. Klein, has come as a result of the growing trend toward employment of married women outside the home. In the past, it was generally believed that the personal development of women ended with marriage. There were few choices and the roles of women were relatively clear and simple; socialization was directed mainly toward becoming a wife and mother. Development of the female was conceptualized mainly in terms of family relationships, marked by the transition to wifehood, parenthood, and grandparenthood. That is no longer the only accepted view of female fulfillment.

The theoretical perspective of family development describes the family life cycle as a series of stages, marked by certain critical transitions and developmental tasks. The early stages of development are most demanding: childbearing and childrearing increase the complexity and demands of family relationships. The potential for strain and conflict during this period becomes greater, however, when women work outside the family and bring home needs similar to those of their working husbands. One limitation of the family development model is that it places primary emphasis on the parent-child developmental tasks, and ignores the developmental needs of adults. This was a conceptualization that fit well with the "child-oriented" society of the 1960s.

"The problem with that conceptualization," Ms. Klein said, "is that the values it represented were even then becoming outmoded because of demographic changes in the family—the shortening period of childbearing in contrast to the total family life cycle, decreasing birth rates, and the increasing employment of women outside the home. I do not wish to ignore the importance of children, or the primary role the family must continue to play in socializing its children. But when we talk about the family today, the husband-wife marital relationship seems to be the most important relationship in the viability of the family. It may well be that strengthening this relationship is the best way to help children."

Ms. Klein went on to say that since the conference was considering the needs and problems of people between the ages of 25 and 40, she would look at the family from the perspective of adults in this age group.

"The crucial question in assessing the family from that point of view," she said, "is to what extent does the American family structure contribute to the optimum development of the potential of its members? This is a key question in assessing not only the traditional nuclear family made up of a husband, wife, and one or more children, but also in assessing family variants — the single-parent family and the reconstituted family."

Ms. Klein's assessment of the family from that point of view was not a favorable one. "I suggest," she said, "that the contemporary family may be meeting the needs of its wife/mother least well."

In support of her hypothesis she called attention to three sets of data that suggest marriage may be less healthy for women than for men. In terms of mental health, 1960 Census data show some interesting differences in the rates of mental hospital residence for married men and women. Married people of both sexes were less likely to be admitted to mental hospitals than widowed, divorced, separated, or single people. However, the advantage was much greater for married men as opposed to married women; married women were hospitalized 128 percent as much as men. By

1969 these rates had changed, so that married women had lower rates than married men for mental hospital residency. It was hypothesized that the dramatic increase in divorce rates may have alleviated some of the stress, particularly as women saw alternatives to remaining married.

"Differential health benefits of marital status are also shown by mortality statistics," she said. "Standardized mortality rates for married persons were lower in 1959-61 than for any other marital status. Women always fare better than men in mortality. However, the relative advantage of women who are married is much less than that of women of any other marital status — for instance the ratio of white male to female death rates is 1.9 for married people, as opposed to 3.3 for divorced and 2.4 for single people. In terms of physical health, then, the benefits of marriage may be less for women than for men.

"In terms of satisfaction with marriage, it has been shown by Feldman that wives' satisfaction drops most precipitously from the early stages of parenting through the period of launching of children. Although initially wives were shown to be more satisfied with marriage during the first two stages of the family life cycle, they were less satisfied than their husbands at all subsequent stages until the stage of aging was reached. Thus, marriage may be meeting the needs of wives less well than those of husbands."

Because of the changing norms and increased opportunities for women, Ms. Klein continued, employment may become an alternative to marriage as an avenue to personal fulfillment. Marriage and divorce trends are quite opposite for men as opposed to women with regard to income levels. The 1960 Census reports show that for men, the percent never married and percent divorced were inversely related to income. For women the relationship between income and percent never married was curvilinear, suggesting that nonmarriage becomes viable when the woman's income is over the poverty line. In addition, divorce rates for women increased directly with income. These statistics suggest that greater economic resources offer women more freedom to reject marriage and choose alternative lifestyles.

This data may offer at least a partial answer to why the government's income maintenance experiment failed, said Ms. Klein. Rates of divorce and family dissolution were reported to sharply increase after a new government program guaranteeing a minimum wage to poor families was instituted. This program was probably based on the assumption that divorce rates decrease with income level, and failed to consider that additional resources may give women the freedom to break away from an unhappy marriage.

"What I am suggesting," she continued, "is that the changing role of women — their accelerating drive toward equality — has been crucial in changing the functioning of the family, and that stability in the family of the future may increasingly depend on whether the family structure can meet women's needs.

"Personally, I think there is convincing evidence that marriage and family are strong, quite necessary to society, and are here to stay. In spite of the rising likelihood of divorce, people are still showing great interest in getting married. However, we are living in a time of transition. Some old norms are being challenged. Some of the beliefs and expectations we have had about the family are being called into question."

One of the most important myths now being challenged, she continued, is that the nuclear family is the norm. The long-term trend is for the nuclear family to remain in the plurality, but not in the majority. The most important major variations are dual-work families, single-parent families, and reconstituted families. Each of these should be more carefully studied in terms of its ability to meet the developmental needs of all its individual members and to function as a family unit. Such variant families are characterized by the complexity of roles (that of step-parent, for example), role overload (the mother in a single-parent family must be mother, father, and breadwinner), and the relative lack of appropriate norms, rules, and role models. The single-parent family, she said, is the variant most lacking in structure, most in need of further study, and probably most in need of social intervention.

Ms. Klein went on to point out that the changing role of women has had important effects on family functioning. "In the traditional nuclear family," she said, "the husband was the economic provider and the decision maker. The wife's role was that of emotional manager of the family. But the absolute authority of the husband is no longer taken for granted. He is no longer the head of the family merely because he is the man of the house. Rather, he must win this right by virtue of his own skills and accomplishments — in competition with his wife. Currently, the husband has the advantage in this competition because of his greater financial resources, but this, too, is beginning to change — particularly when the wife is working. There is an increasing trend toward more egalitarian relationships in which husbands and wives share power. Consensus is being replaced by bargaining. Families are less likely to be comfortable places to which one retreats from the pressures of the outside world. In fact, the dominant process within the family is sometimes one of conflict rather than stability. Rising divorce rates are one evidence of the growing conflict, particularly among husbands and wives in the 25-to-40 age group."

What is currently regarded as a "disruption" of family life, Ms. Klein continued, is a reflection of what Ogburn referred to as a "cultural lag" — one part of the social system has changed faster and to a greater extent than other parts of the system. The role of one of the family's traditional gatekeepers, the wife-mother, has changed as she has moved into the work force. At the same time the developmental tasks of parenting have been compacted into a shorter time frame. In addition, families are more mobile and often lack the resources of an extended family network. Thus while the pressures on the family unit are great, societal institutions are lagging in responding to the needs of the contemporary family. Creative approaches to day care, more flexible maternity/paternity leaves and other organizational innovations which could help families deal with competing pressurese have been slow to appear.

The family is one of the most dynamic and responsive institutions in our society. Ms. Klein said she preferred to look upon modern variants of traditional family forms — including divorce and remarriage — as positive changes that will, in fact, lead to new and stronger family forms that are more functional, meaningful, and satisfying.

"In my opinion," she continued, "there are two aspects of the ideology of marriage that have become dysfunctional. The first is the notion of happiness as stability, and the second is the concept of commitment as something static and permanent. Implicit in these notions is that adjustment, stability, and maintenance of the marriage are the most desirable outcomes. However, it is clear at this point that the maintenance of some marriages is not a desirable outcome — either for the children or the parents. Moreover, conflict is just as important to maintaining a viable family as is consensus."

Ms. Klein paused at this point to observe that the Vaillant book, *Adaptation to Life*, offers an amusing comment on the importance of conflict to marital health. She quoted as follows from a reply to a Grant Study questionnaire:

Question	Answer
1. How stable do you think your marriage is?	QUITE STABLE
2. Sexual adjustment is, on the whole . . . ?	VERY SATISFYING

3. Separation or divorce has been considered: Never, casually, seriously?

The multiple choice answers were ignored and replaced by a message written boldly in black ink:
WE'VE NEVER THOUGHT OF DIVORCE, BUT MURDER? YES!!

"Commitment," she continued, "is a dynamic process marked by the on-going ability to deal with change, and the success of a marriage should not necessarily be measured by its duration or permanence. I do not mean to imply that short-term marriages, or divorce, are necessarily good or desirable, but rather that they may be inevitable in a time of transition.

"Adults bring to a marriage a need for warmth, compassion, emotional intimacy, personal understanding, and sexual fulfillment. Marriage is increasingly idealized as a relationship marked by a high degree of interdependence. We now recognize the importance of adult individual growth and development. People change during the course of a lifetime, and their concept of self and of their needs often change. Probably at no other time has society put a greater emphasis on individual growth and choice, and people now have more alternatives available to them.

"We need to know much more than we do now about the new variants of the traditional nuclear family. We know very little, for instance, about the reconstituted family and the way it develops. We need to know more about the particular strengths and weaknesses of variant family forms — particularly the single-parent family — in order to develop appropriate intervention strategies."

Ms. Klein said that caution should be exercised in making value judgments regarding the health of non-traditional family forms. The single-parent family, for instance, suffers from a deficit of members and generally excessive demands on the energies of its one adult member. However, it would be unwise to judge all single-parent families as dysfunctional since individuals vary greatly in their needs. For example, a woman in her forties with two adolescent children may choose to postpone remarriage, feeling that remarriage would disrupt her functioning as a parent and interfere with her own career development. Some people find that total integration is not possible, and they make a series of compromises.

"What I mean to suggest," she said, "is that even the weakest of the variant groups may have some strengths, and we need to identify those strengths. A humane society must help support all families — variant and traditional — in meeting the needs of their individual members."

Ms. Klein's remarks touched off a lively discussion about the history of the family, its utility in modern society, its variant forms, the internal and external pressures that bear upon it, the changing roles of its members, and its prospects for the future. The dialogue went as follows:

Mr. Kelso: The nuclear family as we understand it is a modern, largely Western phenomenon. I think one of the questions that deserves to be asked about it is whether it's the unit that we need to strengthen and support, or whether there are other social units that might do a better job of meeting the needs of the individual.

Ms. Klein: We are not talking about *the* family any more, but a series of variant forms of the family.

Mr. Eurich: There are great differences in families from one country to another. In Iran, for example, Sunday is set aside just for the family — not the immediate family, but what the sociologists would call the extended family. All the relatives get together. No one can break into this family from the outside. It's a closed circle. One result is that in Iran there is a high percentage of intermarriage among cousins. Is that more desirable than the kind of family situation we have in this country?

Dr. Robinson: Well, from the point of view of genetics it's much less desirable. People from Iran and India tend to have a very high rate of recessively inherited diseases, many of which are really deforming. Some of them are fatal.

Dr. Cahill: Well right now, for an American family to do something as a unit very often amounts to the chance of having a meal together.

Dr. Ward: I think we've got to recognize that the family unit as we know it may well become an artifact of the past. There may not be any way to protect it. If you look at the number of women who are working, the strong desire for outside involvement, I'm not sure that the nuclear family is going to be a part of our future.

Dr. Cahill: The Socialist countries have already concluded that the family is passe as a social institution. The children are reared in nursery schools — in an isolated environment — almost from the time of weaning. The state is the father. Children happen to have a biological father and mother, period. I don't think we've had the time to assess whether this is good or bad, as far as the well-being of the children is concerned.

But the family *has* been successful in the past. We've got enough data to support that thesis. We know that when you have mixed-up parents you usually have mixed-up kids. And there's enough data to show that when you have mixed-up families you usually have mixed-up adopted kids. So it isn't all genetic. There's a lot of cultural input.

What I am suggesting is that we should look at the old history of the family. Is what we call the nuclear family the best arrangement for the long pull?

Dr. Visotsky: Before you start talking about whether the family is obsolete, you have to decide whether you can replace it with an alternative arrangement. Whether or not other countries have been able to do this successfully is questionable, because they've gone back to the family and family linkages. In Israel, where they have developed the kibbutz system, and in other countries where children are raised in day nurseries, the parents—the biological parents—are required to spend time with their children. The Russians spend a lot of time instructing parents in the art of parenting. So do the Chinese, even though it's a managed arrangement.

It might be helpful if we took some time to look at the values of family life that we want to preserve. There are certain nurturing qualities of family life that have been delegated to day-care centers, kindergarten, and other institutions. The educational function is being turned over not only to the schools but also to television. We have gradually stripped the family of many of the past support systems that brought families together in terms of values and ideals. We may want to recapture some of these by restructuring the system.

Mr. Frieden: I'm not sure that the values of family life have been lost. I think the evidence is mixed. There's a recent article by Nathan Glazer in *Commentary* dealing with several current books on families written by people who, only a few years ago, had considered themselves critics of the family and society. All of the books concluded that the American family is holding up very well, in spite of the rising divorce rate and the stresses caused by the increasing participation of women in the labor force. One of these books, *Here to Stay,* by Mary Jo Bane, maintains that if you compare the American family today with what it was like in the early years of the twentieth century, children are getting about as much or even more parental attention than they did fifty or sixty years ago. More parents work today, but fewer parents die early. She also points out that far fewer children are being raised in institutions. Many more are being raised by at least one parent, or by close relatives. So it may be premature to hang black crepe over the future of the American family.

Mr. Edmunds: I think it might be worthwhile to trace the development of the family itself, because when we talk about the family at conferences like this we make a lot of assumptions about what has existed in the past. For example, I see a lot of mothers today who feel guilty because they can't give to their children what they think their mothers gave to them and their grandmothers gave to their mothers. When I was a boy visiting my grandfather's farm, I am quite certain that the chores for which my grandmother was responsible greatly limited the time she could spend with her children. Maybe the good old days didn't yield as much free time for mothers as we assume.

Mr. Demos: In the light of history, the family seems a remarkably durable institution. The nuclear family has been with us throughout the history of Western civilization, or at least as far back as people kept any sort of records. We used to think that most families in the Western world lived as extended units until the time of the Industrial Revolution and then switched to the nuclear family pattern. That would make the nuclear family only about a couple of centuries old. But now we know that nuclear families were present long before the Industrial Revolution and were, in fact, the norm. Moreover, I think that some sort of a family unit will continue to be the norm for at least the foreseeable future, and that we ought to concentrate on ways of improving family life however it is going to be lived. We certainly don't want to say that there is only one kind of family unit for everybody in the 25-to-40 age group, but I do think we ought to try to make life easier for people who will want to continue to live in families.

One of the most critical threats to the family is the conflict between family life and the work life, and this threat will become even more critical with the increasing participation of women in the labor force. It is particularly acute in families with small children, and most of those families are headed by people in the 25-to-40 age group.

Dr. Visotsky: Communes and other kinds of surrogate families are important indicators of a deficiency in the modern family. The young people who choose to live in them are meeting an unmet need by grouping together into a support system of their own. They either do not receive the support they need in their own family, or they reject what they find. But although this may be a symptom of failure in the family system it does not necessarily mean that the family is obsolete. It just means that the family is not meeting their needs.

Mr. O'Keefe: Economy of scale is a powerful force in persuading people to live together in families, communes, and other groups, because it enables them to pool their resources and share expenses.

Ms. Klein: Economic reasons are important, but emotional considerations are more important.

Dr. Schweppe: One of the very basic questions that we have failed to address is when, in the course of evolution, did man develop a sense of family? When did the sense of family responsibility begin?

Mr. Kelso: It goes way back to the primates. There are family units among gibbons, for example. There are mother-infant bonds, and under certain circumstances fathers take care of their infants. As far as man is concerned, the family is probably as old as the human race. Some kind of a family aggregation is common to all human society, whether it be by residence unit, economic activity, or mutual nurturing and support.

Dr. Cahill: One theory is that the sense of family among humans began to develop when man discovered the importance of having old people around—for example, the grandfather who knew which path led to the water hole, and the grandmother who could stay home and mind the children while the parents were out hunting. We know from our studies of metabolism that the metabolic rate of older people is only about half the rate of young people and that older people require less food. So the theory seems to make sense.

38 Stress and Conflict in Family Life

Parenting is an art that cannot be learned from books. It takes a series of experiences. —Harold Visotsky

Harold Visotsky, Chairman of the Department of Psychiatry and Behavioral Sciences at the Northwestern University Medical School, said that although the family may change in form and content it will continue to be the primary source of support, nurturing, and gratification for people in the 25-to-40 age group.

"It is at once an investment in our future, a mechanism for repairing developmental injuries or defects, and a fulfillment of our expectations," he said. "When it comes to nurturing, it is the only game in town. There is no other comparable system for dealing with the nurturing process. Our government is not a nurturing government. It does what bad parents do—talks a good game but delivers very little."

Dr. Visotsky went on to point out that the family is a durable institution. It has successfully withstood repeated attempts to undermine or replace it.

In 400 B.C., he said, Plato criticized the family on the grounds that it was inefficient and a deterrent to the socialization of children as members of the ideal new Republic he envisioned. "His own experiment in child-rearing followed, and failed," Dr. Visotsky remarked.

In the early days of the Soviet Union, the theories of Karl Marx were used to undermine the family system in the interest of producing ideal Communists. Divorce, illegitimacy, and abortion became common, and the time parents spent with their children was drastically reduced. Children were encouraged to seek their development outside the family, and even to spy on their parents. But by 1935 there was so much delinquency and vandalism among young people that the rules were changed, and it was officially announced that the family was the best system for learning how to be good Communists.

The Chinese Communists, according to Dr. Visotsky, took a different approach. At no time was there any attempt to disrupt or destroy the family. Instead, there was an intensive campaign to replace ancestor worship with a reverence for the state. Mao became the head of the extended family.

The Oneida Community in the United States, which flourished in the latter half of the nineteenth century, represented a third approach to family living. Work, children, and sex were shared on a communal basis. The Oneida system eventually fell apart, but some of its main elements are present in today's communes.

Dr. Visotsky went on to say that although the family remains the basic unit in our social structure, family roles have been modified by social and economic change. One of the major changes that affected family life, he continued, was the separation of home and work.

"In an agricultural society," he said, "children knew what their fathers did because they saw them at work, but today's children have only a fuzzy notion of the kind of work their father does. They may know that he works in an office or in a factory, but that's about all. On the other hand, they do have a very clear idea of what their mother does. Her office and her factory are in the home. The space within the home is usually defined in female terms—the kitchen, the laundry room, and so forth. There may be someplace which is defined as father's room — a den or an office — but home is where the mother works. And it's easy to get an appointment with mother."

Home has become the place for the playing out of emotions, Dr. Visotsky continued. "It's the great disposal center for the tensions of industrial work. If father has had a bad day in the office, or if the train was delayed, or if the bar car ran out of his favorite brand of Scotch, he takes out his anger and frustration on the family. This, in turn, makes him feel guilty because individual happiness has become a criterion of family success."

Another consequence of the separation of home and the work place, Dr. Visotsky continued, is that the very young child is left for extremely long periods of time with a single individual — the mother. This can be a source of stress to the mother as well as to the child.

In the meantime, society changes so rapidly that parents no longer have a point of reference with elements of their own childhood to use in socializing their children. "Today's children have problems that their parents never encountered," he said, "and when they try to ask for help in meeting them the parents have no point of reference — no experience of their own that might be helpful. The parents usually tend to say that it's not relevant, and the result may be confused, frustrated, and rejected children."

The increase in life expectancy, and the gap between male and female life expectancy, also affect family relationships. For the average woman, life expectancy is about 79 or 80, but for a man it is five to eight years less. Consequently, most wives outlive their husbands, and in some cases they actually outlive their older sons. This is particularly true when husband and wife have married early and have had their children early. "Many mothers are functionally younger than their sons," he said.

As a result of all these changes, Dr. Visotsky continued, young adulthood has become a very stressful stage in the human life cycle. "Too much is being compressed into a short period of time. If you add up what students of the family call life-crisis units — the loss of a job, a change in jobs, marriage, child-rearing, divorce, and even death — you find that their number is very high between the ages of 25 and 40. And so, if you believe the theory that the number of these units is related to illness and disease, you can begin to see that the seeds are being sown for subsequent break-downs. Stress is insidious, in that it builds up over a long period of time. It sets the pattern, and once the pattern has been set it goes into effect automatically and keeps repeat-ing itself."

Turning next to parenting as a signficant role in the lives of people from 25 to 40, Dr. Visotsky said some interesting insights into the consequences of good and bad parenting have been developed by Diana Baumrind, a research psy-chologist at the University of California at Berkeley who has conducted a series of major studies of how various types of parental control affect the behavior of children. Among her findings, he cited the following:

1. The most assertive, self-reliant, and self-controlled children had parents who were controlling, demanding, communicative, and loving. Rather than ridiculing or fright-ening their children, or withdrawing affection, these parents were ready, if necessary, to use corporal punishment. Gen-erally, though, they rewarded their children for behaving well rather than punishing them for behaving badly.

2. Children who are discontented, withdrawn, and dis-trustful had parents who were relatively controlling, but detached. "Detachment," said Dr. Visotsky, "is a way of avoiding problems or decisions. In parents, it is usually a

sign of immaturity. On the other hand, it takes a great deal of maturity to be a loving parent — to be able to listen to your children, give them advice, and even tolerate a certain amount of deviance."

3. Children who had little self-control or self-reliance, or tended to retreat from new experiences, had parents who were relatively warm, but not controlling and not demanding. "I think it's very important," Dr. Visotsky said, "to teach children at a very early age the art of coping—how to deal with stress and how to cope with new experiences."

Among the indicators of strength or health in a family Dr. Visotsky listed the following:

- The ability to provide for the physical, emotional, and spiritual needs of family members, and to be sensitive to their distinctive needs.
- The ability to communicate effectively — to express a wide range of feelings and emotions as well as ideas, concepts, beliefs, and values. This includes the ability to share deeply-felt interpersonal experiences, such as the birth of a child, the death of a parent, special achievement or failure.
- The ability to provide support, security, and encouragement.
- The ability to initiate and maintain growth-producing relationships and experiences both within and outside of the family.
- The ability to create and maintain constructive and responsible relationships in the community.
- The ability to grow with and through children.
- The ability for self-help, and the ability to seek outside help when it is needed.
- The ability to perform family roles flexibly.
- Mutual respect for the individuality of family members.
- A concern for family unity, loyalty, and interfamily cooperation.

• The ability to use a crisis or seemingly injurious experience as a means of growth.

A recent study of hospital admissions in Chicago, Dr. Visotsky reported, showed that the greatest incidence of psychological disorders tended to be among people in the 25-to-40 age group. Women were found to have higher rates of incidence than men, and single persons had a much higher rate than married persons. "Married persons complained a lot about emotional distress," he said, "but it was not as significant among them as it was among unlinked individuals. They had far fewer breakdowns than single persons. You might say that single people have a higher rate of breakdowns because they can't get married, or because of their status, but it also may be true that some of them are undesirable marriage partners. Their symptoms may make them unlovable or unlinkable. Separated persons, as distinct from persons who have never been married, are still going through the trauma of being rejected or thinking that they have been rejected. Their self-esteem seems to be at its lowest ebb."

The study showed that schizophrenia is the most common psychological disorder among people in the 25-to-40 age group, followed by alcoholism, anxiety and depressive neuroses, and a variety of psychoses. Drug dependence ranked close to the bottom of the list.

The study also showed that there is a close correlation between low socio-economic status and high rates of psychological disorders, and that genetic defects produce a high degree of vulnerability to mental problems. Downtrends in the economy seem to result in a higher incidence of mental disorders. "Alcoholism seems to be a working-class response to economic distress," Dr. Visotsky said, "and suicide seems to be a response in higher socio-economic groups. But much of the suicide in higher economic groups is also related to alcoholism."

Emotional distress often leads to child abuse and wife-beating, Dr. Visotsky reported. "It doesn't take a genius to figure out that when someone is under stress the stress is going to be manifested in a variety of interpersonal relationships," he remarked. "A lot of those people start hitting." Young children are the principal victims, and the guilty parents are most likely to be those in the 25-to-40 age group. "There isn't any uniform definition of child abuse," he continued, "but the number of children who are killed by their parents or guardians ranges anywhere from 700 to 5,000 a year."

What can the federal government do to help strengthen the family?

In answer to that question, Dr. Visotsky cited three recommendations he had made in the Spring of 1977 at the annual meeting of the American Orthopsychiatric Association:

1. Establishment of an income transfer or maintenance program for *all* families, guaranteeing a basic level of economic support and stability.

2. Development of a national system of high-quality maternal and child health care services, including prenatal and obstetrical care, well-baby clinics, and adequate follow-up and treatment resources.

3. Special programs to assure vital family and child supports, such as maternity leave, sick leave for parents who must attend to seriously ill children, quality day care, visiting housekeeper services, constructive after-school programs for children, and job programs for youth.

"Continuing study of the family and its problems is essential," he said, "but it is not sufficient. We have available already, and have had for some time, the basic information for the establishment of a meaningful family policy. What is needed is the commitment to put the policy into practice without delay, and to maintain it at a meaningful level."

Dr. Visotsky said his top priority recommendations for the conference would be:

- To persuade industry to take a parental responsibility for the health and well-being of its employees.
- To persuade organized labor to play a more important role — quite apart from its economic role — in improving the quality of life of its members.
- To persuade educators to begin to teach children, at a very early age, the art of coping with stress, and the importance of cooperation rather than competition.
- To persuade medical schools to pay more attention to the quality of life in training physicians. "Medicine should be concerned not only with treating sick people," he said, "but also with helping to make well people 'weller'. We need new kinds of careers in medicine."

In the discussion that followed, Susan Klein was the first speaker. Referring to Dr. Visotsky's comments about the need for educating young people to cope with the stresses they will face when they reach young adulthood, she asked if anything were being done to meet that need. "There's not much being done in terms of formal education," Dr. Visotsky replied, "but there are programs like Outward Bound that help young people learn to deal with stress and how to live with others in stressful situations. Programs like that don't solve every problem, but they do teach self-reliance and where to go to get some kind of help in facing problems."

Arthur Robinson said he was delighted to hear Dr. Visotsky talk about child-raising problems. "Actually," he said, "one of the most important activities of the people in the age group we are discussing is having and raising children. Maybe we ought to consider some sort of licensing system for people who want to become parents. They now have to pass certain tests in order to get married, so why not tests in what Dr. Visotsky called parenting? It might be one ap-

proach to the problem of child abuse. And maybe we ought
to do more than that. Maybe we ought to train young people
in high school and college not so much for marriage, but for
parenthood. That, too, might help to prevent child abuse."

Dr. Pellegrino, harking back to Dr. Visotsky's remarks
about the lack of a nurturing government, said he would be
concerned if the government carried that notion to the point
where it intruded on the individual's right to decide what
constituted the good life.

Dr. Visotsky said he did not intend to imply such an intru-
sion. "The function of a nurturing government," he said, "is
to provide the widest range of choices for its citizens, and
then let the individual choose from among them."

Dr. Schweppe agreed. He said the entire welfare system
needs a radical overhaul. "As it presently operates," he said,
"it does not help individuals or families at all. It's really a
police system, spying on people and telling them what they
can't do instead of helping them to help themselves. The
emphasis should be on rehabilitation rather than on family
management."

The discussion then turned to the nature of child abuse
and the question of whether it was a relatively recent phenomenon in American society.

John Demos said that in his studies of the family in Colonial times, he encountered very few references to child
abuse, and was inclined to think that its increasing frequency in recent years may be a measure of new kinds of
stress in family life. Harry Ward suggested that child abuse
in Colonial times may have taken the form of child exploi-
tation. Alec Kelso said it is almost unknown among primates. George Cahill pointed out that caning and birching,
common in the nineteenth century, were a form of child
abuse, and reminded the conferees that Huck Finn was
beaten by a drunken father. "So maybe it's not so recent
after all," he remarked.

Dr. Visotsky said the definition of child abuse should not be limited to physical abuse, because there are some forms of psychological abuse that are even more destructive. "Children just can't cope with rejection," he pointed out, "and neither can they accommodate to alternating periods of affection and sudden outbursts of unexplainable anger." He went on to say that he did not regard mild corporal punishment as child abuse. "I see it as a form of communication," he said. "All it should do is indicate to the child that the parent is displeased with a certain kind of behavior."

Dr. Schweppe wondered whether some children, perhaps for genetic reasons, might be more prone than others to provoke their parents to physical violence. "I've heard it said," he remarked, "that there are some kids who are little bastards from the day they were born."

When the laughter had subsided, Robert Aldrich observed that children do indeed have basic response patterns that are developed very early in life. "These response patterns are quite evident even in the nursery," he continued. "There are some babies who are quite docile when you pick them up by the heels, but there are others who will squirm and scream and turn red and cuss you out. They are the high-powered types. I call them executive infants, because they are the ones who organize the entire nursery. They have some kind of basic quality that they carry with them for the rest of their lives."

The discussion then turned to financial pressures as a cause of stress within the family. Thomas Edmunds said that the stress was generated not so much by the relative size of family income, but by arguments over how it should be spent. "People need better education in how to manage family finances," he continued. "Poorly managed family budgets often lead to emotional instability and distress."

Bernard Frieden agreed. "There's something more than

poverty involved," he said. "There are differences in what poverty means to people. Some observers have distinguished between slums of hope and slums of despair. If you look at some low-income squatter settlements in developing countries closely enough, you will find that the people there aren't too unhappy, because they expect their lives to improve. But if you look at most of the urban slums in this country, you will see nothing but misery and despair. The people in our slums have a higher standard of living than the squatters in Third World countries, but their expectations aren't being met, and you get resentment and despair — a feeling of inequity."

Dr. Pellegrino said Mr. Frieden's remarks raised some significant questions about the whole matter of values. "Even in an affluent society there is a gap between expectations and actuality," he remarked. "Maybe we ought to pay some attention to the question of reorganizing the priority of our values — the whole question of what, in fact, is the good life. Some people, even those in the higher income brackets, are never going to get out of the predicament they're in now because their wants and expectations are simply too high to be fulfilled."

Dr. Aldrich pointed out that the American notion of the good life is quite different from that in some of the Third World countries, and he suggested that in order for professional people in the United States to acquire a better understanding of other philosophies it might be a good idea for the professional schools to require a year of residence in a Third World country. "A lot of students do this on their own," he said, "but the practice ought to be much more widespread."

Mr. Eurich said he would welcome more specificity in the discussion about values. "The question of values has come up over and over again in our discussions," he said, "but we haven't been very specific. We've talked about values in a

general way, and we've implied that they are very important, but what values are we talking about? What are the
values that ought to be preserved? Are there values in our
society that need to be changed? And what about the values
that differ from one country to another?"

Dr. Pellegrino said he had no specific answers to the questions Mr. Eurich had raised, but would like to make a few
observations. "It seems to me," he said, "that we have been
dealing with what *is* rather than what *might be,* and with
means rather than ends. I am particularly concerned about
values, because the transaction between physician and patient is really an exchange of values. Their values may be
different, but they have a common goal — to improve the
health of the patient."

Dr. Aldrich mentioned love and trust as two specific values that ought to be preserved. "They are the earliest values
a child learns," he said, "and they are crucial in building a
healthy parent-child relationship."

Dr. Kelso said he would like to address Mr. Eurich's
questions from the perspective of people who lived 25,000
to 50,000 years ago. "We can learn a lot about values from
their experience," he said. "They lived in groups that had
face-to-face contact. They were by no means without conflict, but in some ways their life was idyllic as we look at it
today because it did have a lot of elements of nurture and
intimacy that we now consider desirable. The great lesson
they taught us, it seems to me, is that human beings are
social animals — that we have evolved from a long lineage
of social organisms, going back to the primates. Aristotle
emphasized that. Yet the distressing thing that is going on
in our present social system — distressing to me as a father
and husband as well as a professor of anthropology — is the
dispersal of people into individualized cubicles. That is to
say, we have placed a very high value on *individual* achievement. We have developed an ethic that isolates us as indi-

viduals, and inhibits us from assisting one another in the achievement of *common* tasks. The effect of this is expressed in a kind of loneliness and a kind of indifference to the value of other human beings in our lives. The real bottom line is that we have in some cultural sense repudiated our very nature as social beings."

"And when we talk about value differences," Mr. Kelso continued, "we tend to disperse people into different value categories. That bothers me because it further isolates human beings as individuals. Ultimately, I can see human beings dispersed into more and more isolated categories of individual value systems. The market place serves as a great propellant for this dispersal. We are differentiated by the clothing we wear, and the cars we drive. We may say these are superficial, but in a sense they are ways of identifying us as consumers. In my view we're paying one hell of a price as primates for living in a consuming society, and it's reflected in the kinds of stresses we have been talking about."

One of the conferees asked Mr. Kelso how he would compare the life of a young adult in the Stone Age, in terms of values, with the life of a young adult in today's society.

"To begin with," Mr. Kelso replied, "young adults in the Stone Age were the elderly. The life span was much shorter. But values were common from one generation to the next. There wasn't a perfect match, but by and large the value system was continuous and stable. People may have been differentiated by occupation or activity, but not on value grounds alone. The elders were respected, the social system was stable, and in some ways I suppose we would regard the lifestyle as very dull, but generational differences were not very great. You would not have heard what we heard just a few years ago, that anyone over thirty was not to be trusted. Now it's probably anyone over 27.8. We're paying a high price for change."

Thomas Edmunds said he agreed with Mr. Kelso. "We have lost a lot of our understanding of what it is to be a human being," he remarked. "And maybe the church, in its own way, has aided and abetted the trend toward rank individualism."

Mr. Kelso said that our educational system could be indicted on the same count, because of the emphasis on individual grades.

Dr. Ward questioned Mr. Kelso's description of man as an essentially social being. "I'm not sure that we know yet what man's essential nature is," he said. "Even in Stone Age society man was a hunter as well as a sharer. I think a case could be made for the proposition that man's essential nature is 'I' rather that 'we,' and I'm not sure that we can base value systems on our understanding of man's essential nature until we know for sure what it is."

Mr. Kelso conceded that hunting was part of man's cultural heritage from the early primates, but he emphasized that social organization was also a part of that heritage. "Among primates," he continued, "there is no such thing as a solitary individual. A solitary primate is a dead primate."

Arthur Robinson entered the argument with a different viewpoint. "I think that when we talk about the essential nature of man we are fighting windmills," he said. "The only essential nature of a biological organism that I know of is the desire and the ability to reproduce itself. This long-term self-interest is present in every individual, and it could well be the basis for an ethic. I think it's a mistake to generalize about the essential nature of modern man and to compare it to the nature of Stone Age people. As I recall, Stone Age people did not live in communities of a million or more."

Dr. Schweppe offered the opinion that some values change as society grows more complex, but others remain the same. "When you move from the individual to the family, and then to the group, the tribe, the nation, and the family of nations, you can no longer talk about a single set of values," he said, "and the problem of establishing priorities among values becomes almost insoluble."

Dr. Pellegrino said he thought the basic conflict between value systems stemmed from the fact that man is an individual and, at the same time, a member of a society that is getting larger and more complex, as Dr. Schweppe had pointed out. "The real problem," he said, "is how to put these two essential sets of values together."

John Demos suggested that the work ethic — the notion that every individual should work long hours with a minimum of rest and relaxation — was in direct conflict with the values of family life that had been discussed earlier, and that women were becoming victims of the work ethic as well as men. Working husbands and wives have very little time to spend together or with their children, he pointed out, and this time is growing shorter and shorter. One result of this trend is to increase the feeling of isolation among family members.

Merrell Clark said that society could help to alleviate work-related stress by creating more meaningful occupations for more people for more of their lives. "People of all ages have to have the opportunity to feel that they can make a useful contribution to society," he said. "The feeling that life is worthwhile is fundamental to human happiness. We cannot have people who believe that they are useless." He went on to say that there were three requirements for creating meaningful occupations:

1. Government policies that foster steady economic growth and encourage investment.

2. An educational system that provides opportunities for lifelong learning and encourages people of all ages to learn new skills.

3. Enlightened employers, who recognize that their employees are their most important resource and make a special effort to help them develop their full potential.

39

Lifestyle Modifications: The Ethics of Intervention

Any effort to improve the lives of people in the 25-to-40 age group must include the three essential components of effective intervention—a sound plan, a capable manager and the necessary resources.—Merrell Clark

By the fourth day of the conference, many of the participants had offered suggestions for helping people in the 25-to-40 age group cope with their special problems, but there was a growing awareness that intervention in the lives of young adults is a matter of great delicacy, involving such questions as what forms of intervention were acceptable in the ethical sense, how far they should go, what individuals and organizations should undertake them, and finally, whether in some cases any intervention at all might be completely unwarranted.

The caution with which the conferees approached the subject of intervention stemmed in part from a parable re-

415

lated by Dr. Visotsky. It concerned a blind man who was walking down a busy street with his seeing-eye dog. When they arrived at an intersection, the dog dutifully stopped and so did the master. But before the light changed, the dog lifted his leg and wet the blind man's trousers. Whereupon, the blind man reached in his pocket, produced a dog biscuit, and gave it to the dog. A behavioral psychologist who had witnessed the incident was horrified. He tapped the blind man on the shoulder and said: "I beg your pardon, sir, but what you have done is completely wrong. You have rewarded your dog for bad behavior." The blind man was indignant. "I wish you would mind your own business," he said. "I was merely trying to find out which end was which so I could kick him in the place where it would do the most good."

Merrell M. Clark, Executive Vice President of the Academy for Educational Development, was the principal speaker in the conference session devoted to lifestyle modifications and the ethics of intervention. He did not attempt to present a grand strategy for improving the lives of people in the 25-to-40 age group. Rather, he sketched on the blackboard the conceptual framework for such a strategy, emphasizing a number of considerations that he said were essential to any effective intervention effort.

He began with what he called the traditional Sunday-school definition of a human being as a combination of mind, body, and spirit — in community.

"That's a remarkably durable definition of human beings," he said. "Most arguments regarding the essence of human nature wind up somewhere in that ball park. The discussion regarding the relationship between human beings and the other primates is reflected there, in the fact that primates share with human beings the characteristics of body, mind, and community. The word spirit seems to describe a quality that is unique among human beings. It is not observable among the other primates."

He then went on to say that the major determinants of lifestyle are the ego, the innate and developed capacities of each individual, and the way the individual allocates his principal resources — time and money — during a day, a week, or a lifetime, in the environment of all available possibilities.

"The marketer of goods or services is well aware of these ingredients," Mr. Clark said, "and anyone who attempts to change the lifestyle or values of the individual, the family, or society as a whole also should keep them in mind. The environment of all available possibilities includes the necessary, or survival elements, but it also includes a whole host of other possibilities — including savings and leisure — and it is in that area where the range of choices has expanded so dramatically in recent years."

The effort to position oneself favorably with respect to others is probably the most stressful element of lifestyle, Mr. Clark continued. "People do allocate time and money to affiliations with churches, clubs, and other organizations, and they do try to gain access to better jobs and better neighborhoods. Part of the trauma of suburban families is the high price they have paid for access."

Recent advances in science and technology have resulted in major modifications of the human lifestyle, he said. By way of illustration, he cited automobiles, airplanes, television, computers, and electronic communications devices. He also cited labor-saving machinery in industry, agriculture, and in the home. "All of these have significantly changed the way we live," he pointed out, "but their greatest significance is that they have dramatically increased the range of choices available to us. Less and less of our time and money are being spent for what used to be necessary for survival, and more and more time and money are being spent in pursuit of other possibilities."

He singled out television for special attention. "Television," he said, "has opened up a whole new realm of lifestyle possibilities for the average family. It also has dramatically increased the desire for more favorable positioning with respect to other individuals and families in our society."

Mr. Clark went on to review some of the principal value systems developed by philosophers and theologians over the years, and pointed out that all of them have had profound effects on the way people live. The concepts of tradition and duty in Judeo-Christian thought have been the most fundamental conservators of the values approved by society, the concept of hope in Utopian thought has been the principal motivating force in persuading people to seek a better life, and the concepts of responsiveness and creativity, which are emphasized in Existentialist thought, help the individual come to terms with the stresses and strains of life.

The capitalistic notion of value as a return on investment is the harshest of all value systems, Mr. Clark continued, because it measures worth in terms of input and output. It is also the value system that permeates the marketing system.

He then turned to what he called some of the "essential dualities" of human existence that must be taken into account in any effort to modify the lifestyles of young adults. He listed the following:

- life and freedom
- equity and initiative
- creature and co-creator
- the individual and society
- investment and return

"These dualities," he said, "represent a way of looking at some of the major problems we have been discussing at this conference. It's almost assured that any intervention that neglects one aspect of any of these dualities will ultimately fail."

Turning to the ethics of intervention, Mr. Clark said that any intervention effort must respect the rights of the individual, and that social intervention must not be allowed to proceed unchecked. "Intervention with great power must especially be disciplined," he said. "That is the essence of our constitutional document. Moreover, interventions that limit individual freedom are the most costly, lease productive, and least durable forms of intervention — particularly in the United States."

Any effort to improve the lifestyle of people in the 25-to-40 age group must take all these considerations into account, Mr. Clark continued, and must also include the three essential components of effective intervention — a sound plan, a capable manager, and the necessary resources. He concluded by saying that any intervention effort directed at people in the 25-to-40 age group should be designed to meet one or more of their principal needs — their need to solidify their own independence, their need to assume trusteeship roles in their families and in society, and their need to cope with threats to their survival.

Thomas Edmunds added one more suggestion to those offered by Mr. Clark. "It seems to me," he said, "that one reason interventions often fail is that they try for a summit breakthrough — one that starts at the top and works its way down to the bottom — instead of starting at the grass roots level and working upward."

Dr. Pellegrino said he would like to stress the importance of education in any effort to improve the lives of people in the 25-to-40 age group. "At that time of life," he said, "people are passing from a period of indoctrination in the traditional values to a period of exposure to a whole series of new values which they have not yet assimilated. They are confronted with a new range of choices — ethical and moral as well as practical. They are moving from an anxiety of possibilities, which they experienced in adolescence, to an anxiety of probabilities, which in some ways is even worse. They begin to assume responsibilities, and the choices become more immediate and more concrete. They go through what I would call a metaphysical crisis. They also come into confrontation with their own mortality, probably for the first time. They begin to see the concreteness of illness and death. They begin to be aware of the fact that the seeds of their own destruction are present within them. They also begin to wonder who they are, and where they are going. I believe, therefore, that education must be at the center of any effort to improve their lives—the kind of education that will prepare them to deal with the problems they face and the choices they must make."

With respect to medical intervention, Dr. Pellegrino maintained that the transaction between a physician and a patient is moral as well as scientific. "I think we need to take a fresh look at the morality of the transaction," he said. "Up to now we have had a kind of paternalism, stemming from the historical roots of the profession, in which the physician decides for the patient. We need a new ethic of medical intervention — one that recognizes the transaction as a participatory relationship, gives the patient a voice in whatever decisions are made, and insures that conflicts between the physician and the patient are handled in a justifiable way. The Hippocratic oath no longer serves as a unifying professional ethic in medicine. Every one of its precepts is violated by physicians, in good conscience, today."

A new ethic is also needed in the field of social intervention, Dr. Pellegrino continued. It is not enough, he said, to decide whether a proposed policy is feasible. It is also necessary to determine whether it is worthwhile. "We need two kinds of advisory bodies in making decisions that affect the lives of a lot of people—one to address itself to the question 'Can we do it?' and the other to address itself to the question 'Should we do it?' "

Dr. Aldrich suggested that Dr. Pellegrino pass along that idea to some of the major foundations. "Some of them are scrambling around like a bunch of chipmunks trying to figure out what's worthwhile," he said. "It's really a kind of a maze." Dr. Pellegrino agreed. "I think foundations have lost their capacity to take chances," he remarked.

The discussion then turned to the question of how much a physician should tell a patient in cases where a certain course of action on the part of the patient could have disastrous consequences.

"It's always a problem," Dr. Robinson said. "Sometimes people are traumatized by information of that kind. And there's always the possibility that the physician could be wrong. I think a lot depends on how the information is given, but in general I believe that the doctor should respect the patient's right to know."

Dr. Visotsky agreed. "I think you have to recognize that when you give a patient information you're in a contractual relationship with that patient, and that the patient has a right to make his own decisions. But you also have to recognize that some people simply cannot handle the information, so you have to help them understand it and also understand that they have a choice and that you will respect their choice."

Dr. Robinson pointed out that in some cases, particularly in families that have a history of genetic defects, the right *not* to know might be involved. "What do you do," he asked, "when a patient with a chromosomal lesion tells you his sister ought to know about it because she might be at a high risk of producing a defective child? Do you tell the sister? Suppose you do, and she says she would rather not have known? Giving people information that they might not want to have can be a serious ethical problem, and maybe a legal problem, too."

No one offered a solution to Dr. Robinson's dilemma.

Dr. King then raised the question of what kind of an impact the conferees could have on some of the problems they had been discussing. "How does a group of this size have any influence on the lives of people in the 25-to-40 age group?" he asked. "What actually can such a diverse group of people — from so many different disciplines and so many different institutions — hope to accomplish? Maybe what is needed is some sort of an action group that would meet regularly — some sort of a school, perhaps — to alert young people about the problems we've been discussing. By the time people reach 35 or 40 they don't change their lives very much. But if you could reach them while they were still in their twenties you might produce tremendous change. Colleges and universities don't attempt anything of that kind. It's too broad and too interdisciplinary. You would need some sort of a special school."

Dr. Aldrich carried Dr. King's suggestion one step further. He recommended that a new science be developed — a multidisciplinary study of what it means to be a human being. "What I have in mind," he said, "is the study of human beings in all cultures and nations, with a pronounced emphasis on social, cultural, environmental, biological, and cognitive similarities rather than differences. This knowledge could be put together into an integrated whole and presented in a skillful and comprehensive way to students at all levels of education, beginning in the elementary grades."

This new multidisciplinary science, he continued, would have to have the same kind of reward system and the same kind of prestige capabilities as the existing academic and scientific disciplines, and might have to be developed in a new kind of educational institution because most colleges and universities are structured along disciplinary lines and are unwilling or unable to start over again with an interdisciplinary point of view.

Dr. Aldrich went on to say that the kind of institution he

had in mind could well be established by the growing number of first-rate scholars and scientists who are being forced to retire while still in their most productive years. Moreover, he continued, the institution should be international in scope and should foster an international dialogue among the world's top-ranking scholars and scientists. He said that in connection with some work he had been doing for the United Nations he had been in touch with a number of research scholars and scientists from various parts of the world and had found them "literally hungry" to work together on common problems.

Mr. Eurich said he agreed with Dr. Aldrich on the importance of establishing new institutions as a means of encouraging change. "Established institutions aren't interested in change," he said. "Our system of higher education in the United States is a prime example of that. It grew out of the development of new institutions — from the liberal arts college of Colonial days, to the land-grant college of the nineteenth century, and down to the present-day community college. All of these were created to meet new problems and challenges that the established institutions could not, or would not, handle."

Dr. Schweppe also endorsed Dr. Aldrich's recommendation, and expressed the hope that such an interdisciplinary group of research scholars and scientists might be able to come up with a set of guidelines that would help people in the 25-to-40 age group cope more effectively with their stresses, strains, and problems. "For example," he said, "it might be possible to identify those factors that lead to a happy family life and those that lead to dissonance and conflict. We certainly need some sort of guidelines that will help to minimize the social, financial, emotional, and environmental strains that weigh upon individuals and families."

Dr. Aldrich also recommended that the communications media here and abroad pay more serious attention to the human life span as a totality and to human development as a continuous process. "They pick up sensational items dealing with childhood, adolescence, or old age," he said, "but they pay no attention to the serious efforts that are being made around the world to synthesize what we know about the human life span and human development, and how events at one stage in the life cycle lead to other events at later stages."

Dr. Aldrich went on to recommend a world-wide study of values and the principles that people live by, with particular attention to those that might be exportable. "This might be very helpful in reducing the amount of tension, hostility, and mistrust that exist in the world today," he said, "because such a study might well find that there are certain values and certain principles that appear to be common to the species of man."

Dr. Pellegrino said the conference report might well have some impact on the lives of younger members of the 25-to-40 age group. "Our discussions have been valuable in the sense that they have been spontaneous, candid, and provocative," he said. "My own style is somewhat more dialectical — presenting a thesis, arguing it, laying out the issues, and coming to conclusions—but although we have not done that I think we have highlighted some of the critical problems and offered suggestions for dealing with them. I hope the report will become the basis for a public discussion of the problems and issues we have identified."

Bernard Frieden said the conferees themselves might be able to have an impact on policies dealing with people in the 25-to-40 age group, not only in the institutions and disciplines they represented but also in Washington. "Many of us belong to networks that are called upon by Congressional committees for opinion and advice on legislation that is being considered," he said.

William Oriol agreed. "John Demos is currently a member of the Family Impact Seminar, sponsored by the George Washington University's Institute for Educational Leadership, that will help to shape federal policy in that area," he pointed out, "and there may well be other seminars or forums which could profit by some of the ideas and suggestions advanced at this conference." Mr. Oriol went on to suggest that the same kind of participatory relationship that Dr. Pellegrino recommended between physicians and their patients ought also to apply between government agencies and the people who are likely to be most directly affected by policy decisions. "The age of the unilateral decision ought to come to an end," he said. One step toward establishing the kind of participatory relationship he had in mind, Mr. Oriol suggested, could be a national poll of people in the 25-to-40 age group, aimed at finding out their attitudes toward federal policies that affect their lives.

Mr. Eurich urged the conferees not to underestimate the ability of the individual to bring about change. He cited the example of the late Beardsley Ruml, who originated the idea of putting the federal income tax on a pay-as-you-go basis in the early years of World War II. Ruml, then treasurer of Macy's department store in New York, could not get anyone in the Roosevelt administration interested in the idea, but he persisted. He took a year's leave of absence from his job, spent the year waging an intensive educational campaign directed at key people in business, labor, the professions, and the federal government, and finally succeeded.

40 Priorities for the Future

Education must be at the center of any effort to improve the lives of young adults.—Edmund Pellegrino

It has been said of some conferences that they need not have been held at all, because the recommendations were written in advance and most of the discussion dealt with how they should be worded.

This was not true of the Aspen conference on young adulthood. The recommendations were not prepared in advance by the steering committee that planned the conference. They were made by the participants as individuals, and they grew out of the conference discussions. Most of them were spontaneous, but all of them were thoughtful and insightful. Some dealt with research, others with programs of action. They extended over the full range of the conference agenda. Most of them have been reported in earlier sections of this report, in the context within which they were made, but for the sake of convenience all of them are summarized here according to the areas of concern to which they were addressed.

427

Demographic Data

Mr. Norton, the Census Bureau's specialist on marriage and family statistics, said that the available socio-economic information about people in the 25-to-40 age groups is far from adequate, and needs to be supplemented by detailed information on the dynamic interaction between parents and children, wives and husbands, and members of peer groups. Especially needed, he said, is better information about the impact of interior and exterior forces on family life.

The Human Life Cycle

Several of the conferees made the point that much remains to be learned about the human life cycle as an entity, the special biomedical and behavioral aspects of each of its stages, and the relationships between stages.

Dr. Robinson made the point that no study of the human life cycle can be complete without some consideration of the periods of conception and gestation, because genetic "errors" at the very beginning of the life cycle have profound effects on subsequent health and behavior. He and Dr. Aldrich stressed the need for more research about the prenatal stage of the life cycle.

Dr. Aldrich said that very little information was available on the biological changes that take place in people at various stages of the life cycle, and suggested the creation of a new branch of medicine that might be called "mediatrics," which would be concerned with making follow-up studies of age-group cohorts as they move from childhood through adolescence to young adulthood. Such a discipline, he said, would gradually build up a new body of knowledge comparable to the one developed by pediatrics. He also recommended the development of ethical as well as medical standards that might serve as guidelines for physicians who are being besieged with requests for artificial insemination and laboratory fertilization.

Dr. Schweppe emphasized the need for a better understanding of the relationship between the biological and environmental factors that affect human behavior at all stages of the life cycle.

Health Problems

Several of the conferees stressed the need for more widespread and more effective genetic education and genetic counseling as a means of forewarning young adults about genetic defects or susceptibilities that might be transmitted to their offspring. It was emphasized, however, that genetic counseling should be descriptive rather than prescriptive — that the counselor should not attempt to interfere in the choice of marriage partners or in the decision of married couples whether or not to have children. Nonetheless, Dr. Robinson did raise a question about the morality of permitting the preventable birth of a severely defective child.

Dr. Robinson was not alone in contending that federal funding of medical research is insufficient and capricious. Several of his colleagues, speaking out of their own experience, seconded his recommendation for a major overhaul of current procedures. Sudden and unexpected interruptions in the flow of funds are especially damaging to long-term research, they said.

One of the conferees recommended a new type of multiple screening system that would make possible the early identification of social, psychological, and environmental stresses to which particular individuals seem highly susceptible, similar to the screening tests that are now used in the medical profession to identify genetic deficiencies and congential diseases. Dr. Robinson, pointing out that there are genetic subgroups that are more susceptible to social and environmental stresses than others, emphasized the need for additional research in the whole area of genetic susceptibility.

During the discussion of health care, Dr. Schweppe said

that one of the principal reasons for the increase in hospital costs has been a rise in the ratio of support staff to professional staff, and he recommended that hospitals make a special effort to hold down that ratio.

The abuse of drugs and alcohol were cited as particular hazards to the health of men and women in the 25-to-40 age group, and several of the conferees urged a more intensive educational effort to warn young people about these hazards. A few of the conferees questioned the effectiveness of educational programs, pointing to the large number of people who are still smoking, but the consensus seemed to be that past failures should not preclude continuing efforts. Dr. Marine suggested that communications specialists might be able to devise a more effective educational campaign than members of the medical profession.

Dr. Cahill recommended two simple preventive measures that people in the 25-to-40 age group might take to improve their changes of survival — pap smear tests for women, and blood pressure checkups for both men and women.

Regular, moderate, and appropriate exercise was also recommended for people in the 25-to-40 age group, but several of the physicians who participated in the conference warned that jogging, a popular pastime among young adults, can be dangerous for some people.

Although it seemed to be the consensus among the physicians that annual physical examinations are advisable but not absolutely necessary, Dr. Pellegrino did suggest an occasional lifestyle review — a review of the patient's habits, attitudes, family situation, and other factors that can affect health.

The physicians repeatedly stressed the importance of treating their patients as unique individuals, rather than as members of any particular age group. They agreed that the patient's right to know what was wrong with him, and the possible consequences, took precedence over the physician's natural inclination to spare the patient undue worry or anxiety, but Dr. Robinson posed a dilemma that went unanswered: What about the right of other members of the family not to be informed?

Mr. Edmunds suggested that the medical profession develop special courses in biology — to be taught as early as high school — that would provide a better understanding of the body systems, how they function, and what can be done to keep them in good working order.

Susan Klein said she thought health education should begin early and should emphasize the need for developing a sense of personal responsibility for one's health.

Dr. Cahill said that medical science has gone about as far as it can currently go in helping more people to live longer, and should now pay more attention to improving the quality of life.

Work Force Participation

As a means of alleviating job frustration and dissatisfaction among young adults, Mr. O'Keefe recommended work sharing, sabbatical leaves, a further shortening of the work week, and an increase in the compensation rate for overtime work. "But our best bet," he said, "is to keep on creating new kinds of jobs that require higher and higher levels of skill." He called attention to the need for extensive retraining programs to help workers upgrade their skills, and recommended better job training and a more sympathetic response to the special needs of working women as ways of making it easier for women to participate in the labor force.

Mr. Oriol recommended that the federal government, as the nation's largest employer, take the lead in improving job satisfaction among people who work in offices and factories.

Dr. Kelso suggested that one way of helping people in the 25-to-40 age group advance faster in the professions would be to allow distinguished members of the professions to begin a "half-time" working arrangement when they reached the age of 50, under which they would devote about 50 per cent of their time for the next fifteen years to their regular work and the remainder to special pursuits that they never have time for because of the time-consuming trivialities of their profession. "That wouldn't work for everybody," he said, "but it would be a boon to a lot of our most creative people. Special annuities or federal fellowships could help offset the reduction in income."

Housing

Mr. Frieden made a number of recommendations for alleviating the major housing problems among people in the 25-to-40 age group. Among them were the following:

1. A revival of the federal program for encouraging home ownership for the working poor.

2. Modification and simplification of the present maze of municipal building codes, zoning ordinances, and environmental regulations that discourage the construction of new homes and add to housing costs.

3. A restructuring of the present system of financing home ownership to match the changing pattern of family income.

4. Special help for people who want to buy and rehabilitate older homes in the central cities.

5. Stabilizing the housing industry so as to eliminate the peaks and valleys in new housing starts.

Family Problems

Since the conferees devoted more time and attention to the family and its problems than to any other subject discussed at the conference, it is not surprising that they came up with a long list of suggestions and recommendations for alleviating the stresses and strains of family life. The list is summarized here in its barest outlines.

Susan Klein recommended a new conceptualization of the family and the roles of its members, and additional research about new variants of the traditional nuclear family.

Dr. Visotsky made two sets of recommendations, one addressed to the federal government and the other to the conference participants. To the federal government he recommended:

1. Establishment of an income transfer or maintenance program for all families, guaranteeing a basic level of economic support and stability.

2. Development of a national system of high-quality maternal and child health care services, including prenatal and obstetrical care, well-baby clinics, and adequate follow-up and treatment centers.

3. Special programs to assure vital family and child supports, such as maternity leave, sick leave for parents who must attend to seriously ill children, quality day care, visiting housekeeper services, constructive after-school programs for children, and job programs for youth.

His recommendations to the conference participants were:

1. To persuade industry to take a parental responsibility for the health and well-being of its employees.

2. To persuade organized labor to play a more important role in improving the quality of life of its members.

3. To persuade educators to begin to teach children, at a very early age, the art of coping with stress, and the importance of cooperation rather than competition.

4. To persuade medical schools to pay more attention to the quality of life in training physicians.

The following recommendations were made by other conference participants:

1. A complete overhaul of the welfare system, to make it less dictatorial and more responsive to the needs of families.

2. Changes in the tax laws — such as raising the allowance for spouses and dependents — that would help poorer families.

3. More convenient arrangements for providing banking, business, and professional services to members of families who work during the week.

4. A greater effort by churches to help young adults cope with the problems of marriage and family living.

5. An intensive effort by churches to supplement their religious work with more specialized services, such as health care, social work, and psychiatric assistance.

Intervention

As indicated earlier in this report, the conference participants approached the question of intervention in the lives of young adults with a great deal of circumspection. They were wary of too much intervention, whether it be by government, medicine, social agencies, or religious groups.

Dr. Schweppe summed up the feelings of his colleagues with respect to medical intervention when he said, during the discussion of genetic counseling: "There's no question about its value in preventive medicine, but we can't let it go too far. I think it would be mighty dangerous to let a doctor decide whether two people would be compatible marriage partners purely on the basis of a damned computer test."

Dr. Visotsky sounded a note of caution about government intervention when he said, in his remarks on stress and conflict in family life: "I'd hate to see the nurturing functions of the family turned over to HEW or any other federal bureaucracy."

Nevertheless, the conference participants did make some recommendations about strategies of intervention.

Dr. Pellegrino said that there ought to be two kinds of advisory bodies involved in the formation of public policy affecting individuals or families — one to decide whether what was being proposed *could* be done, and the other to decide whether it *should* be done.

Mr. Clark said that any intervention should be planned and carried out within the context of the principal values respected by our society.

Mr. Edmunds said that intervention should begin at the grass-roots level and work upward, rather than beginning at the top and percolating downward.

Dr. Aldrich suggested a new type of interdisciplinary study that would focus on what it means to be a human being. It would embrace people in all cultures and nations, and would emphasize similarities rather than differences. He also recommended a world-wide study of values and the principles that people live by, with particular attention to those that might be exportable.

Pointing out that young adults constitute the "takeover generation," the generation that is about to assume the top leadership positions in government, business, and the professions, Dr. Aldrich also recommended that in the interests of world peace and harmony it might be well for members of the "takeover generation" in the United States to get better acquainted with their counterparts in other countries — especially China and the Soviet Union — through an international exchange of one-year internships funded by a foundation or by the countries involved in the exchange.

And finally, Mr. Frieden suggested that the conference participants themselves could help to improve the lives of young adults through their own research and publication and through their affiliation with scholarly networks that are frequently called upon to advise Congress and the federal agencies on matters of public policy.

Selected
Bibliography

Books:

Aries, Philippe. *Centuries of Childhood*. New York: Alfred A. Knopf, 1962.

Bane, Mary Jo. *Here To Stay: American Families in the Twentieth Century*. New York: Basic Books, 1978.

Hechinger, Fred M. and Grace. *Growing Up in America*. New York: McGraw-Hill, 1975.

Levinson, Daniel J. *The Seasons of a Man's Life*. New York: Alfred A. Knopf, 1977.

Keniston, Kenneth and the Carnegie Council on Children. *All Our Children: The American Family Under Pressure.* New York: Harcourt, Brace and Jovanovich, 1977.

Vaughan, Victor C., III, and Brazelton, T. Berry (eds). *The Family: Can It Be Saved?* Chicago: Year Book Medical Publishers, 1976.

Reports and Monographs:

Clark, Robert. *Adjusting Hours to Increase Jobs*. Washington, D.C.: National Commission for Manpower Policy, 1977.

The Family, a special report by *Daedalus*, the journal of the American Academy of Arts and Sciences, Spring 1977.

Glick, Paul C., and Norton, Arthur J. *Marrying, Divorcing, and Living Together in the U.S. Today*. Washington, D.C.: Population Reference Bureau, Inc., October 1977.

Lalonde, Marc, Minister of National Health and Welfare, Government of Canada. *A New Perspective on the Health of Canadians*. Ottawa, 1974.

National Research Council, National Academy of Sciences. *Toward a National Policy for Children and Families,* 1976.

Spierer, Howard. *Major Transitions in the Human Life Cycle.* New York: Academy for Educational Development, 1977.

U.S. Bureau of the Census. *Current Popluation Reports,* Series P-25, No. 721, "Estimates of the Population of the United States by Age, Sex, and Race: 1970 to 1977." Washington, D.C.: U. S. Government Printing Office, 1978.

Articles:

Campbell, Donald T. "On the Conflicts between Biological and Social Evolution and between Psychology and Moral Tradition," *American Psychologist,* December 1975.

Breslow, L., and Somers, Anne R. "The Lifetime Health Monitoring Program: A Practical Approach to Preventive Medicine," *New England Journal of Medicine,* 296, No. 11, March 17, 1977.

Demos, John. "The Changing Shape of the Life Cycle in American History." Paper prepared for the Smithsonian Institution's Sixth International Symposium, "Kin and Communities: The Peopling of America," June 14-17, 1977.

Frieden, Bernard J. "The New Housing-Cost Problem," *The Public Interest,* Fall 1977.

"The Lessons of the Grant Study," *Psychology Today,* September 1977.

"Men and Women," a series of four articles published in the *New York Times,* November 27-30, 1977.

"Multiphasic Health Testing," a series of four articles with an introduction by Morris F. Collen, *Preventive Medicine,* June 1973.

Norton, Arthur J., and Glick, Paul C. "Marital Instability: Past, Present and Future," *Journal of Social Issues,* Vol. 32, No. 1, 1976.

"Saving the Family," *Newsweek,* May 15, 1978.

Thomas, Caroline B. "The Precursors Study," Unpublished manuscript, Johns Hopkins University.

Vaillant, George E. "How the Best and the Brightest Came of Age," *Psychology Today,* September 1977.

Visotsky, Harold M. "Job-Related Stress," *Medical and Health Annual, Encyclopaedia Britannica,* 1978.

Part IV
The Middle Years:
A Multidisciplinary
View

Joan Waring

41

Introduction to Part IV

A wise society will develop the most effective means possible to ensure the continued health, capability, commitment, and positive perspective towards life's opportunities and responsibilities for people in their middle years.—Donald H. Ford

Middle-aged adults are America's most vital resource. They are called upon to govern the nation, anchor families, teach the young, direct industry, lead citizen groups, and at the very least work to support the dependency of the young and the old. Even as they discharge these responsibilities, the middle-aged inevitably grow older and continue to face changes in themselves and in their lives which are inherently stressful. Coping with these ongoing changes—influenced by their past and bearing on their future—makes this long period of transition from youth to old age the "best of times" for some and the "worst of times" for others. Yet relatively little is known about the midlife experience. Even less is being done to relieve its strains.

From July 31 to August 5, 1977 in Aspen, Colorado, experts representing nineteen academic disciplines and

professional pursuits met to share their specialized knowledge and, together, arrive at deeper and broader under-
standings of the human life course, especially its middle
phase. Convened under the auspices of the Academy
for Educational Development and sponsored by the
Schweppe Research and Education Fund, the Second
Annual Conference on Major Transitions in the Human
Life Course was given a special mission. The conferees
were asked "to focus a multidisciplinary lens" on the
interaction between biomedical and social factors in the
aging process, especially as manifested in the middle
years. But there also was a larger mandate. In his opening
remarks Alvin Eurich, President of AED, charged the
conferees to recommend programs of action that would
improve the quality of life in the middle years and carry
it forward—as intact as possible—to enhance old age as
well. Further, the conferees were asked to specify what
interventions might be taken on behalf of those now
young, or even unborn, so as to assure that their full life
course might be as healthy and satisfying as possible.

The group assembled to confer on these matters included experts in anatomy, physiology, pathology, pre-
ventative medicine, pediatrics, geriatrics, rehabilitation,
biochemistry, genetics, gerontology, human development,
sociology, psychology, economics, theology, industrial
relations, communications, environmental design, and
government. They are listed in Appendix A.

This report describes the outcome of this broad multidisciplinary focus on salient domains of experience in the
middle years: work, health, and relationships. The report does not—indeed could not—include all that was
discussed at the formal sessions and their informal con-
tinuation after the daily meetings were adjourned. Rather
it reviews some of the ideas, concerns, and research findings presented at the scheduled meetings and described

in the papers that served as background reading. The report also contains the conferees' suggestions for ways by which the lives of the middle-aged might be made happier and more productive. The focus is on problems and remedies. To be sure, there were many attempts to accentuate the positive—and there is much positive about midlife—but in this report we accentuate the difficulties in order to enlist readers in the effort to make middle-age a more positive experience now and in the years ahead.

The time is ripe for such an effort. That several books describing the crises of midlife are already major best sellers attests to the interest members of the public have in understanding their own lives and how to manage them more successfully. The increasing frequency of seminars and symposia on the middle years indicates a growing interest on the part of the academic and professional communities. New programs to counsel people with midlife career problems, developed by the private sector and educational institutions, represent response to a need. What is required now is further support for research and scientifically-based programs of action.

Having a conference focused on "The Interaction Between Biomedical and Social Factors in the Human Life Course with Emphasis on the Middle Years" was itself a move in this direction. Too little is known about mature adulthood because of past failures to view the life course as a whole. Thus, the conference tried to reduce the knowledge gap created by the earlier almost exclusive attention to young children and, more recently, to older people. Further, by examining middle age from a life course perspective, the conference sought to find crucial linkages between all life stages so that intervention strategies to reduce stress and prevent problems could come into focus. But the conference did more. It underscored the necessity for specialists to move beyond the narrow boundaries of their disciplines and demonstrated the rewards for the study of aging that follow from collaboration among disciplines.

As preparations now begin for the White House Conference on Aging in 1981 and for the World Assembly on Aging possibly the following year, the organizers of these meetings might well attempt to emulate the approach used at the Aspen Conference. As planners look ahead to what old age may be in the future they must also look backward and "get to know" the middle-aged of the present. Plans for the "new old age" must relate to the characteristics and needs of those now middle-aged and not the unique experience of those already old. But these plans must also be made and implemented on the basis of a multidisciplinary understanding of the aging process. Growing older is not simply a biological or a social or a psychological process but a complex interaction of all three. To focus on one to the exclusion of others is to distort the way aging proceeds.

In the more immediate future are plans for the Third Annual Conference on Major Transitions in the Human Life Course scheduled for the summer of 1978. This conference will continue the forward-looking approach that has become the hallmark of Schweppe Research and Education Fund Projects: a multidisciplinary and life course perspective on critical life course transitions. The emphasis will be on Early Adulthood: Family, Health, Work, and Leisure. Again the objectives will be to identify common stresses and recommend ways to facilitate coping and improve the quality of life.

All of the conferences in the Schweppe Fund series are designed to advance the accumulation of knowledge about the human life course and its transition points and then to disseminate this knowledge. Sharing the same distinctive approach and objectives but differing in emphasis, each of the annual conferences builds upon and extends the work of previous ones. The first conference, held in 1976, focused on ways to study life course transitions and the stresses they generate. That conference was itself an outgrowth of an earlier meeting of experts in the social and biological sciences to consider the importance and implications of sociobiology for the human experience.

42 The Middle Years: A Life Course Perspective

Understanding the middle years—and designing interventions to improve them—must be based on understanding of individual development from birth to death . . . and on an understanding of how the social structure shapes that development anew for each generation.—Matilda Riley

Middle age, like prenativity, childhood, youth, or old age is merely an episode in the life-long experience of growing older. It is a phase of the aging process, an event in the sequence of events making up the continuum called the human life course. It is also a social invention. Studied in and of itself, severed from linkages to past and future, middle age can not be adequately understood. What individuals are like and feel and do during their middle years, said sociologist Matilda Riley, has been shaped by myriad influences from the past and has manifold portents for the future. Whether the focus is on physical capacities, mental acuity, health, interest in career change, or emotional crises, the situation at midlife must be viewed from the perspective of the full life course.

449

Like all other phases of the life course, middle age must be seen as part of an ongoing aging process involving the interplay of biological, psychological, and social events that make individuals different from one time to the next. As months and years go by, for example, more of the genetic program laid down at conception is expressed; more information is stored in the nervous system. Physical capacities develop and decline. Attitudes and values are learned and revised. Roles are taken on and relinquished. To be sure, the precise scheduling of these changes varies widely among individuals. Nonetheless, all these age-related events intersect in ways that create the stresses and satisfactions of growing older. In the interaction of these processes lie the origins of many problems that beset people in midlife and old age. By identifying these antecedents, efforts can be made to prevent or minimize the problems they cause.

But it is not enough to recognize that aging changes individuals. It must also be recognized that aging processes themselves can be changed. Aging is inevitable, but not immutable. The ways in which people age—biologically, psychologically and socially—change in response to changes in the physical and social environment. People born in different times and different places age in different ways. Indeed, Matilda Riley suggested, environmental factors so impinge on peoples' lives that it is difficult to imagine a pure aging process.

Consider, for example, how changes in degree of exposure to noxious substances, changes in norms affecting education and work life, changes in attitudes toward marriage and parenthood, changes in nutritional adequacy, changes in medical care have served to make the aging experience different from one cohort (people born at the same time) to the next. Consider also how cohorts born a decade apart, as well as a century apart, necessarily face a unique sequence of historical events—war or peace, prosperity or depression, natural catastrophe or new legislation—that affects what is learned, what is done and what is valued over a lifetime. Consider further how these "changed" people might alter the environment again so that later cohorts age in still other ways. The lesson for the study of middle age is clear: its content and context can undergo rapid change. The problems facing the middle-aged at one point in history are likely to be different at another point. Programs designed for one generation of the middle-aged may be irrelevant for the next.

Scientific investigations of aging processes thus must include not only observations of individuals over the full life course but also observations of the way successive cohorts age, Matilda Riley said. The interdependence of life course patterns and social conditions must be recognized in considering relationships between biomedical and social factors in the middle years and in recommending ways to resolve the problems that occur.

Much as the middle years are not usefully isolated for intensive study, neither are debates on their age-boundaries very fruitful. The boundaries of any life-stage are entirely arbitrary, set by prevailing social custom and subject to rapid change. For example, abolishing mandatory retirement may extend the upper limit of middle age to seventy, much as youthful protesters of the last decade attempted to lower its entry point to a day over thirty. In addition, subjective experience further modifies life-stage boundaries. One 50-year-old man, for example, may feel himself old because of crippling disease while another born the same day may feel young because of a new romance. There is great variation in characteristics within an age category. The boundaries used at the conference predictably varied by participant, but centered roughly around the age range of 40-65.

The Middle-Aged of Contemporary America

The middle-aged of the here and now are different from their predecessors. Their demographic characteristics as well as the circumstances of their birth, childhood, youth, and early adulthood are unique. While the same could be said of all new cohorts occupying the middle-aged slots in a society, the middle-aged of contemporary America have charted an unusual number of new pathways in the process of growing up and growing older. As they move toward old age in the decades hence, they seem likely to continue their innovative ways.

Demographically the middle-aged are different because they belong to small birth cohorts. Most of the younger members of their ranks are the issue of the low fertility rates prevailing around the time of the Great Depression. At present, there are actually fewer people who are middle-aged than just a decade ago. (See Figure 1). Their spare number relative to other age strata has consequences for the present and the future. For now, it means the middle-aged carry extremely heavy burdens of dependency in supporting the more numerous young and old. They are the ones who must not only pay for the education of the young but who also must make the costly income transfers to support the Social Security system for the old. By contrast, when those now middle-aged reach the current age of eligibility for full Social Security benefits, they will make but a modest increase in the ranks of the old. At that time, however, the ranks of the middle-aged will be swelled because of the aging of the large cohorts that were born in the post-World War II period.

Using age 50 as a point of reference, William Oriol, staff director of the U.S. Senate's Special Committee on Aging, recounted some of the experiences that current cohorts of the middle-aged have lived through—experiences that have shaped their orientations and actions.

Figure 1

Distribution of the Total Population, by Age and Sex: April 1, 1970 and July 1, 1976

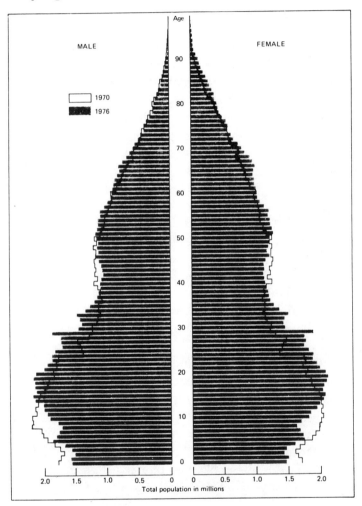

Source: U.S. Bureau of the Census, 1977. *Current Population Reports*, Series P. 25, No. 643, "Estimates of the Population of the United States, by Age, Sex, and Race: July 1, 1974 to 1976."

Oriol pointed out that their childhood was spent in the Depression. World War II was impressed on their youth, with many of the males in this cohort serving in that conflict. They entered college and embraced the responsibilities of adulthood during the Cold War with its "balance of terror" between nuclear nations. They made the transition to adulthood swiftly, finishing school, leaving home, marrying, and becoming parents in practically record time. They bought homes during "the flowering of suburbia" and became the parents of the "baby boom." The women in the cohort nearly all married and made mothering a full-time occupation, but then as children grew older joined the work force in fast accelerating numbers.

The members of this cohort, once fighting for democracy and persuaded of prosperity, then entered their middle years adjusting to campus rebellions, an unpopular foreign war, recession-inflation, and the looming of an energy crisis. Gradually, they saw national attention turn to such "aging" issues as Medicare and Social Security, and began to develop an understanding—and concern—about the far-reaching changes that will occur as the "graying of America" continues.

By reviewing the social biography of members of this cohort we should be better able to anticipate their needs and wants when they become old. What kind of senior citizens will they be? Will they be satisfied with extant programs for the aged? What future changes in society might direct them along new paths?

Attempts to answer these questions raise additional ones. For example, likely to be freed from the provisions of mandatory retirement, will this new and better educated cohort continue the current trend toward early retirement? Or will they fashion a new pattern for old age as they calculate that savings and pensions and Social Security benefits are inadequate to preserve an acceptable standard of living? Will the women continue to work as long as they are able? Will they seek greater educational opportunities for themselves? It is probable, Oriol predicted, that members of this cohort will be "involved in experiments—probably sooner than we expected—that will change the divisions between work and retirement for the better or worse."

What publicly provided services will they seek to enrich their lives? Will the middle-aged when old be attracted to such currently popular programs as Foster Grandparents or Senior Aides or Green Thumb? Will they want federally-underwritten breakfasts and lunches in community facilities? After years of relying on the car, will they be satisfied to wait for senior citizen minibuses? Will they want housing that segregates them from people of other ages? Will health care delivery be adequate as well as relevant to their old age? Will they demand alternatives to hospital and nursing home care?

Oriol pointed out that recent experience with programs for the elderly has revealed a mechanism for keeping public services relevant: organized neighborhood groups that communicate their specific needs to locally-based agencies concerned with old people. (Obviously, these agencies must have enough discretionary power and funds to meet these needs.) If fostered, this device will not only be responsive to continuing and emergent needs of the old, but will also keep them involved, productive, and developing to the end of life.

Here we have emphasized the past in order to speculate about the future of those who are now middle-aged. Next we focus on specific areas in their lives which are important in their day-to-day experience.

43 The Meaning of Work

. . . the fortunate position of the majority of the middle-aged makes no more tolerable the misfortunes of those for whom poor health or involuntary job separation lead to departure from the labor force or slide down the occupational ladder, or, indeed, of those who have never acquired a decent and stable job.—Paul Andrisani and Herbert Parnes

The major fact of midlife is work. Most middle-aged men who are not disabled are in the labor force as full-time, full-year workers. Most have jobs that are stable. Most like their work. And a large majority confess they would continue to work even if they had money enough to live comfortably without doing so.[1] More and

---

[1] Paul J. Andrisani, Eileen Appelbaum, Ross Koppel, and Robert C. Miljus, *Work Attitudes and Labor Market Experience: Evidence From the National Longitudinal Surveys.* Philadelphia, Pa., Center for Labor and Human Resource Studies, May, 1977.

more women are also making work a central activity of their middle years. No longer are many willing or able to preside over an empty nest.

According to the National Longitudinal Surveys (NLS) commissioned by the U.S. Department of Labor, the work situation of the middle-aged male has been generally favorable in recent years. Reporting on the findings of these studies, labor economist Paul Andrisani noted that the men interviewed had on average "moved up the occupational ladder during the course of their careers and regarded their current occupations as the best they have ever had."[1] More than either the young or old, middle-aged men generally hold the kind of work roles that bring high income and opportunities to exercise power and have greater prestige. Not only do middle-aged men (average age 55) claim the majority of executive positions, but even those in lower ranks claim the advantages of seniority and greater job security. As Donald Ford observed, many of these men are expected to supervise subordinates, teach the inexperienced, hire and fire, and make difficult decisions affecting the lives of others. These stresses and their consequences notwithstanding, middle-aged men typically register high degrees of job satisfaction. For those most satisfied with their jobs the intrinsic rewards of their work are liked best—notably the nature of the work itself.

Yet middle age also marks the onset of a range of frustrations and problems involving work. While better-educated workers may still look forward to reaching higher rungs on the occupational ladder, many middle-aged persons cannot entertain such hopes. As one conferee put it, "they can no longer expect recognition for

---

[1] Andrisani and Parnes, 1977. "Five Years in the Work Lives of Middle-Aged Men: Findings from the National Longitudinal Surveys," a paper prepared for the conference.

their future potential." It is a time of "put up or shut up." Some will never realize their career goals. Some, especially those with less schooling, may even experience a slippage in position, either relative or absolute. Many will be forced into reconciling former aspirations with current achievements, a familiar component of the "midlife" crisis.

To be sure, the earnings of middle-aged workers are likely to continue to increase, reflecting trends in the economy as a whole. However, the period of rapid income growth that characterized an earlier period in their work life is likely to come to a halt. If the slower rate of income growth is accompanied by reduced financial responsibility for dependents or the return of the wife to the work force, then the middle-aged may experience a rise in their standard of living. If, however, this reduction in income growth coincides with a need to meet new expenses, such as the costs of higher education for children or of nursing-home care for aged and ailing parents, then financial pressures are added to the other stresses of midlife.

Becoming middle-aged brings other problems in a competitive labor market. Andrisani pointed to several handicaps stemming simply from ongoing changes in the society—changes which leave the middle-aged behind. For example, as a result of a continuing trend toward more years of formal schooling, the middle-aged are likely, on the average, to have more modest educational attainments than the new cohorts entering the labor market. As a result of the introduction of new technologies, the earlier training of the middle-aged may be less relevant to current job requirements than that of the more recently trained. As a result of long tenure in a job, the middle-aged are more likely to be located in older and non-expanding segments of the economy, and hence more susceptible to foreclosed opportunities, lay-offs and plant closings. Should a middle-aged male lose his job, these several disadvantages often combine with evasions of the anti-age discrimination laws to make finding re-employment a difficult and discouraging task.

Unemployment

Although middle-aged workers are less likely to experience unemployment than younger workers, when job loss does occur it may be almost catastrophic in its consequences. The NLS data show that the period of unemployment is often prolonged. A similar finding was reported in a working paper prepared for the U.S. Senate's Special Committee on Aging: "Older workers, during good times and bad, do not find jobs as rapidly as their younger counterparts."[1] The Senate researchers, William Oriol noted, also found a relative scarcity of job openings for men in this age category. They dubbed the older worker the "recession's continuing victim."

When the economy is sluggish, the middle-aged job seeker, lacking good information about what job opportunities exist, is particularly disadvantaged. As a candidate for a new job, he suffers by comparison to younger competitors. Possible educational handicaps are only part of the problem. Often more damaging to his chances is the view held by some employers (however illegal if acted upon) that the older worker is more likely to become ill or disabled, or will be more costly in terms of fringe benefits such as insurance or pension. Further there is the problem of outright prejudice. Some employers simply do not like older people.

---

[1] Marc Rosenblum, "Recession's Continuing Victim: The Older Worker," A Working Paper prepared for use by the Special Committee on Aging. U.S. Senate, U.S. Government Printing Office, Washington, D.C. 1976.

However grim the immediate situation may be, the long-term effects of middle-age unemployment may be still more devastating—even if new employment is found. Summarizing a NLS study that traced the subsequent experiences of job losers, Andrisani said: "Even after they find other work, many individuals continue to suffer the consequences of displacement through less attractive occupational assignments, lower earnings and perhaps some damage to physical and mental well being."[1] Corroborating the latter part of this statement, Matilda Riley referred to the accumulating literature on the relationship between job loss and deleterious changes in health, especially in the absence of adequate social support.

Women, William Oriol reminded the group, are often overlooked when examining the problems of unemployment in the middle years. Yet a growing number are actively seeking work but unable to find it. Many of these women are "displaced homemakers," who, because of divorce, illness or death of the spouse, or loss of family income, seek entry into the labor force as bread winner for the family or sole support of the self. Years of unpaid work in the home, however, have typically left them without unemployment benefits to weather the present, or pensions or Social Security benefits on which to rely in the future. Many lack marketable skills. While programs have sprung up to convert housewifely competencies into labor force assets and to provide retraining or new training for displaced homemakers, it is still too early to tell how successful the programs are in meeting the need.

What can be done about the unemployment problems of the middle-aged? On one point the answer is clear:

---

[1] Paul J. Andrisani and Herbert S. Parnes, 1977. "Five Years in the Work Lives of Middle-Aged Men."

Keeping the economy vigorous provides jobs for people of all ages. Other measures however, are needed when these conditions do not prevail:

- wider dissemination of information about labor market opportunities
- training for those who need and can profit from it
- enforcement of the Age Discrimination in Employment Act
- government as employer of last resort.

Without work and its various rewards, said one conferee, the middle years are likely to become meaningless and miserable.

Poor Health

Increasing age brings increasing risk of chronic illness and disability. Such health problems, said Andrisani, are the source of the most serious labor market problems confronting men in their middle years. Inevitably poor health interferes with work and undermines the quality of life in countless ways.

The NLS data provide some of the richest documentation available on the impact of poor health on the work histories of men in their middle years. By asking respondents to indicate whether or not they had a health problem that somehow limited the amount and kinds of work they do, researchers were able to trace the long-term effects of such problems. At the time of the initial questioning in 1966, when the men in the NLS sample were between the ages of 45-59, approximately 25 percent reported some health-related interference with work capacities. When questioned again five years later, 30 percent of those surviving the interval reported some work-limiting condition. Not unexpectedly, limitations were more pronounced in the older half of the cohort.

Nonetheless, a considerable proportion of all middle-aged workers labor under some physical handicap.

While the NLS data do not provide information on the specific nature of the work limitations, they do contain abundant data on the consequences of such limitations. The chief effect of a health problem, said Andrisani, was to reduce labor force participation. Over the five-year study period, for example, operatives and laborers with health problems were out of work between 76 and 120 weeks more than their healthier peers. Unfortunately, days out of the work force mean days without pay or only partial pay. It was found that those with health limitations earned "substantially less'" than those without. Another finding of the NLS was that blue-collar workers lost more work days than white-collar workers. Yet, unhealthy white-collar workers suffered the greatest absolute dollar loss. Because of their higher income, each non-working day was more costly. Among blacks, however, blue-collar workers with health problems were doubly disadvantaged: losing more days of work and losing more income than their white-collar counterparts.

But temporary withdrawal from the work force was not the only consequence of health limitations. Physical problems also appeared to reduce incentive to undertake a long and thorough job search in the event a former job was lost. Further, some health problems simply hastened the decision to withdraw from the work force entirely. Characteristically, those who retired prematurely for reasons of health had "woefully inadequate" incomes and meager assets to fall back on.

Not surprisingly, those who reported some kind of health problem in the initial questioning were two or three times more likely to die during the research period than those reporting no impairments. Blacks, especially those in the younger half of the cohort, had an appreciably greater risk of dying than whites of similar age and health status. The highest mortality rates among both white and black men were found among service workers and nonfarm laborers. The lowest mortality rates were found among farmers and farm managers.[1]

While some health problems of the middle-aged are related to deleterious biological changes associated with the aging process or are the outcome of a lifetime of circumstances or habits inimical to health, many other problems have their origin in the work place itself. Citing reports from the National Center for Health Statistics, Andrisani said that approximately 10 million workers each year require medical treatment or suffer a measure of restricted activity because of work-related injuries. Referring to another study, he noted that over 14,000 workers died from a work-related injury or disease in 1970. Another 90,000 were permanently disabled, and more than 2.1 million were temporarily disabled. And the situation gets worse. In the seven years since the Occupational Safety and Health Act was passed, it has managed to limit exposure to only 17 of the approximately 1500 carcinogens that scientists believe are present in the work places of the nation.[2]

Several participants in the conference referred to other work hazards of the middle years, such as sedentary jobs, noise levels, the "three martini lunch," and long hours. Also mentioned as undermining both physical

---

[1] Paul J. Andrisani, 1977. "Effects of Health Problems on the Work Experience of Middle-Aged Men." *Industrial Gerontology,* Spring, pp. 97-112.

[2] David Burnham, "Health Hazards in the Work Place," *New York Times,* October 6, 1977.

and mental health were work-place stresses resulting from responsibility for others, the operation of the "Peter principle," and not liking a job but being unable to change it. Still another linkage between the work place and optimal functioning was pointed out. Leslie Libow, in his presentation, observed that a complex and variable day favored longevity. Matilda Riley reported a related association between job complexity and intellectual flexibility. Thus the degree of challenge inherent in the work itself can serve to depress or enhance well being.

What can be done to improve the health of workers or make the consequences of ill health less devastating? The general solution, the conferees agreed, is to encourage good health practices on the part of individuals, to reduce health hazards in the work environment, and to make the best medical care accessible to all people throughout the life course.

Until these conditions are implemented, interim measures were recommended, such as:

- study the health problems that limit labor force participation so that effective interventions might be made
- enforce compliance with the Occupational Health and Safety Act
- change Social Security provisions to provide adequate social and financial support for those who must withdraw from the work because of poor health
- screen workers so that those with family histories of cancer are not exposed to carcinogens and those with family histories of heart disease avoid undue stress
- increase the emphasis on industrial medicine in medical school curricula.

Investments in health care, said the conferees, not only have "salutary effects" on the economic status of the middle-aged but on life satisfaction as well.

Career Change

More and more, middle age is becoming associated
with thoughts of second careers. The timing is appropriate. By then sufficient work time has elapsed for a
fair assessment of the fit between self and current career,
and sufficient lifetime is left to embark on another one.
But it appears that more career changes are contemplated
than carried out in midlife. Although conference participants estimated that a large majority of men seriously
think about making an occupational shift, data from the
NLS indicate that approximately 20 percent of the men
in that sample actually changed employers either volun-
tarily (8%) or involuntarily (12%) over the five-year
period. Altogether, one-third of the sample changed jobs
and about one-fourth crossed the boundaries of a major
occupational group. How much of this job movement
represents planned entry into second careers—as opposed
to promotions or demotions in the same career line—
is not certain.

Those who choose to change jobs, on the average,
profit from doing so, Andrisani reported. Typically,
earnings and job satisfaction both go up. The advantages
attending job change notwithstanding, increasing age is
still a deterrent to job mobility. There are several constraints on the older worker. Fears of fringe-benefit loss
discourage some potential job changers. Fears of en-
dangering the growth of accumulated pension benefits
dissuade others. For many, the major problem is preserving family income during a lengthy period of training
for a new and different career or while waiting for financial returns from a new venture to stabilize. Some, while
motivated to leave their old career, lack direction as to
what new career to pursue. Others are reluctant to ask
their families to make the sacrifices that a career change
might entail. Whatever the most problematic barrier, the

probability of a job change and of improvements in income therefrom are reduced in a sluggish economy. Recession is hostile to taking chances.

Even in the absence of barriers, career changes in midlife should not be undertaken without considerable forethought and planning, warned psychiatrist Eric Pfeiffer. "Don't rock the boat if you can't swim" was his counsel. Observing that people often have lingering attachments to the "dream jobs" of their youth, he suggested these fantasies be re-examined, recognizing that one "is not the same person" many years later. He has found that a frequent outcome of such reassessment is a reaffirmation of commitment to the present career or but a minor change—though sometimes it can lead to a major change.

Good planning as well as the preparation necessary for a second career require professional guidance and opportunities for continuing education, Pfeiffer said. Leslie Libow, a geriatrician, noted that patients often first bring their career problems and work-related illnesses to their medical doctors. Yet, he suggested, many physicians are not well informed as to what alternative occupations are feasible in terms of labor market opportunities. Unaware of the structural barriers to midlife job changing, doctors may inadvertently add to frustrations by encouraging a career change that will be difficult to achieve.

Some of the benefits of a second career may be gained, however, despite the discouraging realities of the labor market. High levels of job satisfaction, for example, can be derived from socially useful work that is not remunerative. Merrell Clark, a foundation executive, observed that programs such as the National Executive Service Corps, Senior Medical Consultants, and National School Volunteers afford many satisfactions to the retired who participate. The middle-aged might find that such participation would enhance their lives as well. William Lea, member of the clergy, suggested that cultural or avocational interests be developed more fully to enrich midlife and to bring meaning in old age.

What can be done to facilitate career change in the middle years, if desired or required? A long-term societal goal would be to break out of the long established sequence of education, work, and retirement as the design of the life course. Frequent entries and exits into these domains could become normative rather than atypical, rewarded rather than punished. Until that time, other measures were proposed, including:

- improve midlife career counseling services
- encourage those making career changes
- expand opportunities for continuing education
- remove barriers to labor market mobility
- protect the accumulated benefits of middle-aged workers during a job change, and also the growth potential of these benefits.

In general, the conferees agreed that changes in career trajectories would increasingly characterize work experience in the middle years.

Retirement

Retirement, said Donald Ford, should be viewed as a "change in circumstances that allows people to grow and develop." But often it is not. As Eric Pfeiffer observed, approximately twenty-five years are given to the transition from play to work, but usually not more than six months are allowed for the transition back from work to play. As a result, retirement often becomes a sudden letdown, "a terminal drop-off," with devastating consequences. Thus the conferees urged that middle-aged people prepare for retirement by developing the capacities that might enhance satisfaction and survival after departure from the work force. Such preparation for retirement should include not only provision for gradual curtailment of work and work commitment, but also efforts to strengthen interpersonal relationships so that they might better sustain the leisure time in retirement, and to improve health habits so as to decrease the infirmities of old age.

Andrisani said that his analysis of the NLS data showed no widespread dread of retirement. On the contrary, many men looked forward to retirement, and fully 40 percent said they intended to retire early. Indeed, he said, the trend toward early retirement appears to be getting stronger. Those who choose to retire early, however, are the comparatively "better off." Typically, they have liberal retirement benefits, no dependents, and often a relatively unfavorable attitude toward their job. How those who choose to retire early fare, however, remains an unanswered question.

Retirement plans, patterns, and trends may change dramatically as a result of recent legislation abolishing mandatory retirement before age 70 and raising the earnings limit for retirees receiving Social Security benefits. Nonetheless, middle age needs to be the time when preparations for old age begin.

Leisure

As counterpoint to the discussion of the problems of work in the middle years, discussion now and again focused on problems of leisure. Many of the conferees noted that middle-aged people often do not have much leisure. The load of responsibilities leaves little time for recreation. Spenser Havlick, an environmentalist, further observed that neither do they often have leisure facilities accessible. While recreation planners and designers make provision for children, young families, and sometimes for old people, they typically neglect the needs of the middle-aged. In many places, for example, there is little opportunity for the middle-aged to get the kind of exercise they desire or require for good health. Moreover, vacation areas—however distant from the city—have become increasingly crowded and urbanized. National parks as well as private resorts afford less and less opportunity to "get away from it all." Respite, for the middle-aged, remains elusive.

Havlick pointed to another problem in the making. As middle-aged singles increase in number, "the lack of a common gathering place becomes an increasingly serious deficiency." Seldom, he added, is there a "plan, program, or design for the middle-aged whose social propensity is high but who prefer not to congregate in a bar or a night club." But there still is a larger problem affecting people of all ages, he indicated. Few communities are designed to facilitate genuine community.

Without leisure and the ability to use it well, middle age is poor preparation for old age.

44 The Importance of Health

*Even though most of us may not wish for immortality, we do wish to maintain our health and vigor to the end—or close to it, at any rate . . .—*Albert Rosenfeld

How the physical organism grows older is better understood than why. As yet there is no universally accepted theory as to what mechanism triggers the senescent changes that seem inevitably to increase the risk of death with increasing age. There is, however, an emerging consensus that more than one process is at work: Nature has an arsenal of fail-safe devices to ensure that people do not live forever. If disease or accident do not prove fatal, old age will. But the future may well be different, suggested science writer and editor Albert Rosenfeld, if researchers soon locate the crucial aging clocks and begin to understand and alter their ways. Yet, doubts remain about extending the human life span. Fears as well.

471

Theories of Biological Aging

At the moment, many hypotheses about biological aging compete for serious attention. Some have relatively long and respected histories and have not yet been rejected. Some are only emerging from laboratory experiments and need further study. Some are reasoned conjectures impossible to prove one way or another. Some await exhaustive tests. Others seek incorporation into a larger theoretical framework. Many are not mutually exclusive. More than one is likely to be true. And few theories contain a promise for "life almost everlasting." Nonetheless, adults in their middle years are continually undergoing time-related biological changes that affect how they feel and how they function. Thus the conference participants discussed some of the theories of aging.

The fact that each species has a characteristic life span, noted Rosenfeld, persuades many that genetic programming is the key to the extent of life. Having long-lived forebears also seems to confer an advantage of additional years. Buttressing this position are the experiments conducted by Hayflick and others that have shown that normal human cells are mortal. When placed in tissue culture, cells taken from the lungs of embryos undergo fifty or so doublings and then deteriorate and die. Similar cells from adults undergo fewer proliferations before they, too, slow down, stop dividing, and die. The predictable death of these cells in the laboratory is viewed by some as evidence of an intrinsic limit to the life span. The cells seem programmed to age and die. Elaborating on this position, Rosenfeld suggested that the later phases of the life course may be as much a part of genetically determined developmental biology as prenatal life or the onset of puberty. And it may be possible, he continued, that aging is a kind of birth defect that at some point in the future can be corrected by modifying the genetic program to prolong life.

James Birren provided a somewhat different perspective on genetic control of senescent change. Since natural selection can favor and pass on to the next generation only those traits that enhance prospects of survival through the reproductive years, the puzzle is to explain those regularly appearing traits that are observed in the post-reproductive years. Birren postulated that the same selective pressures that serve to increase the probability of survival in early life may have counterpart characteristics that are manifested as some of the ordered, predictable and often detrimental biological changes of later life. For example, having the blood pressure elevate in response to stress is adaptive for the young. After years of stress, however, this adaptive response may result in a permanent elevation of blood pressure (hypertension) and thus increase the risk of death.

Other theories of aging focus on the cell. In general, this line of reasoning holds that over time something is bound to go wrong at the cellular level because of chance or the effects of environmental insults. As a result, cell function is in some way impaired and it cannot long survive. For example, various "error" theories suggest that at some point in the complex multistage process of providing genetic instructions for the manufacture of essential enzymes or proteins in the cell, a mistake is made so that the final product is not correct. Even a small mistake in critical cell processes may compromise function; the cell "ages" and in time dies.

The "cross-linking" theory holds that, over time, damaging bonds are formed between two parts of a single molecule or between two different molecules. These irregular linkages alter the physical and chemical properties of the molecules, interfere with cell functioning, and eventually lead to the aging of body tissues. Another theory implicates "free radicals"—pieces of molecules that have somehow become detached during the cell's normal oxidation processes. Unstable, free radicals enter into reactions with other nearby molecules, change the structure and function of cell enzymes and proteins, and so damage cells. Other theories argue that the DNA itself is damaged by hostile elements in the environment. One such theory (though disputed) holds that radiation causes deleterious somatic mutations which accelerate the aging process. Other hypotheses implicate chemicals, infections, or nutritional deprivations as agents of injury to the DNA. Nonetheless, DNA is capable of initiating a self-repair process so that damage to it—unless total—is likely to be temporary.

Several prominent theories of biological aging indict the failing efficiency of body systems. One explanation, for example, focuses on the immune system. With age it not only affords less protection against diseases such as cancer, but also makes errors. That is, it sometimes regards molecules in the body as foreign and dangerous rather than familiar and safe. It then produces antibodies to attack and destroy these molecules. This process, known as autoimmunity, eventually kills normal cells. Furthermore, autoimmunity is the suspected process behind several diseases of adulthood, such as rheumatoid arthritis.

Other theories point to declines in the regulatory and control mechanisms of the body—the nervous and endocrine systems which some scholars view as the major pacemakers of the aging process. Support for this position—which focuses on the total organism rather than the cell or particular organ—is based on the observation that behavioral performances show greater age decrements if they require the coordination of several rather than just a few body systems. Along a similar line is Birren's finding that the more complex a task, the more age decrements are manifested in carrying it out. Impairments of temperature regulation in the body and the declining sensitivity of the pancreas to changes in blood sugar levels are also seen as evidence of control system declines.

Because of its promise for prolongevity, Albert Rosenfeld believes that more attention should be given to Denckla's "death hormone" theory of aging. This theory argues that the aging clock is located in the brain and is hormonal in nature. It focuses on the thyroid, the one gland that has a profound effect on the two systems whose failures most often cause death—the cardiovascular system and the immune system. According to Denckla's theory, the pituitary gland is genetically programmed to release a "death hormone" that prevents body cells from using thyroxin, one of the hormones secreted by the thyroid. Unable to utilize this youth-preserving hormone, cells undergo the destructive changes associated with aging.

Evidence of Aging

Whatever the operative aging processes, one, some, or all serve to put middle-aged persons on a trajectory of increasing statistical probabilities of illness and death. Although rates of the many senescent changes vary widely among individuals and even within the same individual, passage through the middle years often brings with it intimations of old age: some people find that their hair begins to get gray and thin; some note that their hearing becomes less acute, and their vision dimmer; the skin less elastic and more pigmented.

Still other changes begin to occur, according to pathologist Donald West King. The immune system becomes less effective in warding off disease. The healing response begins to decline—broken bones take longer to knit and other injuries longer to repair. Certain enzymes decrease in activity. Hormonal levels change or receptors change in sensitivity to their effects. With menopause, women lose their reproductive capacity, and cease producing estrogen. Men experience some decline in testosterone levels. So far, said physiologist Oscar Hechter, there is no evidence that release of catecholamines changes with age. As age increases, however, it takes longer to return to a normal physiological state after exposure to stress. Indeed some of the biological deficits that accompany growing older are observed only under stressful conditions.

With age, said Michael Gershon, a neuro-anatomist, the brain undergoes numerous changes as it is subjected to increasing years of wear and tear. Neurons die, and since they cannot divide after infancy, they are forever lost. And the loss of specific ones has specific consequences. For example, loss of the dopamine neuron is implicated in creating tremors and Parkinson's disease. In some individuals, the loss of the dopamine neuron may also be responsible for clinical depressions that are chemical in etiology and not a reaction to environmental events. (Gershon noted that in this case the physical problem creates the environmental problem rather than vice versa.) Damage to or the loss of other kinds of neurons, Gershon suggested, might be responsible for other pathologies of aging. As the brain ages and changes it reacts differently to medications than it once did and dosages must be adjusted accordingly. There is, however, much variability among individuals both in the degree of neurological change experienced and in its behavioral manifestations.

All the biological vulnerabilities notwithstanding, the middle-aged are more than likely to survive the period in reasonably good health. Moreover, noted James Birren, some things may get better. With age, people grow in wisdom. They can use the accumulated learnings of experience to compensate for declining capacities and maybe even exceed their former powers. And more, people can also increase in feelings of contentment. For some, as the poet Browning observed, "The best is yet to be."

Squaring the Mortality Curve

In the settlements of Colonial New England, John Riley reported, death had a far different age distribution than now. It is estimated, for example, that up to thirty percent of infants perished in their first year of life and that older children and adolescents were also gravely imperiled by high risks of death. Many women died in their prime because of complications of childbirth. Those managing to survive until what is now considered middle-age probably had a life expectancy not many years shorter than that prevailing in more recent times. Nonetheless, many more died young than old.

As currently drawn, the curve describing age-specific rates of death in the United States resembles a wide J. After the initial rise reflecting infant deaths, the curve slopes downward and stays low, recording the low mortality of children and adolescents, pushing gradually upward in the middle years and then steeply so to document the high mortality of old age. Many now hope to square the mortality curve; to transform the arc of the J into a right angle (_|). Much as past improvements in standard of living, public health, and medicine served to reduce death dramatically among the young and child-bearing women, the aim is now to eliminate those noxious environmental or life-style influences that cause death before the full natural life span has been enjoyed. Essentially, the outcome of squaring the mortality curve

would be that people born together would die together—
of senescence. There would be, perforce, a real "deadline."

Squaring the mortality curve, said John Riley, would
entail making a sharp reduction in the preventable illnesses of midlife and early "old age." Although many of
the major causes of premature death now appear to have
some familial genetic component, a large part of their
etiology is probably environmentally induced. Heart
disease, various cancers, strokes, alcoholism, and many
other conditions fall into this category. Preventing these
diseases, however, will require extensive efforts to persuade young and old members of the public to correct
the behaviors that undermine health and to undertake
the behaviors that enhance it. Further, people would
need to insist that both the public and private sectors
explore and implement other measures to promote health.

Until the promises of prophylactic measures are realized, several problems stand in the way of squaring the
mortality curve. Forestalling death in the interim
requires:

- programs for early detection of disease so as to
  arrest a deadly progression
- support of basic biomedical research, including research on basic aging processes
- support for research on cures for disease
- improving regimens for living with afflictions
- further development of phenotyping procedures to
  identify people with high genetic risks of disease.

"But it is not enough," said Matilda Riley, "to square
the mortality curve. We must also square the morbidity
curve." Little is gained by extending lives already made
miserable by the accumulation of infirmities. Others
concurred. For example, several of the conferees questioned the value of squaring the mortality curve if it

increased the prevalence of senility in the population. Others asked what social goods would have to be foregone to support the costly dependency of ancient ones unable to function. "Any extra years added to life must be good years," said John Riley.

Yet another set of unknowns may vitiate current efforts to square both the mortality and morbidity curves. Authorities responsible for public health scarcely know what threats to long life are now present in the environment. The lethal effects of these hidden threats may not be manifested for years or decades hence. For example, Donald King indicated there are approximately 1300 food additives now in use and we do not know what they do to human cells. Spenser Havlick noted that even the government admits that the effects of various levels of air pollution, water pollution, noise pollution and visual blight remain unclear or unknown. Indeed, he said, "we don't have data on even two generations of pesticide intake."

For the while, however, heart disease is the major obstacle to squaring the mortality curve. Two presentations were made on the subject: one focusing on the relationship between life styles and the development of atherosclerosis and the other on the relationship between psychological factors and coronary heart disease.

Atherosclerosis

"The main cause of the main cause of death in the middle years is atherosclerosis," said Jack Strong, a pathologist and expert on the disease. It is the underlying factor in heart attacks, the leading killer, and sets the stage for strokes, aneurisms of the abdominal aorta, and gangrene of the legs. But the problem of atherosclerosis is not only heightened risk of mortality. The diseases and disabilities it causes are not always quickly fatal. Rather they often involve a long period of chronic illness that undermines the quality of life and incurs high costs to the sufferers and to the society as well. Yet there is reason to be encouraged. Results of investigations under-way around the world indicate that atherosclerosis can, in large measure, be prevented or at least postponed.[1]

Atherosclerosis is but one form of arteriosclerosis, a term used to describe several diseases characterized by hardening and thickening of the arteries. Of these diseases, however, atherosclerosis is the most important clinically. It is most threatening to life. Atherosclerosis involves damage to the intima, or the inner surface, of particular arteries—notably the coronary arteries en-circling the heart and the carotid arteries leading to the brain. The disease is the outcome of a long-term process marked by the deposit first of fats and later of other substances found in the blood on the arterial wall. As these deposits accumulate they narrow the artery and interfere with blood flow.

The buildup of the atherosclerotic lesion is gradual. The first signs appear in childhood and tend to increase in severity with age. While there is an association be-tween severity and age, the development of advanced lesions is not an inevitable aspect of the aging process. Rather the extent of atherosclerosis can vary greatly among individuals of the same age living in the same culture. Moreover, there are marked differenecs in de-gree among cultures by the time middle age is reached. Nonetheless, there does appear to be a "natural history" of atherosclerosis among men in the United States. It follows a fairly regular progression through three stages before the clinical horizon is reached. (See Figure 2).

---

[1] Jack P. Strong, "Atherosclerosis, A Preventable Disease? Autopsy Evidence." Mimeographed.

Figure 2
The Natural History of Atherosclerosis

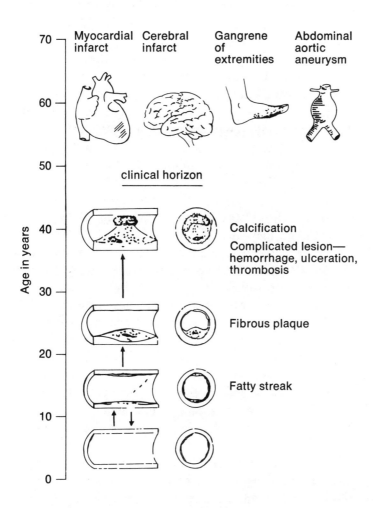

Source: Jack P. Strong, Douglas A. Eggen, Margaret C. Oalmann, Myra L. Richards and Richard E. Tracy, "Pathology and Epidemology of Atherosclerosis." *Journal of The American Dietetic Association* 62 (March 1973), p. 263. Copyright by The American Dietetic Association. Reprinted by permission from the *Journal of the American Dietetic Association*.

The first stage of atherosclerosis is observed surprisingly early in life. It appears that by age three, all children have fatty streaks on the intima of their aorta. However, not much of the surface is involved. By age ten or soon thereafter, fatty streaks begin to appear on the intima of the coronary arteries. Such streaking is now thought to be widespread after age twenty. Nor is prevalence of aortic fatty streaks in children and young adults restricted to the United States. Autopsy evidence gathered from many geographic regions with vastly different cultural patterns shows little difference in the amount of fat deposited on the intima of the aorta through adolescence. From a clinical point of view, fatty streaks are in themselves harmless. They are seldom elevated and so do not narrow the lumen, or opening, of the blood vessels. And they may be reversible.

The second stage is marked by the appearance of fibrous plaques. This is a more serious development in the progress of atherosclerosis. The base of the plaque is made of fat, but in time this becomes covered over by a layer of connective tissue, making the lesion firm and elevated. Narrowing of the arteries results. Unlike fatty streaks, fibrous plaques are not universal but differ in degree and frequency from country to country, and sometimes within subgroups of the same country. Significantly, arterial plaques and the amount of coronary heart disease found in a population tend to be directly related.

In the studies reported on by Strong, for example, male whites in New Orleans and Oslo had as much as three times the intimal surface of the coronaries covered with plaques and more advanced lesions than did men in Guatemala or the Durban Bantu males of South Africa. In the United States, fibrous plaques become evident somewhere between the second and third decade of life.

In the third stage, fibrous plaques enlarge and become the locus of additional problems. By mid-adulthood, hemorrhages may occur within the lesions. Ulceration or thrombosis may be present. Calcium may be deposited on the plaques. The consequence of these accumulating insults is to dangerously narrow the opening of the affected arteries, and set the stage for clinically significant disease and, not infrequently, death. Advanced lesions signal that the clinical horizon has been reached.

Atherosclerosis becomes the underlying cause of disease in several ways. Narrowed arteries deprive vital tissue of the blood supply needed for functioning. An inadequate supply of blood to the heart muscle, for example, is the cause of angina pectoris and an inadequate supply of blood to the brain is the cause of senile confusion. When the blood supply is cut off entirely, tissue in the affected organ dies. Complete occlusion or closing off of an artery can occur when a blood clot or thrombus, unable to gain passage through the narrowed opening, blocks blood flow.

Depending on where the occlusion is located, the result is a myocardial infarction (heart attack), stroke, or eventual gangrene of the leg. And sometimes sudden death. In addition, atherosclerotic lesions themselves can be a source of clots or thrombi. The lesions may also weaken the aortic wall and lead to the out-pouchings called aneurisms.

The problems that atherosclerosis causes are clear. But what causes atherosclerosis? Cross-national similarities in fatty streaks in the early years of life and the wide differences in the severity of lesions in later life, said Strong, implicate environmental factors as important culpable agents. It is clear that the atherosclerotic process begins early in life, but it need not progress to the point where it becomes a factor in coronary heart attacks.

Reporting the findings from the International Atherosclerosis Project, based on studies of coronary arteries at autopsy, Strong pointed out several contributing factors. Obviously age is a factor. The more time that elapses, the greater the opportunity for the buildup of atherosclerotic lesions. And sex appears to be a factor, with men in most cases more susceptible than women (at least, in white populations). But other factors have been found to be related significantly to raised lesions in the coronary arteries in different populations. For example, the higher the serum cholesterol levels, the larger the intimal surface covered with atherosclerotic lesions. The inference here is that diets high in fat are important contributors to the process, although more studies of a longitudinal nature are needed for confirmation. High blood pressure was also found to be associated with the accelerated development of atherosclerotic lesions. Diabetes was found to have a similar effect. Smoking was strongly implicated in the appearance of coronary artery atherosclerosis, with the number of cigarettes smoked daily linked to the severity of the lesions.

All of the factors linked to the development of coronary atherosclerosis are thus also the major risk factors in coronary heart disease and death. These risk factors, however, are more lethal in middle-age than old-age. Many of these factors can be reduced by changes in life style, thus reducing the risk of heart disease. Strong recommended several measures to prevent or retard the development of atherosclerosis:

- Beginning in kindergarten, teach children about risk factors and how to minimize them through diet.
- Educate adults to limit fats and reduce cholesterol in their diets.
- Convince people to stop smoking.
- Promote measures for the early detection and treatment of hypertension.

- Encourage industry to modify foods and food processing for safer food products (such as skim milk), find methods for reducing cholesterol and saturated fats, and develop substitutes for salty snacks.

Stress and Type A Behavior Patterns

Age, sex, serum cholesterol, hypertension, and smoking are indeed important etiologic factors in coronary heart disease, acknowledged psychologist David Glass, but they "still fail to identify most new cases of the disease." In fact, much about the who and why of heart disease still remains to be explained. In order to increase understanding of coronary heart disease, Glass urged that research efforts be broadened to include two "promising" variables: stress and the Type A coronary-prone behavior pattern.

Stresses stemming from work tensions, marital unhappiness, family troubles, and other midlife problems, he pointed out, contribute to the development of atherosclerosis and thus to coronary heart disease. For example, stress raises serum cholesterol and blood pressure, and so in time accelerates damage to the intima of the coronary arteries. According to Glass, studies show that accountants before the April 15th tax deadline have elevated serum cholesterol levels which then fall "sharply" once the deadline passes.[1]

Stress, he said, also sets off a physiological reaction which includes the release of hormonal substances such as adrenalin and noradrenalin, or catecholamines. Catecholamines are believed by some to increase intimal damage and to potentiate the aggregation of blood platelets. Thus, he concludes, life events that trigger the release of catecholamines may be important in the complex

---

[1] David C. Glass, "Stress, Behavior Patterns, and Coronary Disease." In the *American Scientist,* Vol. 65, March-April 1977.

etiology of coronary heart disease. In short, psychological stress may be implicated. But stress alone is not the full answer.

People who respond to stresses in their environment with a Type A behavior pattern, Glass believes, are eminently prone to coronary heart disease. The Type A behavior pattern is characterized by a) competitive achievement striving, b) a sense of time urgency, and c) hostility and agressiveness. It is a response evoked by situations that an individual perceives as threatening to his or her sense of mastery or control—which appear to be most situations having any stress or challenge at all for Type A's.

Much of Glass' research has been directed toward discovering more about these hard-driving individuals. Using questionnaires and interviews to determine who most fits these criteria and who does not (Type B's), he undertook a series of laboratory and field experiments with college students to investigate in what ways the two groups might differ in response to stress. He found that while neither brighter nor more privileged in background than Type B's, Type A students were awarded more college honors and were more likely to have plans for post-graduate education. This success orientation was reflected in the students' approach to the various tasks devised by Glass to study their behavior as well. His findings are consistent in documenting that Type A's do indeed strive hard. And they work just as hard whether or not they have a deadline before them. Type B's, by contrast, tend to work at a more comfortable pace, and press themselves only when given an explicit deadline to meet. Moreover, Type A's tend to suppress—or at least deny—fatigue when it interferes with their performance of a task, and relative to Type B's become agitated and impatient when asked to delay a response.

The various tasks that elicited these differences between Type A's and Type B's were "possible," in the sense that if effort were expended mastery could result. Noting that "Type A's are engaged in a struggle for control, whereas Type B's are relatively free from such concerns," Glass took his experiments a step further. He devised brief tasks that were impossible. No amount of effort would result in mastery. Nor could the stressful situation be escaped. Given such a task, the Type A's initially displayed hyper-responsiveness. They increased their motivation and redoubled their efforts to succeed. When this test was terminated and another—this time a controllable task—confronted them, Type A's made greater efforts than Type B's to re-establish a sense of control over the environment. Type A's enhanced their performances.

In later studies, a crucial difference between Type A's and Type B's emerged. Glass found that after prolonged exposure to uncontrollable and salient stresses (such as a loud noise), Type A's completely gave up on efforts to take charge of their situation. After prolonged but unavailing attempts at mastery, they became dejected and stopped trying altogether. They had learned to be helpless; that is, they were impaired in efforts to learn how to escape or avoid the stress when, subsequently, it was possible to do so. It is this shift from hyper-responsiveness to hypo-responsiveness that Glass thinks may be the important link to coronary heart disease.

Feelings of helplessness and depression, following some adverse life event, have been tied to the onset of a wide range of diseases. Glass thinks, however, that these feelings become the specific precursor of coronary heart disease when the person experiencing the helplessness-inducing event is a Type A responder. He notes that active coping with a stressor increases the level of nor-adrenalin in the blood, while the adrenalin level remains unchanged or may even decline. He hypothesizes that this repeated pattern of over-coping followed by under-coping—which must go on throughout a life course inevitably subject to uncontrollable events—spurs the development of atherosclerosis. In addition, some research suggests that Type A's may be more susceptible to platelet aggregation than Type B's when there are high levels of noradrenalin in the blood. Moreover, Type A's appear to have higher plasma levels of circulating noradrenalin immediately before, during and after the stress than B's.

Since it is virtually impossible to prevent many of life's aversive events, what can be done to reduce or intervene into the Type A pattern so as to make it less damaging or lethal? A revamping of the response pattern would be difficult, says Glass, because there may perhaps be some genetic predisposition that facilitated the learning of this coping response in the first place. Nonetheless, he suggested that relaxation therapies, biofeedback, behavior modification technique, and group therapy are worth trying, even though current evidence evaluating their effectiveness is not persuasive. Pharmacological interventions may also be useful. For the while, however, understanding of both the causes and the cures for coronary heart disease will move forward only if both physiological and psychological factors are considered.

Physician and Patients

While coronary artery disease is the chief cause of mortality and morbidity among middle-aged men and cancer the chief problem for women, these hardly constitute all the complaints for which medical care is sought or required. Yet, said Leslie Libow, "compared to childhood or old age, the middle years are practically uncharted for the medical practitioner." He took note of some of the special problems. Athletic injuries as a result of denying or failing to recognize the fact of growing older; problems with impotence which, however, are rarely related to changing hormone levels; alcoholism which he described as "rampant" in the middle years; obesity; increasing risks of suicide; and the effects of attempts at self medication.

Several of the conferees pointed to the importance of the middle-aged person finding a primary-care physician and then entering into a respectful collaboration with him or her, as a way to optimize present health and reduce future problems. However, the theme that was emphasized again and again was the need for individuals to take responsibility for their own health. "They must become self-regulating," Eric Pfeiffer said. If they have not done so before, by middle age individuals must make a single irreversible decision to take care of themselves— to stop smoking, to stop overeating, to get exercise, to minimize drinking. He posed the Talmudic question "If not now, when?"

45 The Value of Relationships

Social support is defined as information leading the subject to believe that he is cared for and loved, esteemed, and a member of a network of mutual obligations.[1]—
Sidney Cobb

Social support is often the key to surviving stress successfully. Whatever the problem—work pressures, job reversals, illnesses, family troubles, bereavement, signs of aging, or just the event of change itself—social support eases the strains and helps individuals to mobilize the personal resources necessary for coping. Not only does such support facilitate negotiating the transitions of life, but it enables people to flourish in the periods between. And more. It affords some protection against the physical and mental illnesses that stress seems to induce. Social support, said John Riley, helps

---

[1] Sidney Cobb, "Social Support as a Moderator of Life Stress," *Psychosomatic Medicine,* Vol. 38, Sept.-Oct. 1976.

to preserve life itself. Curiously, no one knows for certain how or why the biological, psychological and social processes combine to produce this beneficent effect. But its magic is incontrovertible.

Social support comes from claims on relationships—upon family, friends, neighbors and upon caretaking agencies in the community. Typically, those in the middle years are involved in many of these networks. They are fully engaged in life. Yet, their roles in these networks are often as the providers rather than the receivers of support. Despite the stresses which they themselves face, the middle-aged, as Donald Ford succinctly put it, "give more than they get." The claims for support they are asked to honor add to the overload of midlife responsibilities. Nonetheless, they are expected to take them on. As was frequently noted during the conference, however, the middle-aged also need to receive support while giving it to others.

Family Relationships

The family is often viewed as the quintessential source of social support. Years of intimate interaction are thought to create lasting bonds based on caring and reciprocity. Sociologists have long recognized—and continue to document again and again—the protection against life's outrageous fortunes that family membership confers. For individuals, the first source of social support was the family, and throughout the life course the family often remains the institution of first resort when help is needed.

The middle-aged occupy a strategic position in family relationships. As the middle and anchoring generation, they are the links between younger and older

members of the lineage—sometimes involving two generations of the young and two generations of the old. But
they are more than conduits for communication or the
guardians of roots. The middle-aged are often depended
upon to help both the young and old weather the exigen-
cies of their respective life stages, and successfully manage their own as well.

Robert Aldrich likened the relationship between the
middle-aged couple and other family members to a river
system. The middle-aged pair, in his analogy, is posi-
tioned on the main line of the river; various other family
members are located on the tributary branches upstream.
Much as the river system is subject to unpredictable and
uncontrollable forces such as floods or droughts, so too
is the family. However overwhelming, for example, the
effects of inflation or energy shortages are to family
members, they can do little to correct these problems.
But there are other stresses—no less difficult and not always more predictable—which family members are expected to manage. But they cannot always do so by
themselves. They look to others for support and help.
Such upstream troubles in the lives of parents, children,
siblings and friends as divorce, illness, economic catastrophe, jail terms, etc., suggested Aldrich, often cannot be
contained. They flow downward into the main branch,
in disastrous confluence upon the middle-aged couple.
And more than likely the middle-aged couple already
has troubles of its own. Thus, they are threatened with
being washed away by the flood of accumulating problems. The others, absorbed in their own problems, are
unavailable as sources of support.

By deliberately dramatizing the case, Aldrich said,
the interventions that might obviate the need to provide

disaster relief for the middle-aged couple come into focus. Namely, dams are needed on the upstream branches. The family members so located require some help from other sources to solve their problems and thus reduce the stress on the middle-aged couple.

Participants identified several specific family problems that impinge on the middle-aged and undermine the quality of life. Aldrich, for example, pointed to the problem of "family asynchrony"—the distress experienced by parents when a child takes longer than usual to "grow up" and accept responsibility for self. Others pointed to the difficulty of still having adolescent children to guide and nurture through the transition to adulthood. However, the family problem that elicited the greatest interest and concern was taking on responsibility for elderly parents.

As medical science and technology have extended life and prolonged its terminal phase, many middle-aged individuals are confronted with the task of supervising the final illness and sometimes the mental incompetence of one or more elderly parents or other relatives. The problem is often further complicated by the geographic distance which separates the residence of parent and the adult child. And by the availability of devices to prolong vegetative life.

The illness of parents often forces middle-aged offspring to decide when and where to place them in a nursing home. Because of the stigma attached to nursing homes and often the inadequate care they provide, this is a particularly difficult decision for offspring to make, said Leslie Libow. Yet if they are unable or unwilling to take an ailing parent into the home, and the community lacks home-care services, there is often no alternative. Nonetheless, this decision is often a source of distress for the adult child and, perhaps, also for the parent feeling abandoned by such placement. Undoubtedly, it is made more difficult still by an incipient trend toward the home as the preferred place to die.

With living wills not widely in force—California's Natural Death Act is an exception—whether or not to encourage life-sustaining efforts when there is no hope of recovery is an issue which the offspring need negotiate with the physician in charge on behalf of the parent. John Riley indicated, however, that the changing attitudes of the cohorts now in the middle age may serve to change the ambivalent ambiance in which such decisions are currently made. For themselves and their parents, the middle-aged may seek the "good death." According to a paper prepared for the Hastings Center, a good death:

- occurs after one's life work has been largely accomplished
- occurs after one's responsibilities to others have been substantially discharged
- will not tempt others to rage against fate or the unfairness of life or feel despair
- is without unbearable and degrading pain.

The problems of managing the life and death of a parent are difficult enough when there is a clear medical emergency or when the disabilities unequivocally require professional attention. When the problem is primarily one of senility, solutions are more difficult still. Mental deterioration, for example, is not always a sufficient condition for institutionalization, even though the ability to care for the self or personal affairs is gravely compromised. Nor is all mental deterioration reversible. Nor does mental deterioration necessarily presage a rapid slide to death. Individuals so afflicted may live for several years, with middle-aged offspring strained concomitantly.

Individuals who are mentally incompetent typically lose control over their property and over the direction of their lives. Others must act on their behalf. Leslie Libow suggested that a "penultimate will" would serve to increase individual control of later life. It would also relieve children of decision-making and spare them possible conflict with siblings. A penultimate will is written at a time when mental faculties are intact against a time when they may not be. Such a will can provide guidance to others with regard to: preferred physician; residence, including under what protective conditions the individual would live in the community; whether and where to be institutionalized; whether a return to the community is desired upon improvement; how assets would be expended or conserved.[1] Penultimate wills, however, are not in widespread use.

But what sources of support are available to the middle-aged as they attempt to provide support to parents and children and all the others in their network? What alternative networks might be activated to help the middle-aged cope with the myriad stresses in their lives?

The marital relationship, sometimes improving as adolescent children grow up and leave home, was viewed as an important source of mutual support for the couple. Yet, in some instances, the marital bond weakens in the absence of children. In either case, the cumulative stresses may be overwhelming to both partners. Several conferees observed that the church could be a source of support, but often is not utilized for that purpose. In addition, some of the needed help could be supplied by various public or voluntary agencies assisting kith and kin with their troubles, thus protecting the middle-aged pair from the spillover. These agencies might directly offer aid to the middle-aged as well. John Schweppe

---

[1] Robert Zicklin and Leslie S. Libow, "The Penultimate Will," *New York State Bar Journal,* January 1975.

urged that such help be provided by a multidisciplinary team. Typically, the problems are several, complex, and have many ramifications. A single specialist lacks the expertise necessary for a resolution that will optimize coping.

Robert Aldrich suggested, "We need to study the ecology of small family problems" in order to understand their dynamics. On the basis of this understanding, it is then possible to design interventions to handle problems "locally," i.e., within the family unit. By enabling others to better manage their own problems or by helping the middle generation give support to those in their networks—such as ailing parents or troubled children—some stresses of midlife might be alleviated.

The conferees suggested several measures that would provide support to those in midlife as they try to help significant others in their lives:

- Have easily-accessible multidisciplinary teams available for consultation on problems.
- Increase the number of neighborhood-based home health care services.
- Better regulate nursing homes to ensure adequate patient care.
- Encourage the use of penultimate wills.
- Support legislation for living wills.

Improving Relationships

Although people have been "relating" since birth, they are not necessarily adept at doing so. Through the middle years, Eric Pfeiffer pointed out, relationships tend to be "instrumental." People are brought together and interact to accomplish tasks. They need to get a job done at the work place. They need to collaborate on the rearing of children. They need to organize committees. But as they grow older and such tasks become fewer,

he said, people need to test their competence in relationships of "being" as well as "doing." In the middle years,
people need to test their capacity to relate to others without a task at hand. Can they participate in and enjoy
relationships in which feelings are shared? Selves ex-
plored? Can they satisfactorily resolve interpersonal
conflict?

The emptying of the nest brings to the fore the importance of such skills. But as retirement looms closer,
with its promises of time for intimacy, the need becomes
more urgent. Ideally, said Pfeiffer, such skills should be
part of the person's repertoire from early in life, and it
appears that young people today may be developing such
skills. Nonetheless, those in middle age often have not.
As a result they may find that as they move into old age,
they cannot easily endure the absence of "instrumental"
relationships. Thus, Pfeiffer urged those in their middle
years to acquire whatever interpersonal skills they lack.
Whether through psychiatric training programs, reading
or practice, they need to prepare for the relationships
that are typical of and bring satisfaction in old age.

Relationship to the Self

Until recently, relatively little attention has been given
to psychological development in the middle years. In
part, this was the outcome of a long held but now dis-
credited belief that little development could occur once
adulthood was reached. Patterns set in early childhood
were considered unalterable. Compounding the error
was Freud's dictum not to treat those over age 50. It
discouraged therapists and scientists alike from a focus
on the middle years that might have challenged the
earlier position. Its still-lingering effect, noted several
conferees, is a scarcity of practitioners expert in treating
older clients.

James Birren, a gerontologist, argued that considerable development can and does take place during adulthood. Not only do middle-aged people continue to grow psychologically, but when they sense that their growth has been arrested, they make deliberate efforts to move forward again. One measure often used by educated adults to resume personal growth, he noted, is the life review or life reconciliation. People take stock of where they have been and where they are going. Often serving as "preludes to change," these reflections suggested to Birren that autobiography might be a useful vehicle for enhancing psychological development in the middle years.

For background, Birren said, it is useful to conceptualize the self as having three components. One is the ideal self, constructed out of answers to the question: What do you aspire to? The second is the real self, composed of insights related to the question: Who are you, actually? The third is the social self, derived from responses to the question: What do you think others say about you? As a normal process of maturation, Birren postulates, the distances between these three selves diminish. Aging brings the selves closer together. However, to the degree that these selves are discordant, there is tension (which is common in adolescence), especially in the case of wide differences between the ideal self and the actual self. Psychotherapy, he observed, often attempts to reduce the distance by having the individual scale down the ideal self or aspirations and ambitions, or by a process of enhancing feelings of self worth and thus the perception of the actual self. Autobiography, too, he said, can help put the selves in better alignment and so be therapeutic.

Birren tested out the power of autobiography in a classroom experiment by having students write and then share their personal histories with one another. Students, representing a wide range of ages and backgrounds, were asked to describe their lives in terms of loves and hates, heroes and models, family histories, sex role identities, body images, experiences with death, secret fantasies, religious experiences, and more. The outcome was greater self-knowledge. But there was an additional benefit from the sharing itself. "The developmental exchange," as Birren called it, created "bonding" among the members of the group. There was an appreciation of one another's multifaceted lives. Humble backgrounds were honored. Inner struggles were respected. Feelings of self-esteem increased. Self-deprecatory processes were countered. New directions came into clearer focus. A social support system was created.

Although Birren indicated the experiment was still exploratory, others suggested that the opportunity for "developmental exchange" be made part of the offerings of adult education programs. Further, since life reviews appear to be a normal part of old age, if opportunities for a public accounting for one's life were made available in senior centers, then the self-esteem of old people might be enhanced as well. Merrell Clark observed that variations on this theme were already underway. The new emphasis on finding "roots" and the oral history projects that old people are contributing to confer similar benefits. A few participants, however, expressed concern that such self-revelation might be "damaging." Nonetheless, social support in whatever network it is found provides strength and self-esteem for coping with change.

46 Recommendations and Unresolved Issues

One thing upon which an interventionist in human development can rely is the fact that the people about whom he is concerned will undergo a change . . . The task for the interventionist in human affairs is one of seeking to effect controlled and predictable change—intentional change—change that moves in the direction that people have agreed to seek.—Hugh Urban and William Looft

If a single phrase were required to summarize the Aspen conference, it might well be: Middle Age—a time of change and a time to change. This theme emerged as a link connecting the wide-ranging multidisciplinary presentations and discussions, and it became the "message" of the conference as well. Whatever the perspective on the middle years—psychological, sociological, economic, medical, biological—the emphasis was on changing individuals in a changing situation. Middle-aged individuals were viewed as developing, declining,

501

renewing, failing, and overcoming in response to an array of internal processes and external challenges. Similarly the context which shapes middle-age experience was viewed as undergoing continuing change. New legislation, new environmental threats, new role options, new prohibitions, new scientific discoveries, new attitudes toward the family and toward middle age itself were all seen as working to remake the promises and problems of this phase of life.

In this fluid situation, the conferees saw an opportunity and also a mandate for further change—change directed toward improving the quality of life in the middle years. The time to initiate these changes was seen as now. Over and over again, the conferees urged that individuals immediately adopt those habits, life styles, and patterns of relating to others that research has shown to benefit health and improve the ability to cope with the changes life brings. They also urged that the public and private sectors immediately act to remedy or at least ameliorate those social conditions that create unnecessary stress or compromise health. Yet, concern was expressed that widespread ignorance about midlife might impede the desirable changes. People simply do not recognize that many of the problems of the middle years are of their own making and are correctable. Nor do they recognize that many of these problems have their origin in years gone by and will have consequences for the years ahead. As a result, not only are some hardships unnecessarily endured but opportunities for prevention are also lost.

Recommendations

John Schweppe urged that ways be found to disseminate information on the realities of the middle years widely and effectively. Such information, he said, must be brought to the attention of those now struggling to cope with the changes and problems that undermine their productivity and happiness, and to those who will soon move through this phase of life. Information about midlife also must be brought to the attention of people involved in social-support networks. It must be brought to the attention of educators so that they might pass it on. It must be brought to the attention of those in helping professions who are called upon to render advice and assistance. And it must be brought to the attention of policy makers—along with specific suggestions as to what might be done. But in disseminating information about the middle years, Schweppe warned, we must be careful to put it in perspective. "To educate for midlife," he said, "we need to teach people about human development over the full life course, about the ways in which people age from birth to death." In doing so, he noted, there is an additional benefit: people not only learn about themselves but also about others in the same or different phases of life. Relations between the generations in the family and age strata in the society thereby may be improved.

Throughout the conference numerous recommendations were made regarding measures to improve the quality of life in the middle years both immediately and in the long range. Those given high priority are summarized below under five broad headings.

Education and Public Information

Among the highest priorities, according to the conferees, is creating a more positive attitude toward growing older in the United States. They urged that current media programs and educational efforts be expanded and new ones launched to combat the negative stereotypes about middle age and old age that now exist.

Furthermore, specialists in aging should provide the media with accurate and up-to-date information about the middle years to correct false images and to make programming more relevant to the needs and concerns of persons at midlife. While the conferees recommended that courses in life course development be offered at all instructional levels, such classes were considered especially important for medical students and others preparing for roles in the caretaking agencies. In addition, the conferees urged that physicians avail themselves of course work about aging over the full life course. Also considered necessary are efforts to translate specialized knowledge of life-span development and its problems into a form useful for legislators and other policy makers.

Research and Study

The conferees gave high priority to intensifying the study of the ways biological, psychological, and social processes interact throughout the life course. Considerable support was expressed for expansion of research efforts in all areas of basic biomedical research. In terms of more focused biomedical research, priority was given to investigations centered on preventing hypertension and atherosclerosis. A major emphasis in the recommendations was given to research that would locate and then disseminate "success stories": what demonstration programs worked well in meeting the needs of people in their middle years; what media programs have been effective in producing desirable changes in behavior; what other countries do well for their citizens that might be adapted to the United States. The conferees also thought it important to encourage thorough preparation by government, scholars, and practitioners for the 1981 White House Conference on Aging.

Employment

In addition to recommending an end to mandatory retirement and expansion of job opportunities, the conferees suggested several specific measures to improve the work situation in the middle years. Employers, they felt, should be encouraged to provide flexible work arrangements and to use objective measures of work capacity (not age) to determine access to promotion opportunities, nature of work, and time of retirement. Also considered a very high priority is removing the barriers that inhibit labor force mobility in the middle years. In particular, a mechanism needs to be found to protect the economic security and accumulated benefits of people so that they will not be penalized by a job change.

Health

The first priority in health was seen as devising ways to make people self-monitoring so they might maintain high levels of well being into advanced old age. In addition to seeking greater discussion of a national system of health insurance, the conferees' recommendations focused primarily on ways of improving the quality and kind of health care delivery in the local community. They urged that programs for upgrading nursing homes be put into force and that alternatives to hospital and nursing home care, i.e., home health care, be developed. The emphasis was on community-based services.

Special Programs and Services

Recommendations for special programs and services fell into two major categories. One focused on establishing programs for the retired and elderly that might enrich their lives and provide opportunities for them to contribute their talents to society as well. The other emphasized support for the middle-aged: counseling centers for emotional problems, guidance for those interested in career changes, expansion of opportunities for continuing education. The conferees also urged that "sensitivity" programs be set up to help all people deal with dying and death.

Unresolved Issues

In concluding this report on a conference focused on documenting human needs and recommending measures to meet them, it seems important to mention a theme that surfaced in every discussion session. Priorities. Inevitably, programs of action that intervene in the lives of others raise questions of ethics. Is the social good prior to individual liberty? Inevitably, programs of action that make claims upon scarce public resources raise questions of distributive justice. Who deserves what? Much as the prevailing value system provides little sure guidance on these matters, the questions raised here remained largely unanswered, though not less compelling.

On one issue, however, there was a clear consensus. Evidence, not expedience, must determine the choice of a social program. But beyond the need for excellent data to inform policy is the thornier issue of how complete a data base must be before it should serve as a basis for action. In making medical interventions, for example, how certain must a physician be that the proposed treatment will help the patient and cause no harm? How sure should the government be that a substance is dangerous to health before removing it from the market place? In implementing social programs, how sure must we be that they will not waste public resources that might be used efficiently in another program?

Other questions were even more difficult. Should consideration of quality of life be prior to considerations of quantity of life? Suppose stopping the "aging clock" or even squaring the mortality curve results in increasing the number of hopelessly incompetent old people who consume scarce resources in their progressively costly dependency. Should research and action to this end be continued? What are the social or medical responsibilities toward the old who are now hopelessly ill or mentally incompetent? What responsibilities do individuals themselves have in view of the possibility of their own future helplessness? To what degree and under what conditions should individuals be allowed to decide their own fate?

Still another issue was how public funds should be spent. Since a single discovery from basic biomedical research often does more to improve the general welfare than a multitude of service programs, should such research be supported even when no such discovery can be promised in the immediate future—or at all? What kind of accountability should be asked of scientists? Should certain kinds of research problems (e.g. particular diseases) be studied to the exclusion of others? What responsibilities do scientists have to the public that supports them? Should private industry be rewarded by the public for developing new ways to improve the health and welfare of people or to protect workers from harm? Should drugs be withheld from some for fear of possible abuse by them or others?

Should the current needs of the middle-aged be met if such action deprives children or old people of essential services? Should efforts to prevent problems in the future take precedence over ameliorating the problems that face people now? Should attempts be made to intervene in family life?

A future conference might well address some or all of these questions.

Selected
Bibliography

1. Andrisani, Paul J., and Herbert S. Parners, 1977. "Five Years in the Work Lives of Middle-Aged Men: Findings From the National Longitudinal Surveys." Mimeographed.

2. Andrisani, Paul J., 1977. "Effects of Health Problems on the Work Experience of Middle-Aged Men." *Industrial Gerontology,* Spring, pp. 97-112.

3. Andrisani, Paul J., Eileen Appelbaum, Ross Koppel, and Robert C. Miljus, 1977. *Work Attitudes and Labor Market Experience: Evidence From the National Longitudinal Surveys.* Philadelphia, Pa., Center for Labor and Human Resource Studies, May.

4. Bane, Mary Jo, 1976. *Here To Stay: American Families in the Twentieth Century,* Basic Books, Inc.

5. Binstock, Robert and Ethel Shanas, (editors) 1976. *Handbook of Aging and the Social Sciences,* New York: Van Nostrand Reinhold Co.

6. Birren, James E., and V. Jayne Renner, 1976. "Developments in Research on the Biological and Behavioral Aspects of Aging and Their Implications." In A. W. Exton-Smith and J. G. Evans, (editors) *Care of the Elderly,* New York: Green and Stratton, pp. 194-224.

7. Birren, James E. and V. Jayne Renner, 1976. "Stress & Aging: Psychobiological Theory & Research." Presented at Gerontological Society Meetings, New York, October.

8. Birren, James E. and Werner K. Schaie (editors), 1976. *Handbook of the Psychology of Aging.* New York: Van Nostrand Reinhold Co.

9. Cheiw, Peter, 1976. *The Inner World of the Middle-Aged Men.* New York: Grossman-Viking.

10. Glass, David C. 1977. "Stress, Behavior Patterns, and Coronary Disease." *American Scientist,* Vol. 65 March-April.

11. "The Graying of America," 1977. *Newsweek.* February 28.

12. Havlick, Spenser, 1974. *The Urban Organism.* New York: Macmillan.

13. Levine, James A., 1976. *Who Will Raise the Children?* New York: J. B. Lippincott.

14. Renner, V. Jayne and James E. Birren, 1977. "Objectives and New Directions in the Study of Stress, Disease and Aging." Presented at Symposium No. 5: Society, Stress and Disease-Aging and Old Age, Stockholm, Sweden, May.

15. Riley, Matilda White, 1976. "Age Strata in Social Systems" In *Handbook of Aging and the Social Sciences.* Edited by Robert H. Binstock and Ethel Shanas. New York: Van Nostrand and Reinhold Company.

16. Riley, Matilda White, and Joan Waring, 1976. "Age and Aging." In *Contemporary Social Problems,* 4th ed. Edited by Robert K. Merton and Robert Nisbet. New York: Harcourt Brace Jovanovich, Inc.

17. Rosenblum, Marc. "Recession's Continuing Victim: The Older Worker," A Working Paper prepared for use by the Special Committee on Aging, United States Senate, U.S. Government Printing Office, Wash., D.C. 1976.

18. Rosenfeld, Albert, 1976. *Prolongevity.* New York: Alfred Knopf.

19. Rosenfeld, Albert, 1976. "Are We Programmed to Die?" *Saturday Review,* October.

20. Rosenfeld, Albert, 1976. "The Willy Loman Complex" *Saturday Review,* August.

21. Sheehy, Gail, 1974. *Passages.* New York: E. P. Dutton.

22. Strong, Jack P. "Atherosclerosis, A Preventable Disease? Autopsy Evidence." (Mimeo)

23. Strong, Jack P., Douglas A. Eggen, Margaret C. Oalmann, Myra L. Richards and Richard E. Tracy, 1973. "Pathology and Epidemiology of Atherosclerosis." *Journal of the American Dietetic Association,* Vol. 62, March.

24. Troll, Lillian, 1975. *Early and Middle Adulthood.* Belmont, California: Wadsworth Company.

25. Urban, Hugh B. and William R. Looft, 1973. "Issues in Human Development Intervention." Paper presented at Conference: Applied Human Development: Issues in Intervention. College of Human Development, The Pennsylvania State University, May 30-June 2.

Index

List of Contributors

Rev. Ernest J. Bartell, C.S.C., Department of Economics, University of Notre Dame, Notre Dame, Indiana

Philip Blumstein, Associate Professor of Sociology, University of Washington, Seattle, Washington

John Cannon, Rector, St. John's in the Village, New York, New York

June Jackson Christmas, Medical Professor and Director, Program in the Behavioral Sciences, School of Bio-Medical Education, City College, New York, New York

John G. Darley, Professor Emeritus, Psychology Department, University of Minnesota, Minneapolis, Minnesota

John Demos, Professor of History, Brandeis University, Waltham, Massachusetts

Leonard Duhl, Professor of Public Health and City Planning, University of California at Berkeley, Berkeley, California

Aklilu Habte, Director, Education Department, The World Bank, Washington, D.C.

Alfred J. Kahn, Professor, Social Policy and Planning, and Codirector, Cross-National Studies Research Program, The Columbia University School of Social Work, New York, New York

Sheila B. Kamerman, Associate Professor, Social Policy and Social Planning, and Codirector, Cross-National Studies Research Program, The Columbia University School of Social Work, New York, New York

A.J. Kelso, Professor, Department of Anthropology, University of Colorado, Boulder, Colorado

Donald West King, Delafield Professor of Pathology, Columbia University College of Physicians and Surgeons, New York, New York

Arthur J. Norton, Assistant Chief, Population Division, Bureau of the Census, U.S. Department of Commerce, Washington, D.C.

John Scanlon, Editor, Academy for Educational Development, New York, New York

Edmund Pellegrino, President, Catholic University of America, Washington, D.C.

Howard Spierer, Associate Director for Development, Mt. Sinai Medical Center, New York, New York

William E. Threlkeld, Division Chief, Juvenile Division, Denver Police Department, Denver, Colorado

Harold Visotsky, Professor and Chairman, Department of Psychiatry and Behavioral Sciences, Northwestern University Medical School, Chicago, Illinois

Joan Waring, Director of Corporate Research Services, The Equitable Life Assurance Society of the United States, New York

Helen R. Washburn, President, American Personnel & Guidance Association (as of 1 July 1981), Boise, Idaho

Sidney Werkman, Professor, Department of Psychiatry, University of Colorado School of Medicine, Denver, Colorado

Willard Wirtz, former Secretary, Department of Labor, Washington, D.C.

About the Editor

Alvin C. Eurich was professor of educational psychology successively at the University of Minnesota; Northwestern University; and Stanford University, where he also served as vice-president and acting president; and he was the first chancellor of the State University of New York. Dr. Eurich was a member of President Truman's Commission on Higher Education, President Kennedy's Task Force on Education, and Visiting Fellow at Clare College of the University of Cambridge. Dr. Eurich has written five books, including *Reforming American Education,* his most recent, and numerous articles that have been published in professional and general periodicals.

49345

HQ
799.95
M34

Major transitions
in the human life
cycle

DATE DUE

JUN 7 '85			
AUG 21 '86			
JA 04 '90			
JA18'93			
NO17'95			
JE21'96			
AG2⁻'96			
AP25'02			